Electromagnetic
Anisotropy and
Bianisotropy

A Field Guide

Electromagnetic Anisotropy and Bianisotropy

A Field Guide

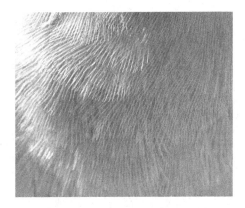

Tom G Mackay

University of Edinburgh, UK and Pennsylvania State University, USA

Akhlesh Lakhtakia

Pennsylvania State University, USA

 **World Scientific**

NEW JERSEY · LONDON · SINGAPORE · BEIJING · SHANGHAI · HONG KONG · TAIPEI · CHENNAI

Published by

World Scientific Publishing Co. Pte. Ltd.

5 Toh Tuck Link, Singapore 596224

USA office: 27 Warren Street, Suite 401-402, Hackensack, NJ 07601

UK office: 57 Shelton Street, Covent Garden, London WC2H 9HE

British Library Cataloguing-in-Publication Data
A catalogue record for this book is available from the British Library.

ELECTROMAGNETIC ANISOTROPY AND BIANISOTROPHY: A FIELD GUIDE

ISBN-13 978-981-4289-61-0
ISBN-10 981-4289-61-2

Printed in Singapore by B & Jo Enterprise Pte Ltd

James Clerk Maxwell in his native city of Edinburgh.[1]

---

[1]Photograph courtesy of Gary Doak and *The Royal Society of Edinburgh*.

Prologue

Σα βγεις στον πηγαιμό για την Ιθάκη,
να εύχεσαι νάναι μακρύς ο δρόμος,
γεμάτος περιπέτειες, γεμάτος γνώσεις.
– Κωνσταντίνος Π. Καβάφης (1911)

As you set out for Ithaka
hope the voyage is a long one,
full of adventure, full of discovery.
– Constantine P. Cavafy (1911)[2]

---

[2]Translated by Edmund Keeley and Philip Sherrard.

With love to our children

Freya Mackay
Magnus Mackay
Natalya Sheetal Lakhtakia

Preface

The focus of this field guide is largely on electromagnetic fields in linear materials. Even with the exclusion of nonlinearity (except in the last chapter), the panorama of electromagnetic properties within the ambit of this monograph is vast — a view which has been widely appreciated for the last 10 years.

An isotropic medium has electromagnetic properties which are the same in all directions. Such mediums provide the setting for introductory courses on electromagnetic theory, as encountered at high school or in early undergraduate classes. But isotropy is an abstraction which requires qualification when applied to real materials. For examples, liquids and random particulate composite mediums may be isotropic on a statistical basis, while cubic crystals are isotropic when viewed at macroscopic length–scales. In the frequency domain, electromagnetically isotropic mediums are characterized simply by scalar constitutive parameters which relate the induction field phasors $\underline{D}$ and $\underline{H}$ to the primitive field phasors $\underline{E}$ and $\underline{B}$.

Often, naturally occurring materials and artificially constructed mediums are more accurately described as anisotropic rather than isotropic. Anisotropic mediums exhibit directionally dependent electromagnetic properties, such that $\underline{D}$ is not aligned with $\underline{E}$ or $\underline{H}$ is not aligned with $\underline{B}$. Dyadics (i.e., second–rank Cartesian tensors) are needed to relate the primitive and the induction field phasors in anisotropic mediums. No wonder, these mediums exhibit a much more diverse range of phenomenons than isotropic mediums do. A visit to the mineralogical section of the local museum should impress upon the reader the array of dazzling optical effects that may be attributed to anisotropy in crystals.

Bianisotropy represents the natural generalization of anisotropy. In the electromagnetic description of a bianisotropic medium, *both* $\underline{D}$ and $\underline{H}$ are anisotropically coupled to *both* $\underline{E}$ and $\underline{B}$. Hence, in general, a linear bian-isotropic medium is characterized by four 3×3 constitutive dyadics. Though seldom described in standard textbooks, bianisotropy is in fact a ubiquitous phenomenon. Suppose a certain medium is characterized as an isotropic di-electric medium by an observer in an inertial reference frame Σ. The same medium generally exhibits bianisotropic properties when viewed by an ob-server in another reference frame that translates at uniform velocity with respect to Σ. While naturally occurring materials with easily appreciable bianisotropic properties in normal environmental conditions are relatively rare (viewed from a stationary reference frame), bianisotropic mediums may be readily conceptualized through the process of homogenization of a com-posite of two or more constituent mediums. Bianisotropy looks set to play an increasingly important role in the rapidly burgeoning fields relating to complex mediums and metamaterials.

Our aim in this field guide is to extend and update the standard treat-ments of crystal optics found in classical textbooks such as Born & Wolf's *Principles of Optics*[3] and Nye's *Physical Properties of Crystals*[4]. We pro-vide a broad overview of electromagnetic anisotropy and bianisotropy. The topics covered are constitutive relations (Chap. 1), examples of anisotropy and bianisotropy (Chap. 2), space–time symmetries (Chap. 3), planewave propagation (Chap. 4), dyadic Green functions including depolarization dyadics (Chap. 5), homogenization formalisms (Chap. 6), and nonlinear aspects (Chap. 7). The target audience comprises graduate students and researchers seeking an introductory survey of the electromagnetic theory of complex mediums. A familiarity with basic electromagnetic theory, and a commensurate level of mathematical expertise, is assumed. An appendix is provided to acquaint the reader with dyadic notation and algebra. SI units are adopted throughout.

Tom G. Mackay, Edinburgh, Scotland
Akhlesh Lakhtakia, University Park, PA
July 2009

---

[3]M. Born and E. Wolf, *Principles of optics, 7th (expanded) ed*, Cambridge University Press, Cambridge, UK, 1999.
[4]J.F. Nye, *Physical properties of crystals*, Oxford University Press, Oxford, UK, 1985.

Acknowledgments

Over the years, we have had many discussions on matters electromagnetic with colleagues worldwide. Most notably, extensive dialogues with Craig F. Bohren and (late) Werner S. Weiglhofer have assisted in shaping our conceptions of electromagnetic fields in complex mediums. We also owe debts of gratitude to: Ioannis M. Besieris, Ricardo A. Depine, Dale M. Grimes, Magdy F. Iskander, Dwight L. Jaggard, Martin W. McCall, Bernhard Michel, J. Cesar Monzon, S. Anantha Ramakrishna, Sandi Setiawan, Ari H. Sihvola, Miguel A. Solano, and Vasundara V. Varadan.

For partial financial support, Tom thanks *The Royal Society of Edinburgh/Scottish Executive* and Akhlesh thanks the *Charles Godfrey Binder Endowment* at the Pennsylvania State University. We thank our families for supporting our collaboration on this book, and the editorial staff of World Scientific Publishing Co. Pte. Ltd. for their assistance.

Contents

Acronyms and Principal Symbols

Acronyms

CLC	cholesteric liquid crystal
CSTF	chiral sculptured thin film
DGF	dyadic Green function
FCM	Faraday chiral medium
HBM	helicoidal bianisotropic medium
HCM	homogenized composite medium
LSPR	localized surface plasmon resonance
NPV	negative phase velocity
PPV	positive phase velocity
QED	quantum electrodynamics
SPFT	strong–property–fluctuation theory

Sets

$\mathbb{C}$	complex numbers
$\mathbb{R}$	real numbers
$\mathcal{W}$	angular frequencies

Scalars

c_0	speed of light in free space
$\tilde{c}$	microscopic charge density (time domain)
f_ℓ	volume fraction of constituent ℓ
g^{FF}	local field factor
i	$\sqrt{-1}$
$\tilde{j}$	microscopic current density (time domain)
k	wavenumber
L_{cor}	correlation length
q	point charge
t	time
v	magnitude of velocity
$\tilde{W}$	total energy density (time domain)
Z	impedance
ϵ_0	permittivity of free space
μ_0	permeability of free space
ρ	average linear particle size
$\tilde{\rho}_{\mathrm{e}}$	externally impressed electric charge density (time domain)
$\tilde{\rho}_{\mathrm{m}}$	externally impressed magnetic charge density (time domain)
ω	angular frequency

3–vectors

$\underline{\tilde{B}}, \underline{B}$	primitive magnetic field (time, frequency domain)
$\underline{\tilde{D}}, \underline{D}$	induction electric field (time, frequency domain)
$\underline{\tilde{E}}, \underline{E}$	primitive electric field (time, frequency domain)
$\underline{\tilde{H}}, \underline{H}$	induction magnetic field (time, frequency domain)
$\underline{\tilde{J}}_{\mathrm{e}}, \underline{J}_{\mathrm{e}}$	externally impressed electric current density (time, frequency domain)

$\tilde{\underline{J}}_{\mathrm{m}}, \underline{J}_{\mathrm{m}}$	externally impressed magnetic current density (time, frequency domain)
$\tilde{\underline{M}}, \underline{M}$	magnetization (time, frequency domain)
$\tilde{\underline{P}}, \underline{P}$	polarization (time, frequency domain)
$\underline{\mathbb{Q}}$	Beltrami field (frequency domain)
$\underline{k}$	wavevector
$\underline{r}$	position vector
$\tilde{\underline{S}}, \underline{S}$	Poynting vector (time, frequency domain)
$\underline{v}$	velocity vector

6–vectors

$\mathbf{C}$	induction electric–primitive magnetic field phasor (frequency domain)
$\mathbf{F}$	primitive electric–induction magnetic field phasor (frequency domain)
$\mathbf{Q}$	source current density phasor (frequency domain)
$\mathbf{N}$	polarization–magnetization phasor (frequency domain)

3×3 dyadics

$\underline{\underline{\alpha}}$	polarizability density dyadic (frequency domain)
$\underline{\underline{D}}$	depolarization dyadic (frequency domain)
$\underline{\underline{G}}$	dyadic Green function (frequency domain)
$\underline{\underline{I}}$	identity dyadic
$\tilde{\underline{\underline{T}}}$	Maxwell stress tensor (time domain)
$\underline{\underline{U}}$	shape dyadic
$\tilde{\underline{\underline{\epsilon}}}_{\mathrm{EB}}, \underline{\underline{\epsilon}}_{\mathrm{EB,EH}}$	permittivity constitutive dyadic (time, frequency domain)
$\tilde{\underline{\underline{\zeta}}}_{\mathrm{EB}}, \underline{\underline{\zeta}}_{\mathrm{EB,EH}}$	magnetoelectric constitutive dyadic (time, frequency domain)
$\underline{\underline{\mu}}_{\mathrm{EH}}$	permeability constitutive dyadic (frequency domain)
$\tilde{\underline{\underline{\nu}}}_{\mathrm{EB}}, \underline{\underline{\nu}}_{\mathrm{EB}}$	impermeability constitutive dyadic (time, frequency domain)

$\tilde{\underline{\underline{\xi}}}_{\mathrm{EB}}, \underline{\underline{\xi}}_{\mathrm{EB,EH}}$ magnetoelectric constitutive dyadic (time, frequency domain)

$\underline{\underline{\Sigma}}^{[n]}$ nth–order mass operator (frequency domain)

6×6 dyadics

$\underline{\underline{\alpha}}$ polarizability density dyadic (frequency domain)

$\underline{\underline{D}}$ depolarization dyadic (frequency domain)

$\underline{\underline{G}}$ dyadic Green function (frequency domain)

$\underline{\underline{I}}$ identity dyadic

$\underline{\underline{K}}_{\mathrm{EB,EH}}$ constitutive dyadic (frequency domain)

$\underline{\underline{\Sigma}}^{[n]}$ nth–order mass operator (frequency domain)

Operators and functions

$[\underline{\underline{A}}]_{m\ell}$ entry of matrix A in row m, column ℓ

$\underline{\underline{A}}^{\mathrm{T}}, \underline{A}^{\mathrm{T}}$ transpose of matrix or vector A

$\underline{\underline{A}}^{-1}$ inverse of matrix A

$\underline{\underline{A}}^{\dagger}$ conjugate transpose of matrix A

adj $\underline{\underline{A}}$ adjoint of matrix A

det $\underline{\underline{A}}$ determinant of matrix A

$\mathcal{C}\{\cdot\}$ conjugation transformation

b^* complex conjugate of b

$\mathrm{Im}\{\cdot\}$ imaginary part

$\underline{\underline{L}}(\cdot)$ linear differential operator (6×6 dyadic)

$\hat{\underline{p}}$ unit vector in direction of vector $\underline{p}$

$\mathcal{P}\{\cdot\}$ spatial inversion

$\mathrm{P}\int\ldots$ principal–value integration

$\mathcal{R}_{\pi/2}\{\cdot\}$ duality transformation

$\mathrm{Re}\{\cdot\}$ real part

$\mathcal{T}\{\cdot\}$	time reversal
$\mathcal{T}_{\mathrm{w}}\{\cdot\}$	Wigner time reversal
tr $\underline{\underline{A}}$	trace of matrix A
$\underline{\underline{\tau}}\{\cdot\}$	transforms $\underline{\underline{\mathbf{K}}}_{\mathrm{EH}}$ to $\underline{\underline{\mathbf{K}}}_{\mathrm{EB}}$
$\delta(\cdot)$	Dirac delta function
$\langle\cdot\rangle_{\mathrm{e}}$	ensemble average
$\langle\cdot\rangle_{\mathrm{t}}$	time average
$\langle\langle\mathrm{p},\,\mathrm{q}\rangle\rangle$	reaction of 'p' sources on 'q' fields
$s(\cdot,\cdot)$	switching function

Chapter 1

The Maxwell Postulates and Constitutive Relations

The action of complex arrangements of matter on electromagnetic waves lies at the heart of this book. Herein, we generally regard matter from a macroscopic perspective, which means that electromagnetic wavelengths are assumed to be large compared with interatomic distances — the atomic (and sub–atomic) nature of matter does not directly concern us. Accordingly, 'mediums' and 'materials' are referred to in this book, rather than 'assemblies of atoms and molecules'. The theoretical basis is provided by the Maxwell postulates for macroscopic fields combined with constitutive relations. The fundamental features of macroscopic electromagnetic theory, which underpin the remainder of the book, are presented in this chapter.

1.1 From microscopic to macroscopic

The electromagnetic properties of materials are conveniently characterized in macroscopic terms, wherein wavelengths are much larger than atomic length–scales (approximately 10^{-10} m). However, the macroscopic electromagnetic viewpoint is build upon a microscopic foundation [1].

Within the realm of microscopic electromagnetism, only two fields are involved: the electric field $\tilde{\underline{e}}(\underline{r}, t)$ and the magnetic field $\tilde{\underline{b}}(\underline{r}, t)$. Both of these fields vary extremely rapidly as functions of position $\underline{r}$ and time t: spatial variations occur over distances $\lesssim 10^{-10}$ m while temporal variations occur on timescales ranging from $\lesssim 10^{-13}$ s for nuclear vibrations to $\lesssim 10^{-17}$ s for electronic orbital motion [2]. The fields $\tilde{\underline{e}}(\underline{r}, t)$ and $\tilde{\underline{b}}(\underline{r}, t)$ develop due to point charges q_ℓ positioned at $\underline{r}_\ell(t)$ and moving with velocity $\underline{v}_\ell(t)$. The microscopic charge density

$$\tilde{c}(\underline{r}, t) = \sum_\ell q_\ell \, \delta \left[\underline{r} - \underline{r}_\ell(t) \right] \tag{1.1}$$

1

and the microscopic current density

$$\tilde{\underline{j}}(\underline{r}, t) = \sum_\ell q_\ell \, \underline{v}_\ell \, \delta\left[\underline{r} - \underline{r}_\ell(t)\right] \tag{1.2}$$

involve the Dirac delta function $\delta(\cdot)$.[1]

The relations between the fields $\tilde{\underline{e}}(\underline{r}, t)$ and $\tilde{\underline{b}}(\underline{r}, t)$ on the one hand and the source densities $\tilde{c}(\underline{r}, t)$ and $\tilde{\underline{j}}(\underline{r}, t)$ on the other hand are provided by the microscopic Maxwell postulates [2]

$$\left.\begin{aligned} &\nabla \times \tilde{\underline{e}}(\underline{r}, t) + \frac{\partial}{\partial t}\, \tilde{\underline{b}}(\underline{r}, t) = \underline{0} \\[2mm] &\nabla \times \tilde{\underline{b}}(\underline{r}, t) - \epsilon_0 \mu_0 \frac{\partial}{\partial t}\, \tilde{\underline{e}}(\underline{r}, t) = \mu_0 \tilde{\underline{j}}(\underline{r}, t) \\[2mm] &\nabla \bullet \tilde{\underline{e}}(\underline{r}, t) = \frac{1}{\epsilon_0}\, \tilde{c}(\underline{r}, t) \\[2mm] &\nabla \bullet \tilde{\underline{b}}(\underline{r}, t) = \underline{0} \end{aligned}\right\}, \tag{1.3}$$

where all quantities are in SI units, with $\epsilon_0 = 8.854 \times 10^{-12}$ F m^{-1} and $\mu_0 = 4\pi \times 10^{-7}$ H m^{-1} being the permittivity and permeability of free space, respectively.

When the length–scales of the variation in electromagnetic fields greatly exceed atomic length–scales, the summation index ℓ in Eqs. (1.1) and (1.2) achieves enormous values. Hence, it becomes a practical necessity to consider the spatiotemporal averages of the microscopic quantities in Eqs. (1.3). In fact, spatial averaging alone suffices due to the finite universal speed $c_0 = (\epsilon_0 \mu_0)^{-1/2}$ [2]. The macroscopic counterparts of Eqs. (1.3) thus arise as

$$\left.\begin{aligned} &\nabla \times \tilde{\underline{E}}(\underline{r}, t) + \frac{\partial}{\partial t}\, \tilde{\underline{B}}(\underline{r}, t) = \underline{0} \\[2mm] &\nabla \times \tilde{\underline{B}}(\underline{r}, t) - \epsilon_0 \mu_0 \frac{\partial}{\partial t}\, \tilde{\underline{E}}(\underline{r}, t) = \mu_0 \tilde{\underline{J}}(\underline{r}, t) \\[2mm] &\nabla \bullet \tilde{\underline{E}}(\underline{r}, t) = \frac{1}{\epsilon_0}\, \tilde{\rho}(\underline{r}, t) \\[2mm] &\nabla \bullet \tilde{\underline{B}}(\underline{r}, t) = 0 \end{aligned}\right\}, \tag{1.4}$$

with the macroscopic fields $\tilde{\underline{E}}(\underline{r}, t)$ and $\tilde{\underline{B}}(\underline{r}, t)$ being the spatial averages of $\tilde{\underline{e}}(\underline{r}, t)$ and $\tilde{\underline{b}}(\underline{r}, t)$, respectively, and the macroscopic charge and current densities $\tilde{\rho}(\underline{r}, t)$ and $\tilde{\underline{J}}(\underline{r}, t)$ being similarly related to $\tilde{c}(\underline{r}, t)$ and $\tilde{\underline{j}}(\underline{r}, t)$.

---

[1] The Dirac delta function is defined in Eq. (5.4).

In matter, a distinction can be made between (i) free charges (which are externally impressed) and (ii) bound charges (which arise due to internal mechanisms). Thus, we have the externally impressed source densities

$$\left. \begin{array}{l} \tilde{\rho}_e(\underline{r},t) = \tilde{\rho}(\underline{r},t) + \nabla \bullet \tilde{\underline{P}}(\underline{r},t) \\[3mm] \tilde{\underline{J}}_e(\underline{r},t) = \tilde{\underline{J}}(\underline{r},t) - \dfrac{\partial}{\partial t}\tilde{\underline{P}}(\underline{r},t) - \nabla \times \tilde{\underline{M}}(\underline{r},t) \end{array} \right\}, \tag{1.5}$$

where the polarization $\tilde{\underline{P}}(\underline{r},t)$ and magnetization $\tilde{\underline{M}}(\underline{r},t)$ characterize the bound source densities. Notice that $\tilde{\underline{P}}(\underline{r},t)$ and $\tilde{\underline{M}}(\underline{r},t)$ are not uniquely specified by Eqs. (1.5). That is, if $\tilde{\underline{P}}(\underline{r},t)$ were replaced by $\tilde{\underline{P}}(\underline{r},t) - \nabla \times \tilde{\underline{A}}(\underline{r},t)$ and $\tilde{\underline{M}}(\underline{r},t)$ replaced by $\tilde{\underline{M}}(\underline{r},t) + (\partial/\partial t)\tilde{\underline{A}}(\underline{r},t)$, then Eqs. (1.5) would still be satisfied for any differentiable vector function $\tilde{\underline{A}}(\underline{r},t)$.

The concepts of polarization and magnetization thus give rise to two further macroscopic fields defined as

$$\left. \begin{array}{l} \tilde{\underline{D}}(\underline{r},t) = \epsilon_0 \tilde{\underline{E}}(\underline{r},t) + \tilde{\underline{P}}(\underline{r},t) \\[3mm] \tilde{\underline{H}}(\underline{r},t) = \dfrac{1}{\mu_0} \tilde{\underline{B}}(\underline{r},t) - \tilde{\underline{M}}(\underline{r},t) \end{array} \right\}. \tag{1.6}$$

The basic framework for our description of electromagnetic anisotropy and bianisotropy is constructed in terms of the four macroscopic electromagnetic fields $\tilde{\underline{E}}(\underline{r},t)$, $\tilde{\underline{D}}(\underline{r},t)$, $\tilde{\underline{B}}(\underline{r},t)$ and $\tilde{\underline{H}}(\underline{r},t)$. These are piecewise differentiable vector functions of position $\underline{r}$ and time t which arise as spatial averages of microscopic fields and bound sources. The fields $\tilde{\underline{E}}(\underline{r},t)$ and $\tilde{\underline{B}}(\underline{r},t)$ are directly measurable quantities which produce the Lorentz force [2]

$$\tilde{\underline{F}}_{\rm Lor}(\underline{r},t) = q(\underline{r},t)\left[\tilde{\underline{E}}(\underline{r},t) + \underline{v}(\underline{r},t) \times \tilde{\underline{B}}(\underline{r},t)\right], \tag{1.7}$$

acting on a point charge $q(\underline{r},t)$ travelling at velocity $\underline{v}(\underline{r},t)$. Accordingly, $\tilde{\underline{E}}(\underline{r},t)$ and $\tilde{\underline{B}}(\underline{r},t)$ are viewed as the *primitive* fields. The fields $\tilde{\underline{D}}(\underline{r},t)$ and $\tilde{\underline{H}}(\underline{r},t)$ develop within a medium in response to the primitive fields; hence, they are considered as *induction* fields. Conventionally, $\tilde{\underline{E}}(\underline{r},t)$ and $\tilde{\underline{D}}(\underline{r},t)$ are called the electric field and the dielectric displacement, respectively. The conventional names for $\tilde{\underline{B}}(\underline{r},t)$ and $\tilde{\underline{H}}(\underline{r},t)$ — magnetic induction and magnetic field, respectively — are confusing and are avoided in this book.

The physical principles governing the behaviour of $\tilde{\underline{E}}(\underline{r},t)$, $\tilde{\underline{D}}(\underline{r},t)$, $\tilde{\underline{B}}(\underline{r},t)$ and $\tilde{\underline{H}}(\underline{r},t)$ are encapsulated by the Maxwell postulates, which — after combining Eqs. (1.4)–(1.6) — we write as two curl postulates

$$\left. \begin{array}{l} \nabla \times \tilde{\underline{H}}(\underline{r},t) - \dfrac{\partial}{\partial t}\tilde{\underline{D}}(\underline{r},t) = \tilde{\underline{J}}_e(\underline{r},t) \\[3mm] \nabla \times \tilde{\underline{E}}(\underline{r},t) + \dfrac{\partial}{\partial t}\tilde{\underline{B}}(\underline{r},t) = -\tilde{\underline{J}}_m(\underline{r},t) \end{array} \right\} \tag{1.8}$$

and two divergence postulates

$$\left.\begin{array}{l} \nabla \bullet \underline{\tilde{D}}(\underline{r},t) = \tilde{\rho}_{\mathrm{e}}(\underline{r},t) \\[2mm] \nabla \bullet \underline{\tilde{B}}(\underline{r},t) = \tilde{\rho}_{\mathrm{m}}(\underline{r},t) \end{array}\right\} . \tag{1.9}$$

The terms on the right sides of Eqs. (1.8) and (1.9) represent sources of fields. Whereas $\underline{\tilde{J}}_{\mathrm{e}}(\underline{r},t)$ and $\tilde{\rho}_{\mathrm{e}}(\underline{r},t)$ are the externally impressed electric current and electric charge densities, respectively, the magnetic current and magnetic charge densities — denoted by $\underline{\tilde{J}}_{\mathrm{m}}(\underline{r},t)$ and $\tilde{\rho}_{\mathrm{m}}(\underline{r},t)$ — do not represent physical quantities but are added for mathematical convenience [3]. In consonance with our macroscopic viewpoint, the source terms are also piecewise differentiable and satisfy the continuity relations

$$\left.\begin{array}{l} \nabla \bullet \underline{\tilde{J}}_{\mathrm{e}}(\underline{r},t) + \dfrac{\partial}{\partial t}\,\tilde{\rho}_{\mathrm{e}}(\underline{r},t) = 0 \\[4mm] \nabla \bullet \underline{\tilde{J}}_{\mathrm{m}}(\underline{r},t) + \dfrac{\partial}{\partial t}\,\tilde{\rho}_{\mathrm{m}}(\underline{r},t) = 0 \end{array}\right\} . \tag{1.10}$$

A redundancy is implicit in Eqs. (1.8)–(1.10), from the macroscopic viewpoint. The continuity relations (1.10), when combined with the Maxwell curl postulates (1.8), yield the Maxwell divergence postulates (1.9). Therefore, under the presumption of source continuity, there is no need for us to consider explicitly the divergence postulates (1.9).

1.2   Boundary conditions

The set of differential Eqs. (1.8) and (1.9) apply locally, at each position $\underline{r}$ and time t. Wherever the derivatives do not exist, boundary conditions and initial conditions must be specified in order to derive unique solutions. The boundary conditions may be established by recasting the Maxwell postulates (1.8) and (1.9) in integral forms, as follows [4]: Consider a region V of finite volume, enclosed by the surface ∂V, with $\underline{\hat{s}}(\underline{r},t)$ representing the outward–pointing unit vector normal to ∂V. The application of the vector identities

$$\left.\begin{array}{l} \displaystyle\int_{V} \nabla \bullet \underline{\tilde{A}}(\underline{r},t)\, d^{3}\underline{r} = \int_{\partial V} \underline{\hat{s}}(\underline{r},t) \bullet \underline{\tilde{A}}(\underline{r},t)\, d^{2}\underline{r} \\[5mm] \displaystyle\int_{V} \nabla \times \underline{\tilde{A}}(\underline{r},t)\, d^{3}\underline{r} = \int_{\partial V} \underline{\hat{s}}(\underline{r},t) \times \underline{\tilde{A}}(\underline{r},t)\, d^{2}\underline{r} \end{array}\right\} \tag{1.11}$$

to Eqs. (1.8) and (1.9) delivers

$$\left.\begin{array}{c} \int_{\partial V} \hat{\underline{s}}(\underline{r},t) \times \underline{\tilde{H}}(\underline{r},t)\, d^2\underline{r} - \int_V \dfrac{\partial}{\partial t}\, \underline{\tilde{D}}(\underline{r},t)\, d^3\underline{r} = \int_V \underline{\tilde{J}}_{\rm e}(\underline{r},t)\, d^3\underline{r} \\[2ex] \int_{\partial V} \hat{\underline{s}}(\underline{r},t) \times \underline{\tilde{E}}(\underline{r},t)\, d^2\underline{r} + \int_V \dfrac{\partial}{\partial t}\, \underline{\tilde{B}}(\underline{r},t)\, d^3\underline{r} = - \int_V \underline{\tilde{J}}_{\rm m}(\underline{r},t)\, d^3\underline{r} \end{array}\right\}$$
$$(1.12)$$

and

$$\left.\begin{array}{c} \int_{\partial V} \hat{\underline{s}}(\underline{r},t) \cdot \underline{\tilde{D}}(\underline{r},t)\, d^2\underline{r} = \int_V \tilde{\rho}_{\rm e}(\underline{r},t)\, d^3\underline{r} \\[2ex] \int_{\partial V} \hat{\underline{s}}(\underline{r},t) \cdot \underline{\tilde{B}}(\underline{r},t)\, d^2\underline{r} = \int_V \tilde{\rho}_{\rm m}(\underline{r},t)\, d^3\underline{r} \end{array}\right\},$$
$$(1.13)$$

respectively. If the region V moves at velocity $\underline{v}(\underline{r},t)$ then, by exploiting the vector identity

$$\frac{d}{dt}\int_V \underline{\tilde{A}}(\underline{r},t)\, d^3\underline{r} = \int_V \frac{\partial}{\partial t}\underline{\tilde{A}}(\underline{r},t)\, d^3\underline{r} + \int_{\partial V} [\hat{\underline{s}}(\underline{r},t) \cdot \underline{v}(\underline{r},t)]\, \underline{\tilde{A}}(\underline{r},t)\, d^2\underline{r},$$
$$(1.14)$$

Eqs. (1.12) may be expressed as

$$\left.\begin{array}{c} \int_{\partial V} \left\{ \hat{\underline{s}}(\underline{r},t) \times \underline{\tilde{H}}(\underline{r},t) + [\hat{\underline{s}}(\underline{r},t) \cdot \underline{v}(\underline{r},t)]\, \underline{\tilde{D}}(\underline{r},t) \right\}\, d^2\underline{r} \\[2ex] = \dfrac{d}{dt}\int_V \underline{\tilde{D}}(\underline{r},t)\, d^3\underline{r} + \int_V \underline{\tilde{J}}_{\rm e}(\underline{r},t)\, d^3\underline{r} \\[2ex] \int_{\partial V} \left\{ \hat{\underline{s}}(\underline{r},t) \times \underline{\tilde{E}}(\underline{r},t) - [\hat{\underline{s}}(\underline{r},t) \cdot \underline{v}(\underline{r},t)]\, \underline{\tilde{B}}(\underline{r},t) \right\}\, d^2\underline{r} \\[2ex] = -\dfrac{d}{dt}\int_V \underline{\tilde{B}}(\underline{r},t)\, d^3\underline{r} - \int_V \underline{\tilde{J}}_{\rm m}(\underline{r},t)\, d^3\underline{r} \end{array}\right\}.$$
$$(1.15)$$

Now suppose that the region V has the form of a pillbox of height $\delta_{\rm h}$, with its lower face lying in region I and its upper face lying in region II, as illustrated in Fig. 1.1. Thus, we have the partition $V = V_{\rm I} \cup V_{\rm II}$ where $V_{\rm I}$ lies only in region I and $V_{\rm II}$ lies only in region II. Let $\hat{\underline{s}}(\underline{r},t) = \hat{\underline{n}}(\underline{r},t)$ on the upper pillbox face. In the limit $\delta_{\rm h} \to 0$, the integral forms (1.13) and (1.15) yield the boundary conditions

$$\left.\begin{array}{c} \hat{\underline{n}}(\underline{r},t) \times \left[\underline{\tilde{H}}_{\rm I}(\underline{r},t) - \underline{\tilde{H}}_{\rm II}(\underline{r},t) \right] + [\hat{\underline{n}}(\underline{r},t) \cdot \underline{v}(\underline{r},t)] \left[\underline{\tilde{D}}_{\rm I}(\underline{r},t) - \underline{\tilde{D}}_{\rm II}(\underline{r},t) \right] \\[2ex] = \underline{\tilde{J}}^{\rm s}_{\rm e}(\underline{r},t) \\[2ex] \hat{\underline{n}}(\underline{r},t) \times \left[\underline{\tilde{E}}_{\rm I}(\underline{r},t) - \underline{\tilde{E}}_{\rm II}(\underline{r},t) \right] - [\hat{\underline{n}}(\underline{r},t) \cdot \underline{v}(\underline{r},t)] \left[\underline{\tilde{B}}_{\rm I}(\underline{r},t) - \underline{\tilde{B}}_{\rm II}(\underline{r},t) \right] \\[2ex] = -\underline{\tilde{J}}^{\rm s}_{\rm m}(\underline{r},t) \end{array}\right\}$$
$$(1.16)$$

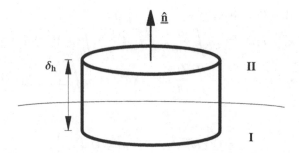

Figure 1.1   A pillbox of height δ_h, with its lower face lying in region I and its upper face lying in region II. The unit vector $\hat{\underline{n}}(\underline{r}, t)$ is normal to the upper face, pointing out of the pillbox.

and

$$\left.\begin{array}{l} \hat{\underline{n}}(\underline{r}, t) \bullet \left[\underline{\tilde{D}}_{\mathrm{I}}(\underline{r}, t) - \underline{\tilde{D}}_{\mathrm{II}}(\underline{r}, t) \right] = \tilde{\rho}_{\mathrm{e}}^{\mathrm{s}}(\underline{r}, t) \\[2mm] \hat{\underline{n}}(\underline{r}, t) \bullet \left[\underline{\tilde{B}}_{\mathrm{I}}(\underline{r}, t) - \underline{\tilde{B}}_{\mathrm{II}}(\underline{r}, t) \right] = \tilde{\rho}_{\mathrm{m}}^{\mathrm{s}}(\underline{r}, t) \end{array}\right\}, \qquad (1.17)$$

respectively. Herein,

$$\left.\begin{array}{l} \underline{\tilde{A}}_{\mathrm{I}}(\underline{r}, t) = \lim_{\delta_h \to 0} \left. \underline{\tilde{A}}(\underline{r}, t) \right|_{\underline{r} \in V_{\mathrm{I}}} \\[3mm] \underline{\tilde{A}}_{\mathrm{II}}(\underline{r}, t) = \lim_{\delta_h \to 0} \left. \underline{\tilde{A}}(\underline{r}, t) \right|_{\underline{r} \in V_{\mathrm{II}}} \end{array}\right\}, \qquad (A = E, B, D, H), \qquad (1.18)$$

while

$$\left.\begin{array}{l} \tilde{\rho}_{\ell}^{\mathrm{s}}(\underline{r}, t) = \lim_{\delta_h \to 0} \tilde{\rho}_{\ell}(\underline{r}, t)\, \delta_h \\[3mm] \underline{\tilde{J}}_{\ell}^{\mathrm{s}}(\underline{r}, t) = \lim_{\delta_h \to 0} \underline{\tilde{J}}_{\ell}(\underline{r}, t)\, \delta_h \end{array}\right\}, \qquad (\ell = \mathrm{e, m}). \qquad (1.19)$$

The surface current densities $\underline{\tilde{J}}_{\mathrm{e,m}}^{\mathrm{s}}(\underline{r}, t)$ and surface charge densities $\tilde{\rho}_{\mathrm{e,m}}^{\mathrm{s}}(\underline{r}, t)$, as defined in Eqs. (1.19), are nonzero only when the corresponding current densities and charge densities are infinite at the boundary between the regions V_{I} and V_{II}. Such an eventuality may arise at the surface of a perfect conductor, for example.

Equations (1.16) and (1.17) imply that, if the boundary moves parallel to the interface, then the boundary conditions are identical to those for a stationary boundary. Furthermore, in the absence of sources, the boundary conditions for a stationary boundary, for which $\hat{\underline{n}}(\underline{r}, t) \equiv \hat{\underline{n}}(\underline{r})$, are given as

$$\left.\begin{aligned} \hat{\underline{n}}(\underline{r}) \times \underline{\tilde{H}}_{\mathrm{I}}(\underline{r}, t) &= \hat{\underline{n}}(\underline{r}) \times \underline{\tilde{H}}_{\mathrm{II}}(\underline{r}, t) \\ \hat{\underline{n}}(\underline{r}) \times \underline{\tilde{E}}_{\mathrm{I}}(\underline{r}, t) &= \hat{\underline{n}}(\underline{r}) \times \underline{\tilde{E}}_{\mathrm{II}}(\underline{r}, t) \end{aligned}\right\} \tag{1.20}$$

and

$$\left.\begin{aligned} \hat{\underline{n}}(\underline{r}) \bullet \underline{\tilde{D}}_{\mathrm{I}}(\underline{r}, t) &= \hat{\underline{n}}(\underline{r}) \bullet \underline{\tilde{D}}_{\mathrm{II}}(\underline{r}, t) \\ \hat{\underline{n}}(\underline{r}) \bullet \underline{\tilde{B}}_{\mathrm{I}}(\underline{r}, t) &= \hat{\underline{n}}(\underline{r}) \bullet \underline{\tilde{B}}_{\mathrm{II}}(\underline{r}, t) \end{aligned}\right\}. \tag{1.21}$$

Thus, the tangential components of $\underline{\tilde{E}}(\underline{r}, t)$ and $\underline{\tilde{H}}(\underline{r}, t)$ are continuous across the boundary, as are the normal components of $\underline{\tilde{D}}(\underline{r}, t)$ and $\underline{\tilde{B}}(\underline{r}, t)$.

Finally, when electromagnetic fields are considered in unbounded mediums, boundary conditions at infinity must be specified in order to derive unique solutions to the Maxwell postulates. These boundary conditions are called *radiation conditions*; they require that at infinity [4]:

- field solutions attenuate no slower than $1/|\underline{r}|$, and
- energy flow is directed outwards.

1.3 Constitutive relations

The Maxwell curl postulates (1.8) represent a system of two linear vector differential equations in terms of the two primitive vector fields $\underline{\tilde{E}}(\underline{r}, t)$ and $\underline{\tilde{B}}(\underline{r}, t)$ and the two induction vector fields $\underline{\tilde{D}}(\underline{r}, t)$ and $\underline{\tilde{H}}(\underline{r}, t)$. In order to solve these differential equations, further information — in the form of constitutive relations relating the induction fields to the primitive fields — is needed. It is these constitutive relations which characterize the electromagnetic response of a medium. The constitutive relations may be naturally expressed in the general form

$$\left.\begin{aligned} \underline{\tilde{D}}(\underline{r}, t) &= \mathcal{F}\left\{\underline{\tilde{E}}(\underline{r}, t), \underline{\tilde{B}}(\underline{r}, t)\right\} \\ \underline{\tilde{H}}(\underline{r}, t) &= \mathcal{G}\left\{\underline{\tilde{E}}(\underline{r}, t), \underline{\tilde{B}}(\underline{r}, t)\right\} \end{aligned}\right\}, \tag{1.22}$$

wherein $\mathcal{F}$ and $\mathcal{G}$ are linear functions of $\underline{\tilde{E}}(\underline{r}, t)$ and $\underline{\tilde{B}}(\underline{r}, t)$ for linear mediums. The case where $\mathcal{F}$ and $\mathcal{G}$ are nonlinear functions of $\underline{\tilde{E}}(\underline{r}, t)$ and $\underline{\tilde{B}}(\underline{r}, t)$ — which is relevant for nonlinear mediums — is taken up in Chap. 7.

In general, the electromagnetic response of a medium is nonlocal with respect to both space and time. Thus, the constitutive relations of a linear medium should be stated as [5]

$$
\left.
\begin{aligned}
\underline{\tilde{D}}(\underline{r},t) &= \int_{t'}\int_{\underline{r}'}\left[\tilde{\underline{\underline{\epsilon}}}_{\text{EB}}(\underline{r}',t')\bullet\underline{\tilde{E}}(\underline{r}-\underline{r}',t-t')\right.\\
&\quad\left.+\tilde{\underline{\underline{\xi}}}_{\text{EB}}(\underline{r}',t')\bullet\underline{\tilde{B}}(\underline{r}-\underline{r}',t-t')\right]d^3\underline{r}'\,dt'\\
\underline{\tilde{H}}(\underline{r},t) &= \int_{t'}\int_{\underline{r}'}\left[\tilde{\underline{\underline{\zeta}}}_{\text{EB}}(\underline{r}',t')\bullet\underline{\tilde{E}}(\underline{r}-\underline{r}',t-t')\right.\\
&\quad\left.+\tilde{\underline{\underline{\nu}}}_{\text{EB}}(\underline{r}',t')\bullet\underline{\tilde{B}}(\underline{r}-\underline{r}',t-t')\right]d^3\underline{r}'\,dt'
\end{aligned}
\right\},
\tag{1.23}
$$

where $\tilde{\underline{\underline{\epsilon}}}_{\text{EB}}(\underline{r},t)$, $\tilde{\underline{\underline{\xi}}}_{\text{EB}}(\underline{r},t)$, $\tilde{\underline{\underline{\zeta}}}_{\text{EB}}(\underline{r},t)$ and $\tilde{\underline{\underline{\nu}}}_{\text{EB}}(\underline{r},t)$ are constitutive dyadics (i.e., second–rank Cartesian tensors) that can be interpreted as 3×3 matrixes. Appendix A provides a guide to dyadic notation and algebra.

Spatial nonlocality can play a significant role when the wavelength is comparable to some characteristic length–scale in the medium [6], but it is commonly neglected and lies outside the scope of our considerations here. In contrast, temporal nonlocality is almost always a matter of central importance, because of the high speeds of electromagnetic signals. Therefore, we focus on linear, spatially local, constitutive relations of the form

$$
\left.
\begin{aligned}
\underline{\tilde{D}}(\underline{r},t) &= \int_{t'}\left[\tilde{\underline{\underline{\epsilon}}}_{\text{EB}}(\underline{r},t')\bullet\underline{\tilde{E}}(\underline{r},t-t')+\tilde{\underline{\underline{\xi}}}_{\text{EB}}(\underline{r},t')\bullet\underline{\tilde{B}}(\underline{r},t-t')\right]dt'\\
\underline{\tilde{H}}(\underline{r},t) &= \int_{t'}\left[\tilde{\underline{\underline{\zeta}}}_{\text{EB}}(\underline{r},t')\bullet\underline{\tilde{E}}(\underline{r},t-t')+\tilde{\underline{\underline{\nu}}}_{\text{EB}}(\underline{r},t')\bullet\underline{\tilde{B}}(\underline{r},t-t')\right]dt'
\end{aligned}
\right\}.
\tag{1.24}
$$

1.4 The frequency domain

Mathematical complexities ensue from the convolution integrals appearing in the constitutive relations (1.24), when those relations are substituted into the Maxwell postulates. In order to circumvent these difficulties (often without loss of essential physics), it is common practice to introduce temporal Fourier transforms as

$$
\mathcal{Z}(\underline{r},\omega) = \int_{-\infty}^{\infty}\tilde{\mathcal{Z}}(\underline{r},t)\,\exp(i\omega t)\,dt\,,
\tag{1.25}
$$

with $\mathcal{Z}$ standing in for $\tilde{\underline{\underline{\epsilon}}}_{\text{EB}}, \tilde{\underline{\underline{\xi}}}_{\text{EB}}, \tilde{\underline{\underline{\zeta}}}_{\text{EB}}, \tilde{\underline{\underline{\nu}}}_{\text{EB}}, \underline{E}, \underline{D}, \underline{B}$ and $\underline{H}$, while ω is called the angular frequency and $i = \sqrt{-1}$. After taking the temporal Fourier transforms of Eqs. (1.24) and implementing the convolution theorem [7], the frequency–domain constitutive relations emerge as

$$\left.\begin{array}{l} \underline{D}(\underline{r},\omega) = \underline{\underline{\epsilon}}_{EB}(\underline{r},\omega) \bullet \underline{E}(\underline{r},\omega) + \underline{\underline{\xi}}_{EB}(\underline{r},\omega) \bullet \underline{B}(\underline{r},\omega) \\ \underline{H}(\underline{r},\omega) = \underline{\underline{\zeta}}_{EB}(\underline{r},\omega) \bullet \underline{E}(\underline{r},\omega) + \underline{\underline{\nu}}_{EB}(\underline{r},\omega) \bullet \underline{B}(\underline{r},\omega) \end{array}\right\}. \qquad (1.26)$$

Often in electromagnetic theory, $\tilde{\underline{E}}(\underline{r},\omega)$ is partnered with $\tilde{\underline{H}}(\underline{r},\omega)$ rather than $\tilde{\underline{B}}(\underline{r},\omega)$; for example, in the formulation of boundary conditions (see Sec. 1.2) and the definition of the time–averaged Poynting vector (cf. Eq. (1.90)) [8][2]. Consequently, frequency–domain constitutive relations may be conveniently expressed as

$$\left.\begin{array}{l} \underline{D}(\underline{r},\omega) = \underline{\underline{\epsilon}}_{EH}(\underline{r},\omega) \bullet \underline{E}(\underline{r},\omega) + \underline{\underline{\xi}}_{EH}(\underline{r},\omega) \bullet \underline{H}(\underline{r},\omega) \\ \underline{B}(\underline{r},\omega) = \underline{\underline{\zeta}}_{EH}(\underline{r},\omega) \bullet \underline{E}(\underline{r},\omega) + \underline{\underline{\mu}}_{EH}(\underline{r},\omega) \bullet \underline{H}(\underline{r},\omega) \end{array}\right\}. \qquad (1.27)$$

Herein $\underline{\underline{\epsilon}}_{EH}(\underline{r},\omega)$, $\underline{\underline{\xi}}_{EH}(\underline{r},\omega)$, $\underline{\underline{\zeta}}_{EH}(\underline{r},\omega)$ and $\underline{\underline{\mu}}_{EH}(\underline{r},\omega)$ are temporal Fourier transforms of $\tilde{\underline{\underline{\epsilon}}}_{EH}(\underline{r},t)$, $\tilde{\underline{\underline{\xi}}}_{EH}(\underline{r},t)$, $\tilde{\underline{\underline{\zeta}}}_{EH}(\underline{r},t)$ and $\tilde{\underline{\underline{\mu}}}_{EH}(\underline{r},t)$, respectively, defined as per Eq. (1.25). The names *Boys–Post* and *Tellegen* are often associated with the constitutive relations (1.26) and (1.27), respectively [9]. A one–to–one correspondence between the Boys–Post representation and the Tellegen representation is straightforwardly established via [5]

$$\left.\begin{array}{l} \underline{\underline{\epsilon}}_{EB}(\underline{r},\omega) = \underline{\underline{\epsilon}}_{EH}(\underline{r},\omega) - \underline{\underline{\xi}}_{EH}(\underline{r},\omega) \bullet \underline{\underline{\mu}}_{EH}^{-1}(\underline{r},\omega) \bullet \underline{\underline{\zeta}}_{EH}(\underline{r},\omega) \\ \underline{\underline{\xi}}_{EB}(\underline{r},\omega) = \underline{\underline{\xi}}_{EH}(\underline{r},\omega) \bullet \underline{\underline{\mu}}_{EH}^{-1}(\underline{r},\omega) \\ \underline{\underline{\zeta}}_{EB}(\underline{r},\omega) = -\underline{\underline{\mu}}_{EH}^{-1}(\underline{r},\omega) \bullet \underline{\underline{\zeta}}_{EH}(\underline{r},\omega) \\ \underline{\underline{\nu}}_{EB}(\underline{r},\omega) = \underline{\underline{\mu}}_{EH}^{-1}(\underline{r},\omega) \end{array}\right\} \qquad (1.28)$$

and

$$\left.\begin{array}{l} \underline{\underline{\epsilon}}_{EH}(\underline{r},\omega) = \underline{\underline{\epsilon}}_{EB}(\underline{r},\omega) - \underline{\underline{\xi}}_{EB}(\underline{r},\omega) \bullet \underline{\underline{\nu}}_{EB}^{-1}(\underline{r},\omega) \bullet \underline{\underline{\zeta}}_{EB}(\underline{r},\omega) \\ \underline{\underline{\xi}}_{EH}(\underline{r},\omega) = \underline{\underline{\xi}}_{EB}(\underline{r},\omega) \bullet \underline{\underline{\nu}}_{EB}^{-1}(\underline{r},\omega) \\ \underline{\underline{\zeta}}_{EH}(\underline{r},\omega) = -\underline{\underline{\nu}}_{EB}^{-1}(\underline{r},\omega) \bullet \underline{\underline{\zeta}}_{EB}(\underline{r},\omega) \\ \underline{\underline{\mu}}_{EH}(\underline{r},\omega) = \underline{\underline{\nu}}_{EB}^{-1}(\underline{r},\omega) \end{array}\right\} \qquad (1.29)$$

wherein the invertibility of $\underline{\underline{\nu}}_{EB}(\underline{r},\omega)$ and $\underline{\underline{\mu}}_{EH}(\underline{r},\omega)$ has been assumed[3]. The Tellegen representation is largely adopted in this book, but with occasional recourse to the Boys–Post representation where appropriate.

---

[2]A notable exception is provided by the Lorentz transformation of fields, wherein $\tilde{\underline{E}}(\underline{r},\omega)$ is partnered with $\tilde{\underline{B}}(\underline{r},\omega)$, as described in Sec. 1.6.3.

[3]The invertibility of constitutive dyadics is an assumption rather than a fact *a priori*. An example of a bianisotropic medium characterized by singular constitutive dyadics is presented in Sec. 2.3.3.

The corresponding frequency–domain Maxwell curl postulates arise as

$$\left.\begin{array}{l} \nabla \times \underline{H}(\underline{r},\omega) + i\omega \underline{D}(\underline{r},\omega) = \underline{J}_{e}(\underline{r},\omega) \\ \nabla \times \underline{E}(\underline{r},\omega) - i\omega \underline{B}(\underline{r},\omega) = -\underline{J}_{m}(\underline{r},\omega) \end{array}\right\}, \tag{1.30}$$

where the source terms $\underline{J}_{e,m}(\underline{r},\omega)$ are the temporal Fourier transforms of $\underline{\tilde{J}}_{e,m}(\underline{r},t)$, defined as in Eq. (1.25) with $\mathcal{Z} = \underline{J}_{e,m}$. The constitutive relations (1.27) — or equally (1.26) — together with the Maxwell curl postulates (1.30), form a self–consistent system into which anisotropy and bianisotropy are incorporated.

The boundary conditions, derived in Sec. 1.2 from the Maxwell postulates in the time domain, carry over to the frequency domain in a straightforward manner. Thus, the frequency–domain counterparts of Eqs. (1.20) and (1.21) are provided as

$$\left.\begin{array}{l} \hat{n}(\underline{r}) \times \underline{H}_{\mathrm{I}}(\underline{r},\omega) = \hat{n}(\underline{r}) \times \underline{H}_{\mathrm{II}}(\underline{r},\omega) \\ \hat{n}(\underline{r}) \times \underline{E}_{\mathrm{I}}(\underline{r},\omega) = \hat{n}(\underline{r}) \times \underline{E}_{\mathrm{II}}(\underline{r},\omega) \end{array}\right\} \tag{1.31}$$

and

$$\left.\begin{array}{l} \hat{n}(\underline{r}) \cdot \underline{D}_{\mathrm{I}}(\underline{r},\omega) = \hat{n}(\underline{r}) \cdot \underline{D}_{\mathrm{II}}(\underline{r},\omega) \\ \hat{n}(\underline{r}) \cdot \underline{B}_{\mathrm{I}}(\underline{r},\omega) = \hat{n}(\underline{r}) \cdot \underline{B}_{\mathrm{II}}(\underline{r},\omega) \end{array}\right\}, \tag{1.32}$$

respectively.

The mathematical simplicity of the frequency–domain formulation in relation to the time–domain formulation is gained at a cost in terms of physical interpretation. The frequency–dependent constitutive dyadics $\underline{\underline{\epsilon}}_{\mathrm{EB,EH}}(\underline{r},\omega)$, $\underline{\underline{\xi}}_{\mathrm{EB,EH}}(\underline{r},\omega)$, $\underline{\underline{\zeta}}_{\mathrm{EB,EH}}(\underline{r},\omega)$, $\underline{\underline{\nu}}_{\mathrm{EB}}(\underline{r},\omega)$ and $\underline{\underline{\mu}}_{\mathrm{EH}}(\underline{r},\omega)$ are complex–valued quantities, and so also are the frequency–dependent field phasors $\underline{E}(\underline{r},\omega)$, $\underline{D}(\underline{r},\omega)$, $\underline{B}(\underline{r},\omega)$, $\underline{H}(\underline{r},\omega)$ and $\underline{J}_{e,m}(\underline{r},\omega)$. The real–valued physical entities they represent surface only indirectly upon subjecting them to the inverse temporal Fourier transform. In this book, phasors are also called fields — in keeping with widespread usage.

Since the inverse temporal Fourier transform

$$\tilde{\mathcal{Z}}(\underline{r},t) = \frac{1}{2\pi} \int_{-\infty}^{\infty} \mathcal{Z}(\underline{r},\omega) \exp(-i\omega t)\, d\omega \tag{1.33}$$

is necessarily real–valued for $\mathcal{Z} \in \{\underline{\underline{\epsilon}}_{\mathrm{EB,EH}}, \underline{\underline{\xi}}_{\mathrm{EB,EH}}, \underline{\underline{\zeta}}_{\mathrm{EB,EH}}, \underline{\underline{\nu}}_{\mathrm{EB}}, \underline{\underline{\mu}}_{\mathrm{EH}}, \underline{E}, \underline{D},$ $\underline{B}, \underline{H}, \underline{J}_{e,m}\}$, the symmetry

$$\mathcal{Z}^{*}(\underline{r},\omega) = \mathcal{Z}(\underline{r},-\omega) \tag{1.34}$$

is imposed, where the superscript $*$ indicates the complex conjugate. Therefore, the frequency–domain quantities represented by $\mathscr{Z}(\underline{r}, \omega)$ are such that

$$
\left.
\begin{array}{l}
\mathrm{Re}\,\{\mathscr{Z}(\underline{r}, \omega)\} = \mathrm{Re}\,\{\mathscr{Z}(\underline{r}, -\omega)\} \\[2mm]
\mathrm{Im}\,\{\mathscr{Z}(\underline{r}, \omega)\} = -\mathrm{Im}\,\{\mathscr{Z}(\underline{r}, -\omega)\}
\end{array}
\right\}, \tag{1.35}
$$

with the operators $\mathrm{Re}\,\{\cdot\}$ and $\mathrm{Im}\,\{\cdot\}$ yielding the real and imaginary parts, respectively.

By virtue of the representation (1.33), the time–domain fields $\tilde{\underline{A}}(\underline{r}, t)$ $(A \in \{\, E,\, B,\, D,\, H,\, J_{\mathrm{e,m}}\})$ may be regarded as continuous sums of time–harmonic components, over all angular frequencies. This leads us to a useful concept in electromagnetics — closely allied to the frequency–domain representation — namely, the representation of monochromatic fields. A monochromatic field, which oscillates at single angular frequency $\omega = \omega_{\mathrm{s}}$, may be represented by the vector

$$
\tilde{\underline{A}}_{\mathrm{mono}}(\underline{r}, t) = \mathrm{Re}\,\left\{ \breve{\underline{A}}(\underline{r}, \omega_{\mathrm{s}})\,\exp\left(-i\omega_{\mathrm{s}} t\right)\right\}, \qquad (A = E, B, D, H, J_{\mathrm{e,m}}), \tag{1.36}
$$

where the amplitude $\breve{\underline{A}}(\underline{r}, \omega_{\mathrm{s}}) \in \mathbb{C}^3$ in general. Taking the temporal Fourier transform of $\tilde{\underline{A}}_{\mathrm{mono}}(\underline{r}, t)$ — which is written as $\underline{A}_{\mathrm{mono}}(\underline{r}, \omega)$, we find that the corresponding frequency–domain representation is

$$
\underline{A}_{\mathrm{mono}}(\underline{r}, \omega) = \frac{1}{2}\,\left[\breve{\underline{A}}(\underline{r}, \omega_{\mathrm{s}})\,\delta(\omega - \omega_{\mathrm{s}}) + \breve{\underline{A}}^{*}(\underline{r}, \omega_{\mathrm{s}})\,\delta(\omega + \omega_{\mathrm{s}})\right]. \tag{1.37}
$$

Notice that the phasor $\underline{A}_{\mathrm{mono}}(\underline{r}, \omega)$ in Eq. (1.37) satisfies the symmetry condition (1.34). The monochromatic field amplitudes $\breve{\underline{A}}(\underline{r}, \omega_{\mathrm{s}})$ satisfy the frequency–domain constitutive relations, Maxwell postulates and boundary conditions in exactly the same way as the frequency–domain phasors $\underline{A}_{\mathrm{mono}}(\underline{r}, \omega)$; that is, $\breve{\underline{A}}(\underline{r}, \omega_{\mathrm{s}})$ can take the place of $\underline{A}_{\mathrm{mono}}(\underline{r}, \omega)$ in Eqs. (1.26), (1.27), (1.30), (1.31) and (1.32) for $A \in \{\, E,\, B,\, D,\, H,\, J_{\mathrm{e,m}}\}$, and with $\omega = \omega_{\mathrm{s}}$ therein.

We close this section with a note of caution. The correspondence between the time and frequency domains may not always be one–to–one: if a time–domain function is not absolutely integrable over the real axis then its Fourier transform does not exist, and therefore the transformation to the frequency domain cannot take place [10]. Further complications can arise from the non–uniqueness of the inverse Fourier transform [11].

1.5   6–vector/6×6 dyadic notation

The use of a 6–vector/6×6 dyadic notation allows the Tellegen constitutive relations (1.27) to be expressed succinctly as

$$\underline{C}(\underline{r},\omega) = \underline{\underline{K}}_{EH}(\underline{r},\omega) \bullet \underline{F}(\underline{r},\omega), \tag{1.38}$$

with the 6–vectors

$$\underline{C}(\underline{r},\omega) = \begin{bmatrix} \underline{D}(\underline{r},\omega) \\ \underline{B}(\underline{r},\omega) \end{bmatrix} \tag{1.39}$$

and

$$\underline{F}(\underline{r},\omega) = \begin{bmatrix} \underline{E}(\underline{r},\omega) \\ \underline{H}(\underline{r},\omega) \end{bmatrix} \tag{1.40}$$

containing components of the electric and magnetic fields, while the 6×6 constitutive dyadic

$$\underline{\underline{K}}_{EH}(\underline{r},\omega) = \begin{bmatrix} \underline{\underline{\epsilon}}_{EH}(\underline{r},\omega) & \underline{\underline{\xi}}_{EH}(\underline{r},\omega) \\ \underline{\underline{\zeta}}_{EH}(\underline{r},\omega) & \underline{\underline{\mu}}_{EH}(\underline{r},\omega) \end{bmatrix}. \tag{1.41}$$

The result of combining the constitutive relations (1.27) with the Maxwell curl postulates (1.30) is thereby compactly expressed as

$$\left[\underline{\underline{L}}(\nabla) + i\omega \underline{\underline{K}}_{EH}(\underline{r},\omega) \right] \bullet \underline{F}(\underline{r},\omega) = \underline{Q}(\underline{r},\omega), \tag{1.42}$$

where the source 6–vector

$$\underline{Q}(\underline{r},\omega) = \begin{bmatrix} \underline{J}_e(\underline{r},\omega) \\ \underline{J}_m(\underline{r},\omega) \end{bmatrix} \tag{1.43}$$

and the linear differential operator

$$\underline{\underline{L}}(\nabla) = \begin{bmatrix} \underline{\underline{0}} & \nabla \times \underline{\underline{I}} \\ -\nabla \times \underline{\underline{I}} & \underline{\underline{0}} \end{bmatrix}, \tag{1.44}$$

with $\underline{\underline{0}}$ and $\underline{\underline{I}}$ being the null and identity 3×3 dyadics, respectively.

In a similar fashion, the four 3×3 dyadics $\underline{\underline{\epsilon}}_{EB}(\underline{r},\omega)$, $\underline{\underline{\xi}}_{EB}(\underline{r},\omega)$, $\underline{\underline{\zeta}}_{EB}(\underline{r},\omega)$ and $\underline{\underline{\nu}}_{EB}(\underline{r},\omega)$, which specify the constitutive properties in the Boys–Post representation (1.26) may be represented by the 6×6 constitutive dyadic

$$\underline{\underline{K}}_{EB}(\underline{r},\omega) = \begin{bmatrix} \underline{\underline{\epsilon}}_{EB}(\underline{r},\omega) & \underline{\underline{\xi}}_{EB}(\underline{r},\omega) \\ \underline{\underline{\zeta}}_{EB}(\underline{r},\omega) & \underline{\underline{\nu}}_{EB}(\underline{r},\omega) \end{bmatrix}. \tag{1.45}$$

The transformations (1.29) and (1.28) may then be expressed in terms of the invertible 6×6 dyadic operator $\underline{\underline{\tau}}$ which we define through the following relationships:[4]

$$
\left.
\begin{aligned}
\underline{\underline{\mathbf{K}}}_{EB}(\underline{r},\omega) &\equiv \underline{\underline{\tau}}\left\{\underline{\underline{\mathbf{K}}}_{EH}(\underline{r},\omega)\right\} \\
&= \begin{bmatrix} \underline{\underline{\epsilon}}_{EH} - \underline{\underline{\xi}}_{EH}\cdot\underline{\underline{\mu}}^{-1}_{EH}\cdot\underline{\underline{\zeta}}_{EH} & \underline{\underline{\xi}}_{EH}\cdot\underline{\underline{\mu}}^{-1}_{EH} \\ -\underline{\underline{\mu}}^{-1}_{EH}\cdot\underline{\underline{\zeta}}_{EH} & \underline{\underline{\mu}}^{-1}_{EH} \end{bmatrix} \\
\underline{\underline{\mathbf{K}}}_{EH}(\underline{r},\omega) &\equiv \underline{\underline{\tau}}^{-1}\left\{\underline{\underline{\mathbf{K}}}_{EB}(\underline{r},\omega)\right\} \\
&= \begin{bmatrix} \underline{\underline{\epsilon}}_{EB} - \underline{\underline{\xi}}_{EB}\cdot\underline{\underline{\nu}}^{-1}_{EB}(\cdot\underline{\underline{\zeta}}_{EB} & \underline{\underline{\xi}}_{EB}\cdot\underline{\underline{\nu}}^{-1}_{EB} \\ -\underline{\underline{\nu}}^{-1}_{EB}(\cdot\underline{\underline{\zeta}}_{EB} & \underline{\underline{\nu}}^{-1}_{EB} \end{bmatrix}
\end{aligned}
\right\}. \quad (1.46)
$$

1.6 Form invariances

Under certain linear transformations of coordinates and fields, the Maxwell postulates retain their form. In this section we describe spatial and temporal invariances as well as spatiotemporal covariance. While spatiotemporal covariance is of immense theoretical importance, invariances with respect to spatial and temporal transformations are commonly applied in many practical situations. Chiral invariance, which captures the nonuniqueness of the Maxwell postulates under linear field transformations, is discussed. A recently discovered invariance of the (frequency–domain) Maxwell postulates to a certain transformation involving complex conjugates is presented. And the implications of various transformations on electromagnetic energy and momentum are also outlined.

1.6.1 *Time reversal*

Let the operation of time reversal be denoted by $\mathcal{T}$, i.e., $\mathcal{T}\{t\} = -t$. Under the presumption that electric and magnetic source densities transform as [12]

$$
\left.
\begin{aligned}
\mathcal{T}\{\tilde{\rho}_e(\underline{r},t)\} &= \tilde{\rho}_e(\underline{r},-t) \\
\mathcal{T}\{\tilde{\rho}_m(\underline{r},t)\} &= -\tilde{\rho}_m(\underline{r},-t)
\end{aligned}
\right\}, \quad (1.47)
$$

---

[4]For compact presentation, the dependency of the 3×3 constitutive dyadics on $\underline{r}$ and ω is omitted from Eqs. (1.46).

the continuity relations (1.10) yield

$$\left.\begin{array}{l} \mathcal{T}\left\{\tilde{\underline{J}}_e(\underline{r},t)\right\} = -\tilde{\underline{J}}_e(\underline{r},-t) \\ \mathcal{T}\left\{\tilde{\underline{J}}_m(\underline{r},t)\right\} = \tilde{\underline{J}}_m(\underline{r},-t) \end{array}\right\}, \qquad (1.48)$$

and the electromagnetic fields are required to transform as

$$\left.\begin{array}{ll} \mathcal{T}\left\{\tilde{\underline{E}}(\underline{r},t)\right\} = \tilde{\underline{E}}(\underline{r},-t), & \mathcal{T}\left\{\tilde{\underline{D}}(\underline{r},t)\right\} = \tilde{\underline{D}}(\underline{r},-t) \\ \mathcal{T}\left\{\tilde{\underline{B}}(\underline{r},t)\right\} = -\tilde{\underline{B}}(\underline{r},-t), & \mathcal{T}\left\{\tilde{\underline{H}}(\underline{r},t)\right\} = -\tilde{\underline{H}}(\underline{r},-t) \end{array}\right\} \qquad (1.49)$$

in order to preserve the form of the Maxwell postulates. From the definition of the temporal Fourier transform (1.25), we see that the frequency–domain counterparts of Eqs. (1.47), (1.48) and (1.49) are as follows:

$$\left.\begin{array}{ll} \mathcal{T}\left\{\rho_e(\underline{r},\omega)\right\} = \rho_e^*(\underline{r},\omega), & \mathcal{T}\left\{\underline{J}_e(\underline{r},\omega)\right\} = -\underline{J}_e^*(\underline{r},\omega) \\ \mathcal{T}\left\{\rho_m(\underline{r},\omega)\right\} = -\rho_m^*(\underline{r},\omega), & \mathcal{T}\left\{\underline{J}_m(\underline{r},\omega)\right\} = \underline{J}_m^*(\underline{r},\omega) \\ \mathcal{T}\left\{\underline{E}(\underline{r},\omega)\right\} = \underline{E}^*(\underline{r},\omega), & \mathcal{T}\left\{\underline{D}(\underline{r},\omega)\right\} = \underline{D}^*(\underline{r},\omega) \\ \mathcal{T}\left\{\underline{B}(\underline{r},\omega)\right\} = -\underline{B}^*(\underline{r},\omega), & \mathcal{T}\left\{\underline{H}(\underline{r},\omega)\right\} = -\underline{H}^*(\underline{r},\omega) \end{array}\right\}. \qquad (1.50)$$

Therefore, under time reversal, the constitutive dyadics transform as

$$\left.\begin{array}{l} \mathcal{T}\left\{\underline{\underline{\epsilon}}_{\mathrm{EB,EH}}(\underline{r},\omega)\right\} = \underline{\underline{\epsilon}}_{\mathrm{EB,EH}}^*(\underline{r},\omega) \\[2mm] \mathcal{T}\left\{\underline{\underline{\xi}}_{\mathrm{EB,EH}}(\underline{r},\omega)\right\} = -\underline{\underline{\xi}}_{\mathrm{EB,EH}}^*(\underline{r},\omega) \\[2mm] \mathcal{T}\left\{\underline{\underline{\zeta}}_{\mathrm{EB,EH}}(\underline{r},\omega)\right\} = -\underline{\underline{\zeta}}_{\mathrm{EB,EH}}^*(\underline{r},\omega) \\[2mm] \mathcal{T}\left\{\underline{\underline{\nu}}_{\mathrm{EB}}(\underline{r},\omega)\right\} = \underline{\underline{\nu}}_{\mathrm{EB}}^*(\underline{r},\omega) \\[2mm] \mathcal{T}\left\{\underline{\underline{\mu}}_{\mathrm{EH}}(\underline{r},\omega)\right\} = \underline{\underline{\mu}}_{\mathrm{EH}}^*(\underline{r},\omega) \end{array}\right\}, \qquad (1.51)$$

by virtue of Eqs. (1.26) and (1.27). The time–reversal asymmetry which is exhibited by the magnetoelectric constitutive dyadics $\underline{\underline{\xi}}_{\mathrm{EB,EH}}(\underline{r},\omega)$ and $\underline{\underline{\zeta}}_{\mathrm{EB,EH}}(\underline{r},\omega)$ originates from irreversible physical processes, such as can develop through the application of quasistatic biasing fields or by means of relative motion [13]. This issue is enlarged upon in Chap. 2 in the context of Faraday chiral mediums and Lorentz–transformed constitutive dyadics.

1.6.2   *Spatial inversion*

We turn now to the inversion of space, denoted by the operator $\mathcal{P}$ as $\mathcal{P}\left\{\underline{r}\right\} = -\underline{r}$. Similarly to the time–reversal scenario presented in Sec. 1.6.1, if it is assumed that the electric and magnetic charge densities transform as [12]

$$\left. \begin{array}{l} \mathcal{P}\left\{\tilde{\rho}_{\mathrm{e}}(\underline{r},t)\right\} = \tilde{\rho}_{\mathrm{e}}(-\underline{r},t) \\ \mathcal{P}\left\{\tilde{\rho}_{\mathrm{m}}(\underline{r},t)\right\} = -\tilde{\rho}_{\mathrm{m}}(-\underline{r},t) \end{array} \right\}, \tag{1.52}$$

then, by virtue of the continuity relations (1.10), we have

$$\left. \begin{array}{l} \mathcal{P}\left\{\underline{\tilde{J}}_{\mathrm{e}}(\underline{r},t)\right\} = -\underline{\tilde{J}}_{\mathrm{e}}(-\underline{r},t) \\ \mathcal{P}\left\{\underline{\tilde{J}}_{\mathrm{m}}(\underline{r},t)\right\} = \underline{\tilde{J}}_{\mathrm{m}}(-\underline{r},t) \end{array} \right\}; \tag{1.53}$$

also, the form invariance of the Maxwell postulates enjoins the relationships

$$\left. \begin{array}{ll} \mathcal{P}\left\{\underline{\tilde{E}}(\underline{r},t)\right\} = -\underline{\tilde{E}}(-\underline{r},t), & \mathcal{P}\left\{\underline{\tilde{D}}(\underline{r},t)\right\} = -\underline{\tilde{D}}(-\underline{r},t) \\ \mathcal{P}\left\{\underline{\tilde{B}}(\underline{r},t)\right\} = \underline{\tilde{B}}(-\underline{r},t), & \mathcal{P}\left\{\underline{\tilde{H}}(\underline{r},t)\right\} = \underline{\tilde{H}}(-\underline{r},t) \end{array} \right\}. \tag{1.54}$$

Switching from the time domain to the frequency domain does not alter the action of the spatial–inversion operator $\mathcal{P}$ on the field quantities. Hence, from the constitutive relations (1.26) and (1.27) we find that

$$\left. \begin{array}{l} \mathcal{P}\left\{\underline{\underline{\epsilon}}_{\mathrm{EB,EH}}(\underline{r},\omega)\right\} = \underline{\underline{\epsilon}}_{\mathrm{EB,EH}}(-\underline{r},\omega) \\[4pt] \mathcal{P}\left\{\underline{\underline{\xi}}_{\mathrm{EB,EH}}(\underline{r},\omega)\right\} = -\underline{\underline{\xi}}_{\mathrm{EB,EH}}(-\underline{r},\omega) \\[4pt] \mathcal{P}\left\{\underline{\underline{\zeta}}_{\mathrm{EB,EH}}(\underline{r},\omega)\right\} = -\underline{\underline{\zeta}}_{\mathrm{EB,EH}}(-\underline{r},\omega) \\[4pt] \mathcal{P}\left\{\underline{\underline{\nu}}_{\mathrm{EB}}(\underline{r},\omega)\right\} = \underline{\underline{\nu}}_{\mathrm{EB}}(-\underline{r},\omega) \\[4pt] \mathcal{P}\left\{\underline{\underline{\mu}}_{\mathrm{EH}}(\underline{r},\omega)\right\} = \underline{\underline{\mu}}_{\mathrm{EH}}(-\underline{r},\omega) \end{array} \right\}. \tag{1.55}$$

1.6.3 *Lorentz covariance*

Suppose that an inertial reference frame Σ' moves with constant velocity $\underline{v} = v\hat{\underline{v}}$ with respect to an inertial reference frame Σ. The spacetime coordinates $(\underline{r}',t')$ in Σ' are related to the spacetime coordinates $(\underline{r},t)$ in Σ by Lorentz transformation [8]

$$\left. \begin{array}{l} \underline{r}' = \underline{\underline{Y}} \bullet \underline{r} - \gamma\,\underline{v}t \\[4pt] t' = \gamma\left(t - \dfrac{\underline{r} \bullet \underline{v}}{c_0^2}\right) \end{array} \right\}, \tag{1.56}$$

with

$$\left. \begin{array}{l} \underline{\underline{Y}} = \underline{\underline{I}} + (\gamma - 1)\,\hat{\underline{v}}\,\hat{\underline{v}} \\[4pt] \gamma = \left(1 - \beta^2\right)^{-\frac{1}{2}} \\[4pt] \beta = \dfrac{v}{c_0} \end{array} \right\}, \tag{1.57}$$

and $c_0 = (\epsilon_0 \mu_0)^{-1/2}$ being the speed of light in free space (i.e., vacuum[5]).

By application of the transformations (1.56), the time–domain fields $\tilde{E}(\underline{r}, t)$, $\tilde{B}(\underline{r}, t)$, $\tilde{D}(\underline{r}, t)$ and $\tilde{H}(\underline{r}, t)$ in the inertial reference frame Σ are found to be related to their counterparts in inertial reference frame Σ', namely $\tilde{E}'(\underline{r}', t')$, $\tilde{B}'(\underline{r}', t')$, $\tilde{D}'(\underline{r}', t')$ and $\tilde{H}'(\underline{r}', t')$ as

$$
\left.
\begin{aligned}
\tilde{E}'(\underline{r}', t') &= \gamma \left[\underline{\underline{Y}}^{-1} \cdot \tilde{E}(\underline{r}, t) + \underline{v} \times \tilde{B}(\underline{r}, t) \right] \\
\tilde{B}'(\underline{r}', t') &= \gamma \left[\underline{\underline{Y}}^{-1} \cdot \tilde{B}(\underline{r}, t) - \frac{1}{c_0^2} \underline{v} \times \tilde{E}(\underline{r}, t) \right] \\
\tilde{D}'(\underline{r}', t') &= \gamma \left[\underline{\underline{Y}}^{-1} \cdot \tilde{D}(\underline{r}, t) + \frac{1}{c_0^2} \underline{v} \times \tilde{H}(\underline{r}, t) \right] \\
\tilde{H}'(\underline{r}', t') &= \gamma \left[\underline{\underline{Y}}^{-1} \cdot \tilde{H}(\underline{r}, t) - \underline{v} \times \tilde{D}(\underline{r}, t) \right]
\end{aligned}
\right\}.
\tag{1.58}
$$

The Maxwell postulates are *Lorentz covariant*, which means that they retain their form under the spatiotemporal transformation (1.56). The Lorentz covariance of the Maxwell postulates has far–reaching implications for the constitutive relations that develop in uniformly moving reference frames, as described in Sec. 2.3.1.

1.6.4 *Chiral invariance*

In addition to being form–invariant under spatial, temporal and spatiotemporal transformations described in Secs. 1.6.1–1.6.3, the Maxwell postulates do not change their form under the following transformation of fields [3]

$$
\left.
\begin{aligned}
\mathcal{R}_\psi \{ \tilde{E}(\underline{r}, t) \} &= \tilde{E}(\underline{r}, t) \cos \psi - Z \tilde{H}(\underline{r}, t) \sin \psi \\
\mathcal{R}_\psi \{ \tilde{H}(\underline{r}, t) \} &= Z^{-1} \tilde{E}(\underline{r}, t) \sin \psi + \tilde{H}(\underline{r}, t) \cos \psi \\
\mathcal{R}_\psi \{ \tilde{B}(\underline{r}, t) \} &= \tilde{B}(\underline{r}, t) \cos \psi + Z \tilde{D}(\underline{r}, t) \sin \psi \\
\mathcal{R}_\psi \{ \tilde{D}(\underline{r}, t) \} &= -Z^{-1} \tilde{B}(\underline{r}, t) \sin \psi + \tilde{D}(\underline{r}, t) \cos \psi
\end{aligned}
\right\}
\tag{1.59}
$$

and source densities

$$
\left.
\begin{aligned}
\mathcal{R}_\psi \{ \tilde{\rho}_e(\underline{r}, t) \} &= \tilde{\rho}_e(\underline{r}, t) \cos \psi - Z^{-1} \tilde{\rho}_m(\underline{r}, t) \sin \psi \\
\mathcal{R}_\psi \{ \tilde{\rho}_m(\underline{r}, t) \} &= Z \tilde{\rho}_e(\underline{r}, t) \sin \psi + \tilde{\rho}_m(\underline{r}, t) \cos \psi \\
\mathcal{R}_\psi \{ \tilde{\underline{J}}_e(\underline{r}, t) \} &= \tilde{\underline{J}}_e(\underline{r}, t) \cos \psi - Z^{-1} \tilde{\underline{J}}_m(\underline{r}, t) \sin \psi \\
\mathcal{R}_\psi \{ \tilde{\underline{J}}_m(\underline{r}, t) \} &= Z \tilde{\underline{J}}_e(\underline{r}, t) \sin \psi + \tilde{\underline{J}}_m(\underline{r}, t) \cos \psi
\end{aligned}
\right\}.
\tag{1.60}
$$

---

[5]The classical electrodynamic approach is adopted here in which free space and vacuum are viewed as being equivalent. The nonclassical representation of vacuum is described in Sec. 7.4.

Herein the scalar Z is an impedance required to maintain dimensional integrity and ψ is a complex–valued angle. If $\psi \in \mathbb{R}$ (i.e., the set of real numbers), then the transformation operator $\mathcal{R}_\psi$ represents a rotation of the fields. For this reason, the Maxwell postulates are said to possess the property of *chiral invariance*.

The special case of $\psi = \pi/2$ is interesting: The electric and magnetic fields, and similarly the electric and magnetic charge densities, interchange under $\mathcal{R}_{\pi/2}$, which is often called the *duality transformation* [2]. By virtue of the duality of the electric charge and the magnetic charge, it is merely a matter of convention whether a particular particle is said to possess an electric charge or a magnetic charge.

Chiral invariance has an important bearing on the existence or nonexistence of magnetic monopoles. In fact, the question of the existence of magnetic monopoles is more fundamentally the question of whether all charged carriers possess the same proportion of electric charge and magnetic charge. If the answer to this question is in the affirmative, then — by applying a $\mathcal{R}_\psi$ transformation with the appropriate choice of ψ — either the magnetic monopole or the electric monopole could be said to not exist. Duality is best considered globally; i.e., for all mediums, at all times, and everywhere. Accordingly, the appropriate choice of ψ is made for physical certainty; however, that choice does not preclude the later application of duality in a local context for mathematical convenience.

The constitutive relations (1.24) retain their form under the transformation of fields (1.59) provided that the constitutive dyadics transform as

$$
\left.
\begin{aligned}
\mathcal{R}_\psi\left\{\tilde{\underline{\underline{\epsilon}}}_{\text{EB}}(\underline{r},t)\right\} &= \cos^2\psi\,\tilde{\underline{\underline{\epsilon}}}_{\text{EB}}(\underline{r},t) + Z^{-2}\sin^2\psi\,\tilde{\underline{\underline{\mu}}}_{\text{EB}}(\underline{r},t) \\
&\quad - Z^{-1}\sin\psi\cos\psi\left(\tilde{\underline{\underline{\xi}}}_{\text{EB}}(\underline{r},t) + \tilde{\underline{\underline{\zeta}}}_{\text{EB}}(\underline{r},t)\right) \\[6pt]
\mathcal{R}_\psi\left\{\tilde{\underline{\underline{\xi}}}_{\text{EB}}(\underline{r},t)\right\} &= \sin\psi\cos\psi\left(Z\,\tilde{\underline{\underline{\epsilon}}}_{\text{EB}}(\underline{r},t) - Z^{-1}\,\tilde{\underline{\underline{\mu}}}_{\text{EB}}(\underline{r},t)\right) \\
&\quad + \cos^2\psi\,\tilde{\underline{\underline{\xi}}}_{\text{EB}}(\underline{r},t) - \sin^2\psi\,\tilde{\underline{\underline{\zeta}}}_{\text{EB}}(\underline{r},t) \\[6pt]
\mathcal{R}_\psi\left\{\tilde{\underline{\underline{\zeta}}}_{\text{EB}}(\underline{r},t)\right\} &= \sin\psi\cos\psi\left(Z\,\tilde{\underline{\underline{\epsilon}}}_{\text{EB}}(\underline{r},t) - Z^{-1}\,\tilde{\underline{\underline{\mu}}}_{\text{EB}}(\underline{r},t)\right) \\
&\quad + \cos^2\psi\,\tilde{\underline{\underline{\zeta}}}_{\text{EB}}(\underline{r},t) - \sin^2\psi\,\tilde{\underline{\underline{\xi}}}_{\text{EB}}(\underline{r},t) \\[6pt]
\mathcal{R}_\psi\left\{\tilde{\underline{\underline{\mu}}}_{\text{EB}}(\underline{r},t)\right\} &= Z^2\sin^2\psi\,\tilde{\underline{\underline{\epsilon}}}_{\text{EB}}(\underline{r},t) + \cos^2\psi\,\tilde{\underline{\underline{\mu}}}_{\text{EB}}(\underline{r},t) \\
&\quad + Z\sin\psi\cos\psi\left(\tilde{\underline{\underline{\xi}}}_{\text{EB}}(\underline{r},t) + \tilde{\underline{\underline{\zeta}}}_{\text{EB}}(\underline{r},t)\right)
\end{aligned}
\right\} . \quad (1.61)
$$

1.6.5   *Conjugate invariance*

The frequency–domain Maxwell curl postulates (1.30) are invariant under a further transformation, namely the conjugate transformation which is effected by the operator $\mathcal{C}$. The conjugate–transformed fields are specified as [14]

$$\left.\begin{aligned}
\mathcal{C}\left\{\underline{E}(\underline{r},\omega)\right\} &= \underline{E}^*(\underline{r},\omega) \\
\mathcal{C}\left\{\underline{H}(\underline{r},\omega)\right\} &= \underline{H}^*(\underline{r},\omega) \\
\mathcal{C}\left\{\underline{D}(\underline{r},\omega)\right\} &= -\underline{D}^*(\underline{r},\omega) \\
\mathcal{C}\left\{\underline{B}(\underline{r},\omega)\right\} &= -\underline{B}^*(\underline{r},\omega)
\end{aligned}\right\}, \qquad (1.62)$$

while the source densities transform as

$$\left.\begin{aligned}
\mathcal{C}\left\{\rho_e(\underline{r},\omega)\right\} &= -\rho_e^*(\underline{r},\omega) \\
\mathcal{C}\left\{\rho_m(\underline{r},\omega)\right\} &= -\rho_m^*(\underline{r},\omega) \\
\mathcal{C}\left\{\underline{J}_e(\underline{r},\omega)\right\} &= \underline{J}_e^*(\underline{r},\omega) \\
\mathcal{C}\left\{\underline{J}_m(\underline{r},\omega)\right\} &= \underline{J}_m^*(\underline{r},\omega)
\end{aligned}\right\}. \qquad (1.63)$$

When applied to linear materials, the Maxwell postulates remain invariant, provided that the 3×3 constitutive dyadics undergo the following transformations:

$$\left.\begin{aligned}
\mathcal{C}\left\{\underline{\underline{\epsilon}}_{EB}(\underline{r},\omega)\right\} &= -\underline{\underline{\epsilon}}_{EB}^*(\underline{r},\omega) \\
\mathcal{C}\left\{\underline{\underline{\xi}}_{EB}(\underline{r},\omega)\right\} &= \underline{\underline{\xi}}_{EB}^*(\underline{r},\omega) \\
\mathcal{C}\left\{\underline{\underline{\zeta}}_{EB}(\underline{r},\omega)\right\} &= \underline{\underline{\zeta}}_{EB}^*(\underline{r},\omega) \\
\mathcal{C}\left\{\underline{\underline{\nu}}_{EB}(\underline{r},\omega)\right\} &= -\underline{\underline{\nu}}_{EB}^*(\underline{r},\omega)
\end{aligned}\right\} \qquad (1.64)$$

and

$$\left.\begin{aligned}
\mathcal{C}\left\{\underline{\underline{\epsilon}}_{EH}(\underline{r},\omega)\right\} &= -\underline{\underline{\epsilon}}_{EH}^*(\underline{r},\omega) \\
\mathcal{C}\left\{\underline{\underline{\xi}}_{EH}(\underline{r},\omega)\right\} &= -\underline{\underline{\xi}}_{EH}^*(\underline{r},\omega) \\
\mathcal{C}\left\{\underline{\underline{\zeta}}_{EH}(\underline{r},\omega)\right\} &= -\underline{\underline{\zeta}}_{EH}^*(\underline{r},\omega) \\
\mathcal{C}\left\{\underline{\underline{\mu}}_{EH}(\underline{r},\omega)\right\} &= -\underline{\underline{\mu}}_{EH}^*(\underline{r},\omega)
\end{aligned}\right\}. \qquad (1.65)$$

The conjugation symmetry represented in Eqs. (1.62)–(1.65) arises as a generalization of the transformation which reverses the sign of the real–valued permittivity and permeability scalars for isotropic dielectric–magnetic mediums [14]. The effect of the conjugate transformation (1.64) would be observable in, for example, planewave propagation through a material slab, and may be fruitfully applied in determining the reflection and transmission characteristics of isotropic dielectric–magnetic materials whose permittivity and permeability scalars have negative real parts. Such materials are considered to be of technological promise because they may support negative refraction, as discussed in Secs. 2.1.2 and 4.8.1 [14].

1.6.6 *Energy and momentum*

We now turn to more practical matters by considering how spatial, temporal and field transformations influence measurable quantities. Let us introduce the energy flow density, as given by the instantaneous Poynting vector

$$\underline{\tilde{S}}(\underline{r},t) = \underline{\tilde{E}}(\underline{r},t) \times \underline{\tilde{H}}(\underline{r},t); \tag{1.66}$$

the total energy density

$$\tilde{W}(\underline{r},t) = \frac{1}{2}\left[\underline{\tilde{D}}(\underline{r},t)\bullet\underline{\tilde{E}}(\underline{r},t) + \underline{\tilde{B}}(\underline{r},t)\bullet\underline{\tilde{H}}(\underline{r},t)\right]; \tag{1.67}$$

and the Maxwell stress tensor

$$\underline{\underline{\tilde{T}}}(\underline{r},t) = -\frac{1}{2}\left[\underline{\tilde{D}}(\underline{r},t)\bullet\underline{\tilde{E}}(\underline{r},t) + \underline{\tilde{B}}(\underline{r},t)\bullet\underline{\tilde{H}}(\underline{r},t)\right]\underline{\underline{I}}$$
$$+\underline{\tilde{D}}(\underline{r},t)\underline{\tilde{E}}(\underline{r},t) + \underline{\tilde{B}}(\underline{r},t)\underline{\tilde{H}}(\underline{r},t). \tag{1.68}$$

A straightforward application of the field transformations (1.49) and (1.54), respectively, reveals that

$$\left.\begin{array}{l} \mathcal{T}\left\{\underline{\tilde{S}}(\underline{r},t)\right\} = -\underline{\tilde{S}}(\underline{r},-t) \\[2mm] \mathcal{T}\left\{\tilde{W}(\underline{r},t)\right\} = \tilde{W}(\underline{r},-t) \\[2mm] \mathcal{T}\left\{\underline{\underline{\tilde{T}}}(\underline{r},t)\right\} = \underline{\underline{\tilde{T}}}(\underline{r},-t) \end{array}\right\} \tag{1.69}$$

and

$$\left.\begin{array}{l} \mathcal{P}\left\{\underline{\tilde{S}}(\underline{r},t)\right\} = -\underline{\tilde{S}}(-\underline{r},t) \\[2mm] \mathcal{P}\left\{\tilde{W}(\underline{r},t)\right\} = \tilde{W}(-\underline{r},t) \\[2mm] \mathcal{P}\left\{\underline{\underline{\tilde{T}}}(\underline{r},t)\right\} = \underline{\underline{\tilde{T}}}(-\underline{r},t) \end{array}\right\}. \tag{1.70}$$

The chiral invariance of the Maxwell postulates carries over to measurable quantities. It follows immediately from Eqs. (1.59) that

$$\left.\begin{array}{l} \mathcal{R}_\psi\left\{\underline{\tilde{S}}(\underline{r},t)\right\} = \underline{\tilde{S}}(\underline{r},t) \\[2mm] \mathcal{R}_\psi\left\{\tilde{W}(\underline{r},t)\right\} = \tilde{W}(\underline{r},t) \\[2mm] \mathcal{R}_\psi\left\{\underline{\underline{\tilde{T}}}(\underline{r},t)\right\} = \underline{\underline{\tilde{T}}}(\underline{r},t) \end{array}\right\}. \tag{1.71}$$

A notable consequence is that electromagnetic fields cannot be thus uniquely determined from measurements of electromagnetic energy and/or momentum.

1.7   Constitutive dyadics

Let us now examine more closely the constitutive dyadics which characterize the electromagnetic response of a medium. In the most general linear scenario, the 6×6 constitutive dyadic $\underline{\underline{\mathbf{K}}}_{EH}(\underline{r}, \omega)$ is assembled from 36 complex–valued scalar parameters. This vast parameter space may be reduced through the imposition of physical constraints which require the constitutive dyadics to satisfy certain symmetries. Also, our attention is often restricted to special cases and idealizations which manifest as symmetries of the constitutive dyadics.

Note that the constitutive dyadics of *homogeneous* mediums are not functions of $\underline{r}$.

1.7.1   *Constraints*

1.7.1.1   *Causality and Kramers–Kronig relations*

The formulations of constitutive relations for any realistic material must conform to the principle of causality; i.e., 'effect' must appear *after* the 'cause'. Hence, neither can a cause and its effect be simultaneous nor can an effect precede its cause. The principle of causality is most transparently implemented in the time domain for constitutive relations of the form given in Eqs. (1.22).

The induced fields $\underline{\tilde{D}}(\underline{r}, t)$ and $\underline{\tilde{H}}(\underline{r}, t)$ develop in response to the primitive fields $\underline{\tilde{E}}(\underline{r}, t)$ and $\underline{\tilde{B}}(\underline{r}, t)$, such that

$$\left.\begin{aligned}\underline{\tilde{D}}(\underline{r}, t) &= \epsilon_0 \underline{\tilde{E}}(\underline{r}, t) + \underline{\tilde{P}}(\underline{r}, t) \\ \underline{\tilde{H}}(\underline{r}, t) &= \frac{1}{\mu_0}\underline{\tilde{B}}(\underline{r}, t) - \underline{\tilde{M}}(\underline{r}, t)\end{aligned}\right\}. \tag{1.72}$$

The polarization $\underline{\tilde{P}}(\underline{r}, t)$ and the magnetization $\underline{\tilde{M}}(\underline{r}, t)$ indicate the electromagnetic response of a medium, and must therefore be causally connected to the primitive fields.[6]

---

[6]For classical vacuum — which is not a material medium — the polarization and magnetization vectors are null–valued. In this case there is no distinction between primitive fields and induction fields. For a description of the nonclassical vacuum, see Sec. 7.4.

With regard to the time–domain linear constitutive relations (1.24), causality dictates that

$$\left.\begin{array}{l} \tilde{\underline{\underline{\epsilon}}}_{EB}(\underline{r},t) - \epsilon_0\delta(\underline{r})\underline{\underline{I}} \equiv \underline{\underline{0}} \\[6pt] \tilde{\underline{\underline{\xi}}}_{EB}(\underline{r},t) \equiv \underline{\underline{0}} \\[6pt] \tilde{\underline{\underline{\zeta}}}_{EB}(\underline{r},t) \equiv \underline{\underline{0}} \\[6pt] \mu_0^{-1}\delta(\underline{r})\underline{\underline{I}} - \tilde{\underline{\underline{\nu}}}_{EB}(\underline{r},t) \equiv \underline{\underline{0}} \end{array}\right\} \quad \text{for} \quad t \le 0. \qquad (1.73)$$

When translated into the frequency domain, the causality requirement (1.73) gives rise to integral relations between the real and imaginary parts of the frequency–dependent constitutive parameters, as we now outline.

Suppose that the scalar function $\tilde{f}(\underline{r},t)$ represents an arbitrary component of a Boys–Post constitutive dyadic; i.e., $\tilde{f}(\underline{r},t)$ is a component of $\tilde{\underline{\underline{\epsilon}}}_{EB}(\underline{r},t) - \epsilon_0\delta(\underline{r})\underline{\underline{I}}, \tilde{\underline{\underline{\xi}}}_{EB}(\underline{r},t), \tilde{\underline{\underline{\zeta}}}_{EB}(\underline{r},t)$ or $\mu_0^{-1}\delta(\underline{r})\underline{\underline{I}} - \tilde{\underline{\underline{\nu}}}_{EB}(\underline{r},t)$. The temporal Fourier transform of $\tilde{f}(\underline{r},t)$ may be expressed as

$$f(\underline{r},\omega) = \int_0^\infty \tilde{f}(\underline{r},t)\,\exp(i\omega t)\,dt, \qquad (1.74)$$

wherein the causality constraint (1.73) has been applied to set the lower limit of integration equal to zero. The analytic continuation of $f(\underline{r},\omega)$ in the upper complex–ω plane is provided by the Cauchy integral formula

$$f(\underline{r},\omega) = \frac{1}{2\pi i}\oint \frac{f(\underline{r},s)}{s-\omega}\,ds, \qquad (1.75)$$

where the integration contour extends around the upper half plane. The integrand in Eq. (1.75) vanishes as $|s| \to \infty$ for $\text{Im}\,\{s\} > 0$ due to the $\exp(i\omega t)$ factor occurring in the integral representation (1.74). Hence, the contour integral specified in Eq. (1.75) reduces to an integral along the real axis. Counting the single pole on the real axis at $\omega = s$ as a half residue, we have

$$f(\underline{r},\omega) = \frac{1}{\pi i}\text{P}\int_{-\infty}^\infty \frac{f(\underline{r},s)}{s-\omega}\,ds, \qquad (1.76)$$

where P indicates the Cauchy principal value. Hence, we have the Hilbert transforms

$$\left.\begin{array}{l} \text{Re}\,\{f(\underline{r},\omega)\} = \dfrac{1}{\pi}\text{P}\displaystyle\int_{-\infty}^\infty \dfrac{\text{Im}\,\{f(\underline{r},s)\}}{s-\omega}\,ds \\[14pt] \text{Im}\,\{f(\underline{r},\omega)\} = -\dfrac{1}{\pi}\text{P}\displaystyle\int_{-\infty}^\infty \dfrac{\text{Re}\,\{f(\underline{r},s)\}}{s-\omega}\,ds \end{array}\right\}. \qquad (1.77)$$

In addition, since $\tilde{f}(\underline{r}, t)$ is real–valued, the symmetry condition (cf. Eq. (1.34))

$$f(\underline{r}, -\omega) = f^*(\underline{r}, \omega) \tag{1.78}$$

relates f to its complex conjugate f^*. Thus, Eqs. (1.77) yield the *Kramers– Kronig* relations [13]

$$\left. \begin{aligned} \text{Re}\left\{f(\underline{r}, \omega)\right\} &= \frac{2}{\pi} \text{P} \int_0^\infty \frac{s\,\text{Im}\left\{f(\underline{r}, s)\right\}}{s^2 - \omega^2}\, ds \\ \text{Im}\left\{f(\underline{r}, \omega)\right\} &= -\frac{2}{\pi} \text{P} \int_0^\infty \frac{\omega\,\text{Re}\left\{f(\underline{r}, s)\right\}}{s^2 - \omega^2}\, ds \end{aligned} \right\}. \tag{1.79}$$

Although the relations (1.79) are presented here for components of the Boys–Post constitutive dyadics, analogous relations hold for components of the Tellegen constitutive dyadics by virtue of Eqs. (1.29).

An alternative approach to the derivation of the Kramers–Kronig relations, exploiting the properties of Herglotz functions, has recently been reported [15].

The Kramers–Kronig relations represent a particular example of *dispersion relations*[7] that apply generally to frequency–dependent, causal, linear systems [16]. Often these are usefully employed in experimental determinations of constitutive parameters [17].

1.7.1.2   *Post constraint*

A structural constraint — called the Post constraint — is available for those linear mediums that exhibit magnetoelectric coupling [13]. This constraint may be expressed in terms of Boys–Post constitutive dyadics as

$$\text{tr}\left[\underline{\underline{\zeta}}_{\text{EB}}(\underline{r}, \omega) - \underline{\underline{\xi}}_{\text{EB}}(\underline{r}, \omega)\right] = 0, \tag{1.80}$$

or, equivalently, in terms of Tellegen constitutive dyadics as

$$\text{tr}\left\{\underline{\underline{\mu}}_{\text{EH}}^{-1}(\underline{r}, \omega) \bullet \left[\underline{\underline{\zeta}}_{\text{EH}}(\underline{r}, \omega) + \underline{\underline{\xi}}_{\text{EH}}(\underline{r}, \omega)\right]\right\} = 0. \tag{1.81}$$

Hence, under the Post constraint, only 35 independent complex–valued parameters are needed to characterize the most general linear medium.

The origins of the Post constraint lie in the microscopic nature of the primitive electromagnetic fields and the Lorentz covariance of the Maxwell equations [18]. While Post established his eponymous constraint more than

---

[7]These causal dispersion relations should be distinguished from the planewave dispersion relations described in Sec. 4.2.

40 years ago [13], two more recent, independent, proofs — one based on a uniqueness requirement [19, 20] and another based on multipole considerations [21] — further secured the standing of the Post constraint. On the other hand, recent experimental evidence that the Post constraint is violated at low frequencies has been reported [22], but there is no microscopic understanding as yet of this evidence [23]. The incorporation of the hitherto–undiscovered axion will lead to a re–evaluation of the Post constraint even for free space [24, 25].

1.7.1.3 *Onsager relations*

The Onsager relations are a set of reciprocity relations that are applicable generally to coupled linear phenomenons at macroscopic length–scales [26–28]. While the Onsager relations were originally established for instantaneous phenomenons, their scope may be extended by means of the fluctuation–dissipation theorem [29] to include time–harmonic phenomenons too [30].

The assumption of microscopic reversibility is central to the Onsager relations. As a consequence, in order to apply the Onsager relations to electromagnetic constitutive relations, the contribution of free space must be excluded because microscopic processes cannot occur in free space. The frequency–dependent vector quantities $\underline{P}(\underline{r}, \omega)$ and $\underline{M}(\underline{r}, \omega)$, which are the temporal Fourier transforms of the polarization $\underline{\tilde{P}}(\underline{r}, t)$ and the magnetization $\underline{\tilde{M}}(\underline{r}, t)$, represent the electromagnetic response of a medium relative to the electromagnetic response of free space. For linear homogeneous mediums, the bianisotropic constitutive relations (1.26) reduce to

$$\left. \begin{aligned} \underline{P}(\underline{r}, \omega) &= \left[\underline{\underline{\epsilon}}_{EB}(\omega) - \epsilon_0 \underline{\underline{I}} \right] \bullet \underline{E}(\underline{r}, \omega) + \underline{\underline{\xi}}_{EB}(\omega) \bullet \underline{B}(\underline{r}, \omega) \\ \underline{M}(\underline{r}, \omega) &= -\underline{\underline{\zeta}}_{EB}(\omega) \bullet \underline{E}(\underline{r}, \omega) + \left[\frac{1}{\mu_0} \underline{\underline{I}} - \underline{\underline{\nu}}_{EB}(\omega) \right] \bullet \underline{B}(\underline{r}, \omega) \end{aligned} \right\}. \quad (1.82)$$

When a linear homogeneous medium is subjected to an external, spatially uniform, magnetostatic field $\underline{B}_{dc}$, application of the Onsager relations to the Boys–Post constitutive relations (1.82) delivers the constraints [31]

$$\left. \begin{aligned} \underline{\underline{\epsilon}}_{EB}(\omega) \Big|_{\underline{B}_{dc}} &= \underline{\underline{\epsilon}}_{EB}^{T}(\omega) \Big|_{-\underline{B}_{dc}} \\ \underline{\underline{\xi}}_{EB}(\omega) \Big|_{\underline{B}_{dc}} &= \underline{\underline{\zeta}}_{EB}^{T}(\omega) \Big|_{-\underline{B}_{dc}} \\ \underline{\underline{\nu}}_{EB}(\omega) \Big|_{\underline{B}_{dc}} &= \underline{\underline{\nu}}_{EB}^{T}(\omega) \Big|_{-\underline{B}_{dc}} \end{aligned} \right\}, \quad (1.83)$$

where the superscript 'T' denotes the transpose operation. The equivalent contraints for the Tellegen constitutive dyadics follow straight from Eqs. (1.28) as

$$
\left.\begin{aligned}
\underline{\underline{\epsilon}}_{\text{EH}}(\omega)\Big|_{\underline{B}_{\text{dc}}} &= \underline{\underline{\epsilon}}^{\text{T}}_{\text{EH}}(\omega)\Big|_{-\underline{B}_{\text{dc}}} \\[4pt]
\underline{\underline{\xi}}_{\text{EH}}(\omega)\Big|_{\underline{B}_{\text{dc}}} &= -\underline{\underline{\zeta}}^{\text{T}}_{\text{EH}}(\omega)\Big|_{-\underline{B}_{\text{dc}}} \\[4pt]
\underline{\underline{\mu}}_{\text{EH}}(\omega)\Big|_{\underline{B}_{\text{dc}}} &= \underline{\underline{\mu}}^{\text{T}}_{\text{EH}}(\omega)\Big|_{-\underline{B}_{\text{dc}}}
\end{aligned}\right\}.
\qquad (1.84)
$$

1.7.2 *Specializations*

1.7.2.1 *Lorentz reciprocity*

Lorentz reciprocity – which is closely related to the topics of time reversal and the Onsager relations — is a topic that frequently crops up in theoretical analyses involving complex mediums [32]. It is often presented in terms of the interchangeability of transmitters and receivers [4, 33].

Let us consider two frequency–domain electric source current densities, namely $\underline{J}^{\text{p}}_{\text{e}}(\underline{r}, \omega)$ and $\underline{J}^{\text{q}}_{\text{e}}(\underline{r}, \omega)$, and two frequency–domain magnetic source current densities, namely $\underline{J}^{\text{p}}_{\text{m}}(\underline{r}, \omega)$ and $\underline{J}^{\text{q}}_{\text{m}}(\underline{r}, \omega)$. The sources labelled 'p' generate fields denoted as $\underline{E}^{\text{p}}(\underline{r}, \omega)$ and $\underline{H}^{\text{p}}(\underline{r}, \omega)$, whereas the sources labelled 'q' generate fields denoted as $\underline{E}^{\text{q}}(\underline{r}, \omega)$ and $\underline{H}^{\text{q}}(\underline{r}, \omega)$. The interaction of the 'p' sources with the fields generated by the 'q' sources is gauged by the *reaction* [4, 33]

$$
\langle\langle \text{p, q} \rangle\rangle = \int_{V_{\text{p}}} \left[\underline{J}^{\text{p}}_{\text{e}}(\underline{r}, \omega) \bullet \underline{E}^{\text{q}}(\underline{r}, \omega) - \underline{J}^{\text{p}}_{\text{m}}(\underline{r}, \omega) \bullet \underline{H}^{\text{q}}(\underline{r}, \omega) \right] d^3\underline{r}, \qquad (1.85)
$$

where the integration region V_{p} contains the 'p' sources. Similarly, the interaction of the 'q' sources with field generated by the 'p' sources is represented by the reaction $\langle\langle \text{q, p} \rangle\rangle$. If the medium which supports $\underline{J}^{\text{p,q}}_{\text{e,m}}(\underline{r}, \omega)$, $\underline{E}^{\text{p,q}}(\underline{r}, \omega)$ and $\underline{H}^{\text{p,q}}(\underline{r}, \omega)$ is such that

$$
\langle\langle \text{p, q} \rangle\rangle = \langle\langle \text{q, p} \rangle\rangle, \qquad (1.86)
$$

then it is called Lorentz–reciprocal.

Combining the Tellegen constitutive relations (1.27) with the Maxwell curl postulates (1.30) and integrating thereafter, we obtain the reaction difference

$$\langle\langle \mathrm{p, q}\rangle\rangle - \langle\langle \mathrm{q, p}\rangle\rangle =$$

$$-i\omega \int_{V_{\mathrm{p}} \cup V_{\mathrm{q}}} \left\{ \underline{E}^{\mathrm{q}}(\underline{r}, \omega) \bullet \left[\underline{\underline{\epsilon}}_{\mathrm{EH}}(\underline{r}, \omega) - \underline{\underline{\epsilon}}^{\mathrm{T}}_{\mathrm{EH}}(\underline{r}, \omega) \right] \bullet \underline{E}^{\mathrm{p}}(\underline{r}, \omega) \right.$$

$$+\underline{H}^{\mathrm{p}}(\underline{r}, \omega) \bullet \left[\underline{\underline{\mu}}_{\mathrm{EH}}(\underline{r}, \omega) - \underline{\underline{\mu}}^{\mathrm{T}}_{\mathrm{EH}}(\underline{r}, \omega) \right] \bullet \underline{H}^{\mathrm{q}}(\underline{r}, \omega)$$

$$+\underline{E}^{\mathrm{q}}(\underline{r}, \omega) \bullet \left[\underline{\underline{\xi}}_{\mathrm{EH}}(\underline{r}, \omega) + \underline{\underline{\zeta}}^{\mathrm{T}}_{\mathrm{EH}}(\underline{r}, \omega) \right] \bullet \underline{H}^{\mathrm{p}}(\underline{r}, \omega)$$

$$+\underline{H}^{\mathrm{p}}(\underline{r}, \omega) \bullet \left[\underline{\underline{\zeta}}_{\mathrm{EH}}(\underline{r}, \omega) + \underline{\underline{\xi}}^{\mathrm{T}}_{\mathrm{EH}}(\underline{r}, \omega) \right] \bullet \underline{E}^{\mathrm{q}}(\underline{r}, \omega) \right\} d^3\underline{r},$$

$$(1.87)$$

where the integration region $V_{\mathrm{p}} \cup V_{\mathrm{q}}$ contains both the sources 'p' and sources 'q'. Thus, Lorentz reciprocity is signalled by [34]

$$\left.\begin{array}{l} \underline{\underline{\epsilon}}_{\mathrm{EH}}(\underline{r}, \omega) = \underline{\underline{\epsilon}}^{\mathrm{T}}_{\mathrm{EH}}(\underline{r}, \omega) \\[4pt] \underline{\underline{\xi}}_{\mathrm{EH}}(\underline{r}, \omega) = -\underline{\underline{\zeta}}^{\mathrm{T}}_{\mathrm{EH}}(\underline{r}, \omega) \\[4pt] \underline{\underline{\mu}}_{\mathrm{EH}}(\underline{r}, \omega) = \underline{\underline{\mu}}^{\mathrm{T}}_{\mathrm{EH}}(\underline{r}, \omega) \end{array}\right\}. \qquad (1.88)$$

The corresponding symmetries for the Boys–Post representation follow immediately from Eqs. (1.29) as

$$\left.\begin{array}{l} \underline{\underline{\epsilon}}_{\mathrm{EB}}(\underline{r}, \omega) = \underline{\underline{\epsilon}}^{\mathrm{T}}_{\mathrm{EB}}(\underline{r}, \omega) \\[4pt] \underline{\underline{\xi}}_{\mathrm{EB}}(\underline{r}, \omega) = \underline{\underline{\zeta}}^{\mathrm{T}}_{\mathrm{EB}}(\underline{r}, \omega) \\[4pt] \underline{\underline{\nu}}_{\mathrm{EB}}(\underline{r}, \omega) = \underline{\underline{\nu}}^{\mathrm{T}}_{\mathrm{EB}}(\underline{r}, \omega) \end{array}\right\}. \qquad (1.89)$$

Notice that the Lorentz–reciprocity conditions (1.88) and (1.89) coincide with the Onsager relations (1.84) and (1.83), respectively, in the absence of a magnetostatic field.

Many frequently encountered anisotropic and bianisotropic mediums satisfy the Lorentz reciprocity conditions. Lorentz–reciprocal mediums arise commonly as dielectric and magnetic crystals, whereas plasmas and mediums moving at uniform velocity are Lorentz–nonreciprocal mediums. These topics are discussed further in Chap. 2 in the context of specific types of anisotropic and bianisotropic mediums.

1.7.2.2 *Dissipative, nondissipative and active mediums*

No passive medium — with the unique exception of free space (which is not a material) — responds instantaneously to an applied electromagnetic field, which characteristic is enshrined as the principle of causality [8, 35]. Dissipation is therefore exhibited by all passive material mediums. However, occasionally it can be expedient to neglect dissipation, especially if attention is confined to a narrow range of angular frequencies wherein dissipation is very small over the length–scales of interest.

In order to concentrate on dissipation, we introduce the time–averaged Poynting vector for monochromatic fields

$$\langle \breve{\underline{S}}(\underline{r},\omega)\rangle_t = \frac{1}{2}\mathrm{Re}\left\{\breve{\underline{E}}(\underline{r},\omega)\times\breve{\underline{H}}^*(\underline{r},\omega)\right\}, \qquad (1.90)$$

which may be interpreted as the time–averaged power per unit area. The complex–valued fields amplitudes $\breve{\underline{E}}(\underline{r},\omega)$ and $\breve{\underline{H}}(\underline{r},\omega)$ were introduced in Eq. (1.36). Of particular relevance to us here is the divergence of Eq. (1.90), representing the time–averaged power density. Using the Maxwell curl postulates (1.30) in the absence of sources (i.e., $\underline{J}_{e,m}(\underline{r},\omega)\equiv\underline{0}$), together with the Tellegen constitutive relations for a bianisotropic medium (1.27), we find that the divergence of Eq. (1.90) yields [4]

$$\langle \nabla\bullet\breve{\underline{S}}(\underline{r},\omega)\rangle_t = \frac{\omega}{4}\left[\breve{\underline{E}}(\underline{r},\omega)\;\breve{\underline{H}}(\underline{r},\omega)\right]^*\bullet\underline{\underline{m}}(\underline{r},\omega)\bullet\begin{bmatrix}\breve{\underline{E}}(\underline{r},\omega)\\[4pt]\breve{\underline{H}}(\underline{r},\omega)\end{bmatrix}, \qquad (1.91)$$

wherein the 6×6 Hermitian dyadic

$$\underline{\underline{m}}(\underline{r},\omega) = i\begin{bmatrix}\underline{\underline{\epsilon}}_{EH}(\underline{r},\omega)-\underline{\underline{\epsilon}}^\dagger_{EH}(\underline{r},\omega) & \underline{\underline{\xi}}_{EH}(\underline{r},\omega)-\underline{\underline{\zeta}}^\dagger_{EH}(\underline{r},\omega)\\[6pt]\underline{\underline{\zeta}}_{EH}(\underline{r},\omega)-\underline{\underline{\xi}}^\dagger_{EH}(\underline{r},\omega) & \underline{\underline{\mu}}_{EH}(\underline{r},\omega)-\underline{\underline{\mu}}^\dagger_{EH}(\underline{r},\omega)\end{bmatrix}, \qquad (1.92)$$

with the superscript † indicating the conjugate transpose.

A medium is nondissipative provided that $\langle\nabla\bullet\breve{\underline{S}}(\underline{r},\omega)\rangle_t = 0$. Thus, dissipation is neglected by enforcing the equalities [8]

$$\left.\begin{aligned}\underline{\underline{\epsilon}}_{EH}(\underline{r},\omega) &= \underline{\underline{\epsilon}}^\dagger_{EH}(\underline{r},\omega)\\[4pt]\underline{\underline{\xi}}_{EH}(\underline{r},\omega) &= \underline{\underline{\zeta}}^\dagger_{EH}(\underline{r},\omega)\\[4pt]\underline{\underline{\mu}}_{EH}(\underline{r},\omega) &= \underline{\underline{\mu}}^\dagger_{EH}(\underline{r},\omega)\end{aligned}\right\}, \qquad (1.93)$$

or, equivalently,

$$\left.\begin{aligned}\underline{\underline{\epsilon}}_{EB}(\underline{r},\omega) &= \underline{\underline{\epsilon}}^\dagger_{EB}(\underline{r},\omega)\\[4pt]\underline{\underline{\xi}}_{EB}(\underline{r},\omega) &= -\underline{\underline{\zeta}}^\dagger_{EB}(\underline{r},\omega)\\[4pt]\underline{\underline{\nu}}_{EB}(\underline{r},\omega) &= \underline{\underline{\nu}}^\dagger_{EB}(\underline{r},\omega)\end{aligned}\right\}. \qquad (1.94)$$

The distinction between the conditions for the neglect of dissipation and for Lorentz reciprocity should be noted. These conditions are summarized

Table 1.1 Conditions imposed by Lorentz reciprocity and neglect of dissipation.

$\underline{\underline{K}}_{EH}$	Lorentz reciprocity	Neglect of dissipation
$\left[\begin{pmatrix} \epsilon & 0 & 0 \\ 0 & \epsilon & 0 \\ 0 & 0 & \epsilon \end{pmatrix} \begin{pmatrix} \xi & 0 & 0 \\ 0 & \xi & 0 \\ 0 & 0 & \xi \end{pmatrix} \\ \begin{pmatrix} \zeta & 0 & 0 \\ 0 & \zeta & 0 \\ 0 & 0 & \zeta \end{pmatrix} \begin{pmatrix} \mu & 0 & 0 \\ 0 & \mu & 0 \\ 0 & 0 & \mu \end{pmatrix} \right]$	$\xi = -\zeta$	$\epsilon = \epsilon^*$ $\xi = \zeta^*$ $\mu = \mu^*$
$\left[\begin{pmatrix} \epsilon_{11} & 0 & 0 \\ 0 & \epsilon_{22} & 0 \\ 0 & 0 & \epsilon_{33} \end{pmatrix} \begin{pmatrix} \xi_{11} & 0 & 0 \\ 0 & \xi_{22} & 0 \\ 0 & 0 & \xi_{33} \end{pmatrix} \\ \begin{pmatrix} \zeta_{11} & 0 & 0 \\ 0 & \zeta_{22} & 0 \\ 0 & 0 & \zeta_{33} \end{pmatrix} \begin{pmatrix} \mu_{11} & 0 & 0 \\ 0 & \mu_{22} & 0 \\ 0 & 0 & \mu_{33} \end{pmatrix} \right]$	$\xi_{\ell\ell} = -\zeta_{\ell\ell}$	$\epsilon_{\ell\ell} = \epsilon_{\ell\ell}^*$ $\xi_{\ell\ell} = \zeta_{\ell\ell}^*$ $\mu_{\ell\ell} = \mu_{\ell\ell}^*$
$\left[\begin{pmatrix} \epsilon_{11} & \epsilon_{12} & \epsilon_{13} \\ \epsilon_{21} & \epsilon_{22} & \epsilon_{23} \\ \epsilon_{31} & \epsilon_{32} & \epsilon_{33} \end{pmatrix} \begin{pmatrix} \xi_{11} & \xi_{12} & \xi_{13} \\ \xi_{21} & \xi_{22} & \xi_{23} \\ \xi_{31} & \xi_{32} & \xi_{33} \end{pmatrix} \\ \begin{pmatrix} \zeta_{11} & \zeta_{12} & \zeta_{13} \\ \zeta_{21} & \zeta_{22} & \zeta_{23} \\ \zeta_{31} & \zeta_{32} & \zeta_{33} \end{pmatrix} \begin{pmatrix} \mu_{11} & \mu_{12} & \mu_{13} \\ \mu_{21} & \mu_{22} & \mu_{23} \\ \mu_{31} & \mu_{32} & \mu_{33} \end{pmatrix} \right]$	$\epsilon_{\ell m} = \epsilon_{m\ell}$ $\xi_{\ell m} = -\zeta_{m\ell}$ $\mu_{\ell m} = \mu_{m\ell}$	$\epsilon_{\ell m} = \epsilon_{m\ell}^*$ $\xi_{\ell m} = \zeta_{m\ell}^*$ $\mu_{\ell m} = \mu_{m\ell}^*$

Three different forms of the 6×6 Tellegen constitutive dyadic $\underline{\underline{K}}_{EH}$ for a passive medium are represented. Notice that for the medium represented in the first example, the Lorentz–reciprocity condition $\xi = -\zeta$ must be satisfied in order to comply with the Post constraint.

in Table 1.1 for three often–encountered forms of the constitutive dyadic $\underline{\underline{K}}_{EH}(\underline{r}, \omega)$.

A medium is dissipative provided that $\langle \nabla \bullet \underline{\breve{S}}(\underline{r}, \omega) \rangle_t < 0$. By inspection of Eq. (1.91), the dyadic $\underline{\underline{m}}(\underline{r}, \omega)$ must be negative definite in order to satisfy this dissipative condition [36]. Accordingly, for a medium to be dissipative, it is necessary and sufficient that the two 3×3 dyadics [37]

$$
\left.
\begin{aligned}
\underline{\underline{m}}_1(\underline{r},\omega) &= i\left[\underline{\underline{\epsilon}}_{\mathrm{EH}}(\underline{r},\omega) - \underline{\underline{\epsilon}}^{\dagger}_{\mathrm{EH}}(\underline{r},\omega)\right] \\
\underline{\underline{m}}_2(\underline{r},\omega) &= i\Big\{\underline{\underline{\mu}}_{\mathrm{EH}}(\underline{r},\omega) - \underline{\underline{\mu}}^{\dagger}_{\mathrm{EH}}(\underline{r},\omega) \\
&\quad - \left[\underline{\underline{\zeta}}_{\mathrm{EH}}(\underline{r},\omega) - \underline{\underline{\xi}}^{\dagger}_{\mathrm{EH}}(\underline{r},\omega)\right] \cdot \underline{\underline{m}}_1^{-1}(\underline{r},\omega) \cdot \left[\underline{\underline{\xi}}_{\mathrm{EH}}(\underline{r},\omega) - \underline{\underline{\zeta}}^{\dagger}_{\mathrm{EH}}(\underline{r},\omega)\right]\Big\}
\end{aligned}
\right\}
$$

$$(1.95)$$

are negative definite. Equivalently, if the two 3×3 dyadics [37]

$$
\left.
\begin{aligned}
\underline{\underline{m}}_3(\underline{r},\omega) &= i\left[\underline{\underline{\mu}}_{\mathrm{EH}}(\underline{r},\omega) - \underline{\underline{\mu}}^{\dagger}_{\mathrm{EH}}(\underline{r},\omega)\right] \\
\underline{\underline{m}}_4(\underline{r},\omega) &= i\Big\{\underline{\underline{\epsilon}}_{\mathrm{EH}}(\underline{r},\omega) - \underline{\underline{\epsilon}}^{\dagger}_{\mathrm{EH}}(\underline{r},\omega) \\
&\quad - \left[\underline{\underline{\xi}}_{\mathrm{EH}}(\underline{r},\omega) - \underline{\underline{\zeta}}^{\dagger}_{\mathrm{EH}}(\underline{r},\omega)\right] \cdot \underline{\underline{m}}_3^{-1}(\underline{r},\omega) \cdot \left[\underline{\underline{\zeta}}_{\mathrm{EH}}(\underline{r},\omega) - \underline{\underline{\xi}}^{\dagger}_{\mathrm{EH}}(\underline{r},\omega)\right]\Big\}
\end{aligned}
\right\}
$$

$$(1.96)$$

are negative definite, then the necessary and sufficient conditions for dissipation are also satisfied.

A medium which is characterized by $\langle \nabla \cdot \underline{S}(\underline{r},\omega)\rangle_t > 0$ is an active medium.

References

[1] J.Z. Buchwald, *From Maxwell to microphysics: Aspects of electromagnetic theory in the last quarter of the nineteenth century*, University of Chicago Press, Chicago, IL, USA, 1985.

[2] J.D. Jackson, *Classical electrodynamics, 3rd ed*, Wiley, New York, NY, USA, 1999.

[3] A. Lakhtakia, Covariances and invariances of the Maxwell postulates, *Advanced electromagnetism: Foundations, theory and applications* (T.W. Barrett and D.M. Grimes, eds), World Scientific, Singapore, 1995, 390–410.

[4] J.A. Kong, *Electromagnetic wave theory*, Wiley, New York, NY, USA, 1986.

[5] W.S. Weiglhofer, Constitutive characterization of simple and complex mediums, *Introduction to complex mediums for optics and electromagnetics* (W.S. Weiglhofer and A. Lakhtakia, eds), SPIE Press, Bellingham, WA, USA, 2003, 27–61.

[6] S. Ponti, C. Oldano and M. Becchi, Bloch wave approach to the optics of crystals, *Phys Rev E* **64** (2001), 021704.

[7] G.B. Arfken and H.J. Weber, *Mathematical methods for physicists, 4th ed*, Academic Press, London, UK, 1995.

[8] H.C. Chen, *Theory of electromagnetic waves*, McGraw–Hill, New York, NY, USA, 1983.

[9] W.S. Weiglhofer, A perspective on bianisotropy and *Bianisotropics '97*, *Int J Appl Electromagn Mech* **9** (1998), 93–101.

[10] D.C. Champeney, *A handbook of Fourier theorems*, Cambridge University Press, Cambridge, UK, 1989.

[11] T. Sarkar, D. Weiner and V. Jain, Some mathematical considerations in dealing with the inverse problem, *IEEE Trans Antennas Propagat* **29** (1981), 373–379.

[12] J.A. Kong, Theorems of bianisotropic media, *Proc IEEE* **60** (1972), 1036–1046.

[13] E.J. Post, *Formal structure of electromagnetics*, Dover Press, New York, NY, USA, 1997.

[14] A. Lakhtakia, A conjugation symmetry in linear electromagnetism in extension of materials with negative real permittivity and permeability scalars, *Microw Opt Technol Lett* **40** (2004), 160–161.

[15] F.W. King, Alternative approach to the derivation of dispersion relations for optical constants, *J Phys A: Math Gen* **39** (2006), 10427–10435.

[16] J. Hilgevoord, *Dispersion relations and causal descriptions*, North–Holland, Amsterdam, The Netherlands, 1962.

[17] C.F. Bohren and D.R. Huffman, *Absorption and scattering of light by small particles*, Wiley, New York, NY, USA, 1983.

[18] A. Lakhtakia, On the genesis of Post constraint in modern electromagnetism, *Optik* **115** (2004), 151–158.

[19] A. Lakhtakia and W.S. Weiglhofer, Lorentz covariance, Occam's razor, and a constraint on linear constitutive relations, *Phys Lett A* **213** (1996), 107–111. Corrections: **222** (1996), 459.

[20] A. Lakhtakia and W.S. Weiglhofer, Constraint on linear, spatiotemporally nonlocal, spatiotemporally nonhomogeneous constitutive relations, *Int J Infrared Millim Waves* **17** (1996), 1867–1878.

[21] O.L. de Lange and R.E. Raab, Post's constraint for electromagnetic constitutive relations, *J Opt A: Pure Appl Opt* **3** (2001), L23–L26.

[22] F.W. Hehl, Y.N. Obukhov, J.–P. Rivera and H. Schmid, Relativistic analysis of magnetoelectric crystals: Extracting a new 4–dimensional P odd and T odd pseudoscalar from Cr_2O_3 data, *Phys Lett A* **372** (2007), 1141–1146.

[23] A. Lakhtakia, Remarks on the current status of the Post constraint, *Optik* **120** (2009), 422–424.

[24] F.W. Hehl and Yu.N. Obukhov, Linear media in classical electrodynamics and the Post constraint, *Phys Lett A* **334**, (2005), 249–259.

[25] A. Lakhtakia, Boundary–value problems and the validity of the Post constraint in modern electromagnetism, *Optik* **117** (2006), 188–192.

[26] L. Onsager, Reciprocal relations in irreversible processes. I, *Phys Rev* **37** (1931), 405–426.

[27] L. Onsager, Reciprocal relations in irreversible processes. II, *Phys Rev* **38** (1931), 2265–2279.

[28] H.B.G. Casimir, On Onsager's principle of microscopic reversibility, *Rev Mod Phys* **17** (1945), 343–350.

[29] H.B. Callen and R.F. Greene, On a theorem of irreversible thermodynamics, *Phys Rev* **86** (1952), 702–710.

[30] H.B. Callen, M.L. Barasch and J.L. Jackson, Statistical mechanics of irreversibility, *Phys Rev* **88** (1952), 1382–1386.

[31] A. Lakhtakia and R.A. Depine, On Onsager relations and linear electromagnetic materials, *Int J Electron Commun (AEÜ)* **59** (2005), 101–104.

[32] C. Altman and K. Suchy, *Reciprocity, spatial mapping and time reversal in electromagnetics*, Kluwer Academic Publishers, Dordrecht, The Netherlands, 1991.

[33] V.H. Rumsey, Reaction concept in electromagnetic theory, *Phys Rev* **94** (1954), 1483–1491. Corrections **95** (1954), 1705.

[34] C.M. Krowne, Electromagnetic theorems for complex anisotropic media, *IEEE Trans Antennas Propagat* **32** (1984), 1224–1230.

[35] W.S. Weiglhofer and A. Lakhtakia, On causality requirements for material media, *Arch Elektron Übertrag* **50** (1996), 389–391.

[36] I.V. Lindell and F.M. Dahl, Conditions for the parameter dyadics of lossy bianisotropic media, *Microw Opt Technol Lett* **29** (2001), 175–178.

[37] E.L. Tan, Reduced conditions for the constitutive parameters of lossy bi-anisotropic media, *Microw Opt Technol Lett* **41** (2004), 133–135.

Chapter 2

Linear Mediums

A simple understanding of the notions of anisotropy and bianisotropy is that the adjectives 'anisotropic' and 'bianisotropic' describe mediums which are not *isotropic*. Accordingly, prior to discussing anisotropic and bianisotropic mediums in this chapter, we first establish our terms of reference by briefly considering isotropic mediums. We then present a survey of the commonly encountered classifications of anisotropy and bianisotropy, as expressed in terms of their constitutive relations and constitutive dyadics. Our attention is restricted in this chapter to linear anisotropic and bianisotropic mediums; a description of the more complex constitutive relations associated with nonlinear anisotropy and bianisotropy is postponed to Chap. 7. Furthermore, we are chiefly concerned with nonactive mediums (cf. Sec. 1.7.2.2).

2.1 Isotropy

The primitive fields and induction fields in an isotropic medium are co–directional. Hence the corresponding constitutive dyadics simply reduce to scalars.

2.1.1 *Free space*

The most fundamental medium in electromagnetics is free space, otherwise referred to as vacuum. It represents the reference medium, relative to which the electromagnetic responses of all material mediums are gauged [1]. From our viewpoint of classical physics, free space is devoid of all matter. Quantum electrodynamical processes, through which energetic fluctuations may become interwoven with free space, are not relevant to us in here; we postpone our consideration of nonclassical vacuum until Sec. 7.4.

In this section our discussion of the electromagnetic properties of free space pertains to flat spacetime exclusively. The topic of free space in generally curved spacetime is pursued in Sec. 2.4.2.

By definition, free space is both isotropic and homogeneous. Its constitutive properties are specified by the scalar permittivity ϵ_0 and scalar permeability μ_0. Free space holds the unique distinction of being the only medium for which there is an exact, spatiotemporally local relationship between the induction fields and the primitive fields. The time–domain and frequency–domain constitutive relations of free space have the same form:

$$\left.\begin{aligned} \tilde{\underline{D}}(\underline{r},t) &= \epsilon_0\, \tilde{\underline{E}}(\underline{r},t) \\ \tilde{\underline{H}}(\underline{r},t) &= \frac{1}{\mu_0}\, \tilde{\underline{B}}(\underline{r},t) \end{aligned}\right\}, \tag{2.1}$$

and

$$\left.\begin{aligned} \underline{D}(\underline{r},\omega) &= \epsilon_0\, \underline{E}(\underline{r},\omega) \\ \underline{H}(\underline{r},\omega) &= \frac{1}{\mu_0}\, \underline{B}(\underline{r},\omega) \end{aligned}\right\}. \tag{2.2}$$

2.1.2 *Dielectric–magnetic mediums*

Isotropic, homogeneous, dielectric–magnetic mediums are the simplest material mediums. Their electromagnetic properties are characterized in terms of the scalar permittivity $\epsilon(\omega)$ and scalar permeability $\mu(\omega)$ by the frequency–domain Tellegen constitutive relations

$$\left.\begin{aligned} \underline{D}(\underline{r},\omega) &= \epsilon(\omega)\, \underline{E}(\underline{r},\omega) \\ \underline{B}(\underline{r},\omega) &= \mu(\omega)\, \underline{H}(\underline{r},\omega) \end{aligned}\right\}. \tag{2.3}$$

For a dissipative medium, both ϵ and μ are ω–dependent, complex–valued parameters with $\text{Im}\,\{\epsilon(\omega)\} > 0$ and $\text{Im}\,\{\mu(\omega)\} > 0$. The placement of $\epsilon(\omega)$ and $\mu(\omega)$ in the upper half complex plane follows from the sign convention for the exponent of temporal Fourier transform kernel in Eq. (1.25).

Parenthetically, although isotropic dielectric–magnetic mediums have been extensively analyzed since the earliest days of electromagnetics, their fundamental properties are still the subject of ongoing research. For example, negative phase velocity — a phenomenon which follows for uniform plane waves as a consequence of the condition [2]

$$\frac{\text{Re}\,\{\epsilon(\omega)\}}{\text{Im}\,\{\epsilon(\omega)\}} + \frac{\text{Re}\,\{\mu(\omega)\}}{\text{Im}\,\{\mu(\omega)\}} < 0 \tag{2.4}$$

being satisfied at a particular angular frequency ω — is a topic of intense current interest [3]; see Sec. 4.8.1 for further details.

2.1.3 *Isotropic chirality*

A more general linear, isotropic, homogeneous medium is described by the frequency–domain Tellegen constitutive relations

$$\left.\begin{array}{l} \underline{D}(\underline{r},\omega) = \epsilon(\omega)\,\underline{E}(\underline{r},\omega) + \chi(\omega)\,\underline{H}(\underline{r},\omega) \\ \underline{B}(\underline{r},\omega) = -\chi(\omega)\,\underline{E}(\underline{r},\omega) + \mu(\omega)\,\underline{H}(\underline{r},\omega) \end{array}\right\}. \tag{2.5}$$

The medium characterized by Eqs. (2.5) is called an isotropic chiral medium. Three complex–valued, ω–dependent, scalars — namely, permittivity ϵ, permeability μ, and the chirality parameter χ — specify the electromagnetic properties of this medium [4]. The mirror conjugate of this medium is described by the constitutive relations

$$\left.\begin{array}{l} \underline{D}(\underline{r},\omega) = \epsilon(\omega)\,\underline{E}(\underline{r},\omega) - \chi(\omega)\,\underline{H}(\underline{r},\omega) \\ \underline{B}(\underline{r},\omega) = \chi(\omega)\,\underline{E}(\underline{r},\omega) + \mu(\omega)\,\underline{H}(\underline{r},\omega) \end{array}\right\}. \tag{2.6}$$

A key property of isotropic chiral mediums is that they exhibit optical activity [5]. That is, they can distinguish between left–handed and right–handed electromagnetic fields. Measures of optical activity due to an isotropic chiral medium and its mirror conjugate differ only in sign. When an isotropic chiral medium and its mirror conjugate are mixed together in equal proportions (and provided that no chemical reactions or molecular conformational changes occur), the resulting *racemic* mixture does not exhibit optical activity; instead, it is described by the constitutive relations (2.3). Since an isotropic chiral medium exhibits the same measures of optical activity in all directions, its optical activity is often designated as *natural* optical activity.

Notice that, in view of the definition presented in Sec. 1.7.2, isotropic chiral mediums are Lorentz–reciprocal. The corresponding nonreciprocal isotropic mediums would be described by the Tellegen constitutive relations

$$\left.\begin{array}{l} \underline{D}(\underline{r},\omega) = \epsilon(\omega)\,\underline{E}(\underline{r},\omega) + \chi_1(\omega)\,\underline{H}(\underline{r},\omega) \\ \underline{B}(\underline{r},\omega) = \chi_2(\omega)\,\underline{E}(\underline{r},\omega) + \mu(\omega)\,\underline{H}(\underline{r},\omega) \end{array}\right\}, \tag{2.7}$$

with $\chi_1(\omega) \neq -\chi_2(\omega)$. Such a medium is called a biisotropic medium.

2.2 Anisotropy

The primitive fields and induction fields in isotropic mediums are co–directional. In contrast, the defining characteristic of anisotropic mediums is that $\underline{E}(\underline{r},\omega)$ is not aligned with $\underline{D}(\underline{r},\omega)$ and/or $\underline{B}(\underline{r},\omega)$ is not aligned with $\underline{H}(\underline{r}, \omega)$. Hence, whereas scalars provide the constitutive characterizations for isotropic mediums, dyadics are needed to describe the constitutive properties of anisotropic mediums.

Dielectric anisotropy is commonplace within the the realm of crystal optics [6, 7]. Chiral nematic and chiral smectic liquid crystals belong to the anisotropic dielectric category, albeit these are also nonhomogeneous mediums [8, 9]. Magnetic anisotropy is an important characteristic of many diamagnetic and paramagnetic substances [7].

2.2.1   *Uniaxial anisotropy*

At macroscopic length–scales, a *uni*axial medium is identified by a single distinguished axis. The distinguished axis may originate from microscopic attributes, such as the underlying crystalline structure or particulate geometry. Let the unit vector $\hat{\underline{u}}$ point in the direction of the distinguished axis. The frequency–domain constitutive relations of a uniaxial homogeneous dielectric medium are then given in the Tellegen representation as

$$\left. \begin{aligned} \underline{D}(\underline{r}, \omega) &= \underline{\underline{\epsilon}}_{\text{uni}}(\omega) \bullet \underline{E}(\underline{r}, \omega) \\ \underline{B}(\underline{r}, \omega) &= \mu_0 \, \underline{H}(\underline{r}, \omega) \end{aligned} \right\}, \tag{2.8}$$

where the ω–dependent permittivity dyadic $\underline{\underline{\epsilon}}_{\text{uni}}$ may be written in the form

$$\underline{\underline{\epsilon}}_{\text{uni}}(\omega) = \epsilon(\omega) \left(\underline{\underline{I}} - \hat{\underline{u}}\,\hat{\underline{u}} \right) + \epsilon_u(\omega)\, \hat{\underline{u}}\,\hat{\underline{u}}. \tag{2.9}$$

The permittivity scalars $\epsilon(\omega)$, $\epsilon_u(\omega) \in \mathbb{C}$ (i.e., the set of complex numbers), with $\text{Im}\{\epsilon(\omega)\} > 0$ and $\text{Im}\{\epsilon_u(\omega)\} > 0$ as dictated by the principle of causality for a dissipative medium.

Without loss of generality, a coordinate system may be chosen with an orientation such that the distinguished direction coincides with the z axis. Thereby, the permittivity dyadic $\underline{\underline{\epsilon}}_{\text{uni}}(\omega)$ acquires the diagonal matrix form

$$\underline{\underline{\epsilon}}_{\text{uni}}(\omega) = \begin{pmatrix} \epsilon(\omega) & 0 & 0 \\ 0 & \epsilon(\omega) & 0 \\ 0 & 0 & \epsilon_u(\omega) \end{pmatrix}. \tag{2.10}$$

Similarly, a uniaxial homogeneous magnetic medium may be described by the Tellegen constitutive relations

$$\left. \begin{aligned} \underline{D}(\underline{r}, \omega) &= \epsilon_0 \, \underline{E}(\underline{r}, \omega) \\ \underline{B}(\underline{r}, \omega) &= \underline{\underline{\mu}}_{\text{uni}}(\omega) \bullet \underline{H}(\underline{r}, \omega) \end{aligned} \right\}, \tag{2.11}$$

with

$$\underline{\underline{\mu}}_{\text{uni}}(\omega) = \mu(\omega) \left(\underline{\underline{I}} - \hat{\underline{u}}\,\hat{\underline{u}} \right) + \mu_u(\omega)\, \hat{\underline{u}}\,\hat{\underline{u}}. \tag{2.12}$$

The ω–dependent parameters μ and μ_u lie in the upper half complex plane for dissipative mediums. The choice $\hat{u} = \hat{z}$ leads to the diagonal–matrix representation

$$\underline{\underline{\mu}}_{\text{uni}}(\omega) = \begin{pmatrix} \mu(\omega) & 0 & 0 \\ 0 & \mu(\omega) & 0 \\ 0 & 0 & \mu_u(\omega) \end{pmatrix}. \tag{2.13}$$

A uniaxial dielectric–magnetic medium is described by the Tellegen constitutive relations

$$\left. \begin{aligned} \underline{D}(\underline{r}, \omega) &= \underline{\underline{\epsilon}}_{\text{uni}}(\omega) \bullet \underline{E}(\underline{r}, \omega) \\ \underline{B}(\underline{r}, \omega) &= \underline{\underline{\mu}}_{\text{uni}}(\omega) \bullet \underline{H}(\underline{r}, \omega) \end{aligned} \right\}, \tag{2.14}$$

which represents the amalgamation of Eqs. (2.8) and (2.11).

2.2.2 *Biaxial anisotropy*

The natural generalization of the uniaxial dielectric medium given by Eqs. (2.8)–(2.10) is the *orthorhombic* biaxial dielectric medium described by the Tellegen constitutive relations

$$\left. \begin{aligned} \underline{D}(\underline{r}, \omega) &= \underline{\underline{\epsilon}}_{\text{bi}}^{\text{ortho}}(\omega) \bullet \underline{E}(\underline{r}, \omega) \\ \underline{B}(\underline{r}, \omega) &= \mu_0 \underline{H}(\underline{r}, \omega) \end{aligned} \right\}, \tag{2.15}$$

where the permittivity dyadic $\underline{\underline{\epsilon}}_{\text{bi}}^{\text{ortho}}(\omega)$ has the diagonal form

$$\underline{\underline{\epsilon}}_{\text{bi}}^{\text{ortho}}(\omega) = \epsilon_x(\omega)\, \hat{x}\hat{x} + \epsilon_y(\omega)\, \hat{y}\hat{y} + \epsilon_z(\omega)\hat{z}\hat{z} \tag{2.16}$$

$$\equiv \begin{pmatrix} \epsilon_x(\omega) & 0 & 0 \\ 0 & \epsilon_y(\omega) & 0 \\ 0 & 0 & \epsilon_z(\omega) \end{pmatrix}. \tag{2.17}$$

By virtue of the principle of causality, the imaginary parts of the ω–dependent permittivity scalars ϵ_x, ϵ_y and ϵ_z are positive–valued for a dissipative medium.

The diagonal–dyadic/matrix representations (2.16) and (2.17) can be mathematically convenient, but are not particularly insightful. The equivalent representation [10]

$$\underline{\underline{\epsilon}}_{\text{bi}}^{\text{ortho}}(\omega) = \epsilon_{\text{p}}(\omega)\underline{\underline{I}} + \epsilon_{\text{q}}(\omega)\,(\hat{u}_m \hat{u}_n + \hat{u}_n \hat{u}_m) \tag{2.18}$$

highlights the *bi*axial symmetry via the two unit vectors $\hat{\underline{u}}_m$ and $\hat{\underline{u}}_n$ [11]. In the absence of dissipation, the ω–dependent permittivity scalars ϵ_p and ϵ_q are real–valued and the unit vectors $\hat{\underline{u}}_m$ and $\hat{\underline{u}}_n$ are aligned with the *optic ray axes*. Generally, electromagnetic radiation may propagate in one of two different modes in a given direction through a biaxial medium. The two modes are distinguished from each other by their different rates of energy flow. However, in the two privileged directions — delineated by the optic ray axes — only one energy velocity is permissible [6].[1] For dissipative mediums, both ϵ_p and ϵ_q are necessarily complex–valued with $\mathrm{Im}\,\{\epsilon_p(\omega), \epsilon_q(\omega)\} > 0$; in which case $\mathrm{Re}\,\{\epsilon_p(\omega), \epsilon_q(\omega)\}$ and $\mathrm{Im}\,\{\epsilon_p(\omega), \epsilon_q(\omega)\}$ are generally associated with different pairs of $(\hat{\underline{u}}_m, \hat{\underline{u}}_n)$ axes [12].

More general biaxial, homogeneous, dielectric mediums are characterized by the Tellegen constitutive relations

$$
\left.
\begin{aligned}
\underline{D}(\underline{r}, \omega) &= \left.
\begin{aligned}
\underline{\underline{\epsilon}}^{\,\mathrm{mono}}_{\mathrm{bi}}(\omega) \\
\underline{\underline{\epsilon}}^{\,\mathrm{tri}}_{\mathrm{bi}}(\omega)
\end{aligned}
\right\} \cdot \underline{E}(\underline{r}, \omega) \\
\underline{B}(\underline{r}, \omega) &= \mu_0\, \underline{H}(\underline{r}, \omega)
\end{aligned}
\right\},
\tag{2.19}
$$

where the 3×3 permittivity dyadics have the symmetric matrix representations

$$
\underline{\underline{\epsilon}}^{\,\mathrm{mono}}_{\mathrm{bi}}(\omega) = \begin{pmatrix} \epsilon_x(\omega) & \epsilon_\alpha(\omega) & 0 \\ \epsilon_\alpha(\omega) & \epsilon_y(\omega) & 0 \\ 0 & 0 & \epsilon_z(\omega) \end{pmatrix}
\tag{2.20}
$$

and

$$
\underline{\underline{\epsilon}}^{\,\mathrm{tri}}_{\mathrm{bi}}(\omega) = \begin{pmatrix} \epsilon_x(\omega) & \epsilon_\alpha(\omega) & \epsilon_\beta(\omega) \\ \epsilon_\alpha(\omega) & \epsilon_y(\omega) & \epsilon_\gamma(\omega) \\ \epsilon_\beta(\omega) & \epsilon_\gamma(\omega) & \epsilon_z(\omega) \end{pmatrix}.
\tag{2.21}
$$

The constitutive permittivity dyadics which specify biaxial anisotropy are summarized in Table 2.1.

As in the orthorhombic scenario, the ω–dependent permittivity scalars ϵ_x, ϵ_y and ϵ_z are complex–valued with causality dictating that $\mathrm{Im}\,\{\epsilon_x(\omega), \epsilon_y(\omega), \epsilon_z(\omega)\} > 0$ for dissipative mediums. The off–diagonal ω–

---

[1]The optic ray axes should be distinguished from the *optic axes* that identify the two privileged directions in which electromagnetic waves may propagate through a biaxial medium with only one phase speed. See Sec. 4.4.2.

Table 2.1 Constitutive permittivity dyadics for three bi-axial dielectric crystal systems.

Crystal system	Constitutive dyadic form	Complex–valued and real–valued scalars
Orthorhombic	$\begin{pmatrix} \epsilon_x & 0 & 0 \\ 0 & \epsilon_y & 0 \\ 0 & 0 & \epsilon_z \end{pmatrix}$	$\epsilon_x, \epsilon_y, \epsilon_z \in \mathbb{C}$
Monoclinic	$\begin{pmatrix} \epsilon_x & \epsilon_\alpha & 0 \\ \epsilon_\alpha & \epsilon_y & 0 \\ 0 & 0 & \epsilon_z \end{pmatrix}$	$\epsilon_x, \epsilon_y, \epsilon_z \in \mathbb{C}$ $\epsilon_\alpha \in \mathbb{R}$
Triclinic	$\begin{pmatrix} \epsilon_x & \epsilon_\alpha & \epsilon_\beta \\ \epsilon_\alpha & \epsilon_y & \epsilon_\gamma \\ \epsilon_\beta & \epsilon_\gamma & \epsilon_z \end{pmatrix}$	$\epsilon_x, \epsilon_y, \epsilon_z \in \mathbb{C}$ $\epsilon_\alpha, \epsilon_\beta, \epsilon_\gamma \in \mathbb{R}$

dependent scalars ϵ_α, ϵ_β and ϵ_γ are real–valued. The permittivity dyadics specified by Eqs. (2.20) and (2.21), characterize biaxial mediums belonging to the *monoclinic* and *triclinic* crystal systems, respectively [7]. The distinction between the three biaxial crystal systems derives from the symmetries of primitive unit cell belonging to the underlying Bravais lattice. If all three basis vectors of the primitive unit cell are orthogonal then the crystal system is orthorhombic; only two basis vectors are mutually orthogonal in the monoclinic system, whereas there are no orthogonal basis vectors for the triclinic crystal system [13].

The concept of biaxiality applies to magnetic mediums in precisely the same way as it does to dielectric mediums. Thus, the Tellegen constitutive relations

$$\left. \begin{array}{l} \underline{D}(\underline{r},\omega) = \epsilon_0\, \underline{E}(\underline{r},\omega) \\[1em] \underline{B}(\underline{r},\omega) = \left. \begin{array}{l} \underline{\underline{\mu}}^{\mathrm{ortho}}_{\mathrm{bi}}(\omega) \\[0.5em] \underline{\underline{\mu}}^{\mathrm{mono}}_{\mathrm{bi}}(\omega) \\[0.5em] \underline{\underline{\mu}}^{\mathrm{tri}}_{\mathrm{bi}}(\omega) \end{array} \right\} \bullet \underline{H}(\underline{r},\omega) \end{array} \right\}, \qquad (2.22)$$

describe orthorhombic, monoclinic and triclinic biaxial magnetic mediums, respectively, where the corresponding permeability dyadics have the symmetric matrix forms

$$\underline{\underline{\mu}}_{\text{bi}}^{\text{ortho}}(\omega) = \begin{pmatrix} \mu_x(\omega) & 0 & 0 \\ 0 & \mu_y(\omega) & 0 \\ 0 & 0 & \mu_z(\omega) \end{pmatrix}, \tag{2.23}$$

$$\underline{\underline{\mu}}_{\text{bi}}^{\text{mono}}(\omega) = \begin{pmatrix} \mu_x(\omega) & \mu_\alpha(\omega) & 0 \\ \mu_\alpha(\omega) & \mu_y(\omega) & 0 \\ 0 & 0 & \mu_z(\omega) \end{pmatrix}, \tag{2.24}$$

$$\underline{\underline{\mu}}_{\text{bi}}^{\text{tri}}(\omega) = \begin{pmatrix} \mu_x(\omega) & \mu_\alpha(\omega) & \mu_\beta(\omega) \\ \mu_\alpha(\omega) & \mu_y(\omega) & \mu_\gamma(\omega) \\ \mu_\beta(\omega) & \mu_\gamma(\omega) & \mu_z(\omega) \end{pmatrix}, \tag{2.25}$$

with $\mu_x(\omega)$, $\mu_y(\omega)$, $\mu_z(\omega) \in \mathbb{C}$ and $\mu_\alpha(\omega)$, $\mu_\beta(\omega)$, $\mu_\gamma(\omega) \in \mathbb{R}$.

A biaxial dielectric–magnetic medium is biaxial with respect to both its dielectric and magnetic properties. Such a medium is described by the Tellegen constitutive relations

$$\left. \begin{array}{l} \underline{D}(\underline{r}, \omega) = \underline{\underline{\epsilon}}_{\text{bi}}(\omega) \bullet \underline{E}(\underline{r}, \omega) \\ \underline{B}(\underline{r}, \omega) = \underline{\underline{\mu}}_{\text{bi}}(\omega) \bullet \underline{H}(\underline{r}, \omega) \end{array} \right\}, \tag{2.26}$$

wherein the symmetric permittivity and permeability dyadics $\underline{\underline{\epsilon}}_{\text{bi}}(\omega)$ and $\underline{\underline{\mu}}_{\text{bi}}(\omega)$, respectively, may be of the orthorhombic, monoclinic or triclinic type.

2.2.3　*Gyrotropy*

The uniaxial and biaxial manifestations of anisotropy introduced in Secs. 2.2.1 and 2.2.2 are characterized mathematically in terms of symmetric permittivity and permeability dyadics. In light of Sec. 1.7.2.1, uniaxial and biaxial mediums are therefore Lorentz–reciprocal. We turn now to a fundamentally different type of anisotropy, namely *gyrotropy*, which is characteristic of mediums that are Lorentz–nonreciprocal.

Realizations of gyrotropy are found within the context of magneto–optic mediums [14, 15]. As an illustrative example, let us consider an incompressible plasma of electrons in thermal motion [16]. The plasma comprises n_{el} electrons per unit volume, and has charge density $-n_{\text{el}}q_{\text{el}}$ and mass density $n_{\text{el}}m_{\text{el}}$, where the electronic charge $q_{\text{el}} = 1.6022 \times 10^{-19}$ C and electronic mass $m_{\text{el}} = 9.1096 \times 10^{-31}$ kg. A spatially uniform, quasi-static, magnetic field $\underline{B}_{\text{qs}} = B_{\text{qs}}\hat{\underline{u}}$ of strength B_{qs} is applied in the direction of the unit vector $\hat{\underline{u}}$. From our macroscopic viewpoint, we assume that the plasma is homogeneously distributed in free space. Also, the plasma density is taken to be sufficiently low so that collisions between electrons may be neglected. The average electron velocity is $\underline{\tilde{v}}(\underline{r}, t)$.

The governing equations are the time–domain Maxwell curl postulates

$$\left. \begin{aligned} \nabla \times \underline{\tilde{H}}(\underline{r}, t) - \epsilon_0 \frac{\partial}{\partial t} \underline{\tilde{E}}(\underline{r}, t) &= -n_{\text{el}} q_{\text{el}} \, \underline{\tilde{v}}(\underline{r}, t) \\ \nabla \times \underline{\tilde{E}}(\underline{r}, t) + \mu_0 \frac{\partial}{\partial t} \underline{\tilde{H}}(\underline{r}, t) &= \underline{0} \end{aligned} \right\}, \tag{2.27}$$

and the equation of motion

$$-n_{\text{el}} q_{\text{el}} \left[\underline{\tilde{E}}(\underline{r}, t) + \underline{\tilde{v}}(\underline{r}, t) \times \underline{B}_{\text{qs}} \right] = n_{\text{el}} m_{\text{el}} \frac{\partial}{\partial t} \underline{\tilde{v}}(\underline{r}, t). \tag{2.28}$$

After using Eq. (2.28) to eliminate $\underline{\tilde{v}}(\underline{r}, t)$ from Eq. $(2.27)_1$ and thereupon comparing with the source–free versions of the Maxwell curl postulates (1.8), the constitutive relations emerge as [1]

$$\left. \begin{aligned} \underline{\tilde{D}}(\underline{r}, t) &= \underline{\underline{\tilde{\epsilon}}}_{\text{gyro}} \left\{ \underline{\tilde{E}}(\underline{r}, t) \right\} \\ \underline{\tilde{H}}(\underline{r}, t) &= \frac{1}{\mu_0} \, \underline{\tilde{B}}(\underline{r}, t) \end{aligned} \right\}. \tag{2.29}$$

The permittivity dyadic operator $\underline{\underline{\tilde{\epsilon}}}_{\text{gyro}} \{\cdot\}$ is defined formally by

$$\underline{\underline{\tilde{\epsilon}}}_{\text{gyro}} = \epsilon_0 \underline{\underline{I}} + \epsilon_0 \omega_{\text{p}}^2 \left[\underline{\underline{I}} \frac{\partial^2}{\partial t^2} - \omega_{\text{c}} \left(\hat{u} \times \underline{\underline{I}} \right) \frac{\partial}{\partial t} \right]^{-1}, \tag{2.30}$$

with

$$\omega_{\text{p}} = q_{\text{el}} \left(\frac{n_{\text{el}}}{m_{\text{el}} \epsilon_0} \right)^{1/2} \tag{2.31}$$

being the *plasma frequency* and

$$\omega_{\text{c}} = q_{\text{el}} \frac{B_{\text{qs}}}{m_{\text{el}}} \tag{2.32}$$

being the *gyrofrequency*; and the inverse of the temporal differential operator may be interpreted as

$$\left(\frac{\partial}{\partial t} \right)^{-1} f(t) \equiv \int_{-\infty}^{t} f(s) \, ds. \tag{2.33}$$

By some straightforward dyadic manipulations, Eq. (2.30) may be expressed in the more convenient form [10]

$$\underline{\underline{\tilde{\epsilon}}}_{\text{gyro}} = \epsilon_0 \left[(\underline{\underline{I}} - \hat{u}\hat{u}) \left(1 + \frac{\omega_{\text{p}}^2}{\omega_{\text{c}}^2 + \partial^2/\partial t^2} \right) + (\hat{u} \times \underline{\underline{I}}) \frac{\omega_{\text{p}}^2 \omega_{\text{c}}}{(\omega_{\text{c}}^2 + \partial^2/\partial t^2) \, \partial/\partial t} \right.$$

$$\left. + \hat{u}\hat{u} \left(1 + \frac{\omega_{\text{p}}^2}{\partial^2/\partial t^2} \right) \right]; \tag{2.34}$$

herein the interpretation

$$g(t) = \frac{1}{\alpha + \partial^2/\partial t^2}\, f(t) \quad \Leftrightarrow \quad \left(\alpha + \frac{\partial^2}{\partial t^2}\right) g(t) = f(t) \qquad (2.35)$$

is implicit.

The partial derivative operators appearing in the representations (2.30) and (2.34) signify temporal dispersion. Upon applying a temporal Fourier transform, the differential operator $\partial/\partial t$ transforms to the multiplicative factor $-i\omega$. Thereby, the time–domain constitutive relations (2.29) transform into the frequency–domain constitutive relations

$$\left.\begin{aligned} \underline{D}(\underline{r},\omega) &= \underline{\underline{\epsilon}}_{\mathrm{gyro}}(\omega) \bullet \underline{E}(\underline{r},\omega) \\ \underline{B}(\underline{r},\omega) &= \mu_0\, \underline{H}(\underline{r},\omega) \end{aligned}\right\} \qquad (2.36)$$

in the Tellegen representation. The ω–dependent permittivity dyadic $\underline{\underline{\epsilon}}_{\mathrm{gyro}}$ is given by

$$\underline{\underline{\epsilon}}_{\mathrm{gyro}}(\omega) = \epsilon_0 \left[\left(1 + \frac{\omega_{\mathrm{p}}^2}{\omega_{\mathrm{c}}^2 - \omega^2}\right)\left(\underline{\underline{I}} - \hat{\underline{u}}\,\hat{\underline{u}}\right) + \frac{i\omega_{\mathrm{p}}^2 \omega_{\mathrm{c}}}{(\omega_{\mathrm{c}}^2 - \omega^2)\,\omega}\, \hat{\underline{u}} \times \underline{\underline{I}} \right.$$

$$\left. + \left(1 - \frac{\omega_{\mathrm{p}}^2}{\omega^2}\right) \hat{\underline{u}}\,\hat{\underline{u}} \right]. \qquad (2.37)$$

Without loss of generality, let us orient our coordinate system in order that the quasistatic biasing magnetic field is directed along the z axis; i.e., $\hat{\underline{u}} \equiv \hat{\underline{z}}$. Thus, the constitutive dyadic representation (2.37) may be reformulated in matrix form as

$$\underline{\underline{\epsilon}}_{\mathrm{gyro}}(\omega) = \epsilon_0 \begin{pmatrix} 1 + \dfrac{\omega_{\mathrm{p}}^2}{\omega_{\mathrm{c}}^2 - \omega^2} & -\dfrac{i\omega_{\mathrm{p}}^2 \omega_{\mathrm{c}}}{(\omega_{\mathrm{c}}^2 - \omega^2)\,\omega} & 0 \\[3ex] \dfrac{i\omega_{\mathrm{p}}^2 \omega_{\mathrm{c}}}{(\omega_{\mathrm{c}}^2 - \omega^2)\,\omega} & 1 + \dfrac{\omega_{\mathrm{p}}^2}{\omega_{\mathrm{c}}^2 - \omega^2} & 0 \\[3ex] 0 & 0 & 1 - \dfrac{\omega_{\mathrm{p}}^2}{\omega^2} \end{pmatrix}. \qquad (2.38)$$

The gyrotropic form (2.38) represents uniaxial anisotropy in the direction of the quasistatic biasing magnetic field (i.e., along $\hat{\underline{z}}$), together with antisymmetric terms associated with the plane perpendicular to that direction (i.e., in the xy plane).

In connection with the results of Sec. 1.7.2.2, we observe that the gyrotropic medium described jointly by Eqs. (2.36) and (2.37) is nondissipative. A more realistic plasma representation may be established through taking electron collisions into account. Formally, this may be achieved via the substitution $\omega \to \omega - i\nu_{\rm dp}$, where $\nu_{\rm dp} \in \mathbb{R}$ is a suitably chosen damping parameter [17].

Finally in this section, we add that gyrotropy can also arise in magnetic mediums, for example, in ferrites [18]. The Tellegen constitutive relations for a gyrotropic magnetic medium may be expressed as

$$\left.\begin{array}{l} \underline{D}(\underline{r},\omega) = \epsilon_0\, \underline{E}(\underline{r},\omega) \\[4pt] \underline{B}(\underline{r},\omega) = \underline{\underline{\mu}}_{\rm gyro}(\omega) \boldsymbol{\cdot} \underline{H}(\underline{r},\omega) \end{array}\right\}, \tag{2.39}$$

where the ω–dependent permeability dyadic

$$\underline{\underline{\mu}}_{\rm gyro}(\omega) = \mu(\omega)\left(\underline{\underline{I}} - \hat{\underline{u}}\,\hat{\underline{u}}\right) + i\mu_g(\omega)\,\hat{\underline{u}} \times \underline{\underline{I}} + \mu_u(\omega)\,\hat{\underline{u}}\,\hat{\underline{u}}, \tag{2.40}$$

with $\mu(\omega)$, $\mu_u(\omega)$, $\mu_g(\omega) \in \mathbb{R}$ for nondissipative mediums and $\mu(\omega)$, $\mu_u(\omega)$, $\mu_g(\omega) \in \mathbb{C}$ for dissipative mediums. When $\hat{\underline{u}} = \hat{\underline{z}}$, the permeability dyadic in Eq. (2.40) may be expressed in matrix notation as

$$\underline{\underline{\mu}}_{\rm gyro}(\omega) = \begin{pmatrix} \mu(\omega) & -i\mu_g(\omega) & 0 \\ i\mu_g(\omega) & \mu(\omega) & 0 \\ 0 & 0 & \mu_u(\omega) \end{pmatrix}. \tag{2.41}$$

2.3 Bianisotropy

Bianisotropy stems from the amalgamation of anisotropy with magnetoelectric coupling. Thus, in a bianisotropic medium, $\underline{D}(\underline{r},t)$ is anisotropically coupled to both $\underline{E}(\underline{r},t)$ and $\underline{B}(\underline{r},t)$, and $\underline{H}(\underline{r},t)$ is anisotropically coupled to both $\underline{E}(\underline{r},t)$ and $\underline{B}(\underline{r},t)$ as well [1, 17]. In general, a linear bianisotropic medium is described by four 3×3 constitutive dyadics.

The situations in which bianisotropy features are varied. Bianisotropic effects are observed at low frequencies and temperatures in a host of naturally occurring minerals [19, 20]. Also, bianisotropy is an increasingly important concept in the rapidly developing disciplines pertaining to composite metamaterials [21], as we discuss later in Sec. 4.8 and Sec. 6.7. Furthermore, manifestations of motion–induced bianisotropy [10] and gravitationally induced bianisotropy [22] arise in relativistic scenarios.

2.3.1 *Mediums moving at constant velocity*

Isotropy and anisotropy are not invariant under the Lorentz transformation (1.56). That is, a medium which is isotropic or anisotropic in one inertial reference frame is generally bianisotropic in all other inertial reference frames [10]. Furthermore, a homogeneous medium that is spatially local, but not temporally local, in one reference frame is generally spatiotemporally nonlocal in another reference frame [23, 24]. This fact confines analysis on physically realistic materials moving at constant velocity to a consideration only of plane waves.

2.3.2   *Uniaxial and biaxial bianisotropy*

The uniaxial and biaxial classifications introduced in Secs. 2.2.1 and 2.2.2, respectively, for dielectric and magnetic mediums extend to bianisotropic mediums in a natural way. Thus, a general biaxial structure for a homogeneous bianisotropic medium may be specified by the Tellegen constitutive relations [25]

$$
\begin{bmatrix} \underline{D}(\underline{r},\omega) \\ \underline{B}(\underline{r},\omega) \end{bmatrix} = \underline{\underline{K}}_{bi}(\omega) \cdot \begin{bmatrix} \underline{E}(\underline{r},\omega) \\ \underline{H}(\underline{r},\omega) \end{bmatrix}, \tag{2.42}
$$

where the 6×6 constitutive dyadic $\underline{\underline{K}}_{bi}(\omega)$ is assembled from four symmetric 3×3 dyadics as

$$
\underline{\underline{K}}_{bi}(\omega) = \begin{bmatrix} \underline{\underline{\epsilon}}_{bi}(\omega) & \underline{\underline{\xi}}_{bi}(\omega) \\ \underline{\underline{\zeta}}_{bi}(\omega) & \underline{\underline{\mu}}_{bi}(\omega) \end{bmatrix} \tag{2.43}
$$

$$
= \begin{bmatrix} \begin{pmatrix} \epsilon_x(\omega) & \epsilon_\alpha(\omega) & \epsilon_\beta(\omega) \\ \epsilon_\alpha(\omega) & \epsilon_y(\omega) & \epsilon_\gamma(\omega) \\ \epsilon_\beta(\omega) & \epsilon_\gamma(\omega) & \epsilon_z(\omega) \end{pmatrix} & \begin{pmatrix} \xi_x(\omega) & \xi_\alpha(\omega) & \xi_\beta(\omega) \\ \xi_\alpha(\omega) & \xi_y(\omega) & \xi_\gamma(\omega) \\ \xi_\beta(\omega) & \xi_\gamma(\omega) & \xi_z(\omega) \end{pmatrix} \\ \begin{pmatrix} \zeta_x(\omega) & \zeta_\alpha(\omega) & \zeta_\beta(\omega) \\ \zeta_\alpha(\omega) & \zeta_y(\omega) & \zeta_\gamma(\omega) \\ \zeta_\beta(\omega) & \zeta_\gamma(\omega) & \zeta_z(\omega) \end{pmatrix} & \begin{pmatrix} \mu_x(\omega) & \mu_\alpha(\omega) & \mu_\beta(\omega) \\ \mu_\alpha(\omega) & \mu_y(\omega) & \mu_\gamma(\omega) \\ \mu_\beta(\omega) & \mu_\gamma(\omega) & \mu_z(\omega) \end{pmatrix} \end{bmatrix},
$$

$$
\tag{2.44}
$$

with the ω–dependent constitutive parameters $\epsilon_{x,y,z}$, $\xi_{x,y,z}$, $\zeta_{x,y,z}$, $\mu_{x,y,z} \in \mathbb{C}$ and $\epsilon_{\alpha,\beta,\gamma}$, $\xi_{\alpha,\beta,\gamma}$, $\zeta_{\alpha,\beta,\gamma}$, $\mu_{\alpha,\beta,\gamma} \in \mathbb{R}$. The orthorhombic bianisotropic specialization follows from setting

$$
\left. \begin{array}{r} \epsilon_{\alpha,\beta,\gamma}(\omega) \\ \xi_{\alpha,\beta,\gamma}(\omega) \\ \zeta_{\alpha,\beta,\gamma}(\omega) \\ \mu_{\alpha,\beta,\gamma}(\omega) \end{array} \right\} = 0. \tag{2.45}
$$

A uniaxial bianisotropic structure is specified by setting

$$\left.\begin{array}{l} \epsilon_x(\omega) = \epsilon_y(\omega) \\ \xi_x(\omega) = \xi_y(\omega) \\ \zeta_x(\omega) = \zeta_y(\omega) \\ \mu_x(\omega) = \mu_y(\omega) \end{array}\right\}, \tag{2.46}$$

for example, in addition to the conditions (2.45).

2.3.3 *Mediums with simultaneous mirror–conjugated and racemic chirality characteristics*

A racemic mixture arises from mixing together, in equal proportions, an isotropic chiral medium and its mirror conjugate, as outlined in Sec. 2.1.3. Mirror–conjugated and racemic characteristics can co–exist in bianisotropic mediums characterized by a certain magnetic space group [26]. Magnetic space groups are discussed in Sec. 3.4. The Tellegen constitutive relations for mediums exhibiting simultaneous mirror–conjugated and racemic chirality characteristics may be expressed as

$$\begin{bmatrix} \underline{D}(\underline{r},\omega) \\ \underline{B}(\underline{r},\omega) \end{bmatrix} = \underline{\underline{K}}_{\mathrm{MCR}}(\omega) \bullet \begin{bmatrix} \underline{E}(\underline{r},\omega) \\ \underline{H}(\underline{r},\omega) \end{bmatrix}, \tag{2.47}$$

where the 6×6 constitutive dyadic $\underline{\underline{K}}_{\mathrm{MCR}}(\omega)$ is assembled from four diagonal 3×3 dyadics as

$$\underline{\underline{K}}_{\mathrm{MCR}}(\omega) = \begin{bmatrix} \underline{\underline{\epsilon}}_{\mathrm{MCR}}(\omega) & \underline{\underline{\xi}}_{\mathrm{MCR}}(\omega) \\ -\underline{\underline{\xi}}_{\mathrm{MCR}}(\omega) & \underline{\underline{\mu}}_{\mathrm{MCR}}(\omega) \end{bmatrix} \tag{2.48}$$

$$= \begin{bmatrix} \begin{pmatrix} \epsilon(\omega) & 0 & 0 \\ 0 & \epsilon(\omega) & 0 \\ 0 & 0 & \epsilon_z(\omega) \end{pmatrix} & \begin{pmatrix} \xi(\omega) & 0 & 0 \\ 0 & -\xi(\omega) & 0 \\ 0 & 0 & 0 \end{pmatrix} \\ \begin{pmatrix} -\xi(\omega) & 0 & 0 \\ 0 & \xi(\omega) & 0 \\ 0 & 0 & 0 \end{pmatrix} & \begin{pmatrix} \mu(\omega) & 0 & 0 \\ 0 & \mu(\omega) & 0 \\ 0 & 0 & \mu_z(\omega) \end{pmatrix} \end{bmatrix}. \tag{2.49}$$

The medium described by these constitutive relations is uniaxial with respect to its dielectric and magnetic properties, but biaxial with respect to its magnetoelectric properties. Chirality of opposite handedness along the x and y coordinate axes is evident from the structure of the magnetoelectric dyadic $\underline{\underline{\xi}}_{\mathrm{MCR}}(\omega)$, whereas the absence of a $\hat{z}\hat{z}$ term indicates achirality along the z coordinate axis. Notice that the magnetoelectric dyadic $\underline{\underline{\xi}}_{\mathrm{MCR}}(\omega)$ is singular.

Mediums exhibiting simultaneous mirror–conjugated and racemic chirality characteristics occur in nature in the form of certain minerals of the type RXO_4, where R is a trivalent rare earth, X is either vanadium or arsenic or phosphorus, and O is oxygen [26, 27]. These chirality characteristics are displayed below the Néel temperatures.

Artificial mediums which exhibit simultaneous mirror–conjugated and racemic chirality characteristics may be conceived by immersing parallel identical springs in a host medium [28, 29]. The structure of the constitutive dyadic $\underline{\underline{\mathbf{K}}}_{\mathrm{MCR}}(\omega)$ may be realized, at wavelengths much longer than the linear spring dimensions, by aligning right– and left–handed springs in equal linear densities along the x and y directions on a square lattice, respectively. Either no springs at all or equal proportions of left– and right–handed springs may be aligned with the z axis in order to simulate achirality and racemic chirality, respectively, with respect to this direction [30]. Other fabrication schemes, involving Ω–shaped inclusions for example, may be envisaged [31].

2.3.4   *Faraday chiral mediums*

Faraday chiral mediums (FCMs) may be conceptualized as homogenized composite mediums [32]. Their bianisotropic nature is acquired through combining

- natural optical activity, as exhibited by isotropic chiral mediums [4], with
- Faraday rotation, as exhibited by gyrotropic mediums [10, 18, 33].

Two examples of FCMs are the *chiroferrite* medium which arises from the homogenization of an isotropic chiral medium with a magnetically biased ferrite [34], and the *chiroplasma* medium which arises from the homogenization an isotropic chiral medium with a magnetically biased plasma [35]. Notice that the FCM 'inherits' Lorentz nonreciprocity from its gyrotropic constituent medium.

The frequency–domain constitutive relations of a homogeneous FCM are rigorously established as [36]

$$
\left.
\begin{aligned}
\underline{D}(\underline{r},\omega) &= \underline{\underline{\epsilon}}_{\mathrm{FCM}}(\omega) \cdot \underline{E}(\underline{r},\omega) + i\underline{\underline{\xi}}_{\mathrm{FCM}}(\omega) \cdot \underline{H}(\underline{r},\omega) \\
\underline{B}(\underline{r},\omega) &= -i\underline{\underline{\xi}}_{\mathrm{FCM}}(\omega) \cdot \underline{E}(\underline{r},\omega) + \underline{\underline{\mu}}_{\mathrm{FCM}}(\omega) \cdot \underline{H}(\underline{r},\omega)
\end{aligned}
\right\}, \qquad (2.50)
$$

in the Tellegen representation, where the 3×3 constitutive dyadics

$$
\left.
\begin{aligned}
\underline{\underline{\epsilon}}_{\mathrm{FCM}}(\omega) &= \epsilon(\omega)\,\underline{\underline{I}} + [\epsilon_u(\omega) - \epsilon(\omega)]\,\hat{\underline{u}}\,\hat{\underline{u}} + i\epsilon_g(\omega)\,\hat{\underline{u}} \times \underline{\underline{I}} \\
\underline{\underline{\xi}}_{\mathrm{FCM}}(\omega) &= \xi(\omega)\,\underline{\underline{I}} + [\xi_u(\omega) - \xi(\omega)]\,\hat{\underline{u}}\,\hat{\underline{u}} + i\xi_g(\omega)\,\hat{\underline{u}} \times \underline{\underline{I}} \\
\underline{\underline{\mu}}_{\mathrm{FCM}}(\omega) &= \mu(\omega)\,\underline{\underline{I}} + [\mu_u(\omega) - \mu(\omega)]\,\hat{\underline{u}}\,\hat{\underline{u}} + i\mu_g(\omega)\,\hat{\underline{u}} \times \underline{\underline{I}}
\end{aligned}
\right\}, \qquad (2.51)
$$

all have the same form. The ω–dependent constitutive parameters ϵ, ϵ_u, ϵ_g, ξ, ξ_g, ξ_u, μ, μ_g and μ_u are complex–valued in general, but are real–valued for a nondissipative FCM.

If the direction of the quasistatic magnetic biasing field is chosen along the z axis, then the Tellegen 6×6 constitutive dyadic of the medium may be written as follows:

$$
\underline{\underline{K}}_{\mathrm{FCM}}(\omega) =
\begin{bmatrix}
\underline{\underline{\epsilon}}_{\mathrm{FCM}}(\omega) & i\underline{\underline{\xi}}_{\mathrm{FCM}}(\omega) \\
-i\underline{\underline{\xi}}_{\mathrm{FCM}}(\omega) & \underline{\underline{\mu}}_{\mathrm{FCM}}(\omega)
\end{bmatrix}
\qquad (2.52)
$$

$$
=
\begin{bmatrix}
\begin{pmatrix} \epsilon(\omega) & -i\epsilon_g(\omega) & 0 \\ i\epsilon_g(\omega) & \epsilon(\omega) & 0 \\ 0 & 0 & \epsilon_u(\omega) \end{pmatrix} & i\begin{pmatrix} \xi(\omega) & -i\xi_g(\omega) & 0 \\ i\xi_g(\omega) & \xi(\omega) & 0 \\ 0 & 0 & \xi_u(\omega) \end{pmatrix} \\
i\begin{pmatrix} -\xi(\omega) & i\xi_g(\omega) & 0 \\ -i\xi_g(\omega) & -\xi(\omega) & 0 \\ 0 & 0 & -\xi_u(\omega) \end{pmatrix} & \begin{pmatrix} \mu(\omega) & -i\mu_g(\omega) & 0 \\ i\mu_g(\omega) & \mu(\omega) & 0 \\ 0 & 0 & \mu_u(\omega) \end{pmatrix}
\end{bmatrix}.
$$

$$
(2.53)
$$

The FCM structure described by Eqs. (2.51) is associated with the homogenization scenarios wherein electrically small spheres of the two constituent mediums are mixed together. More general FCMs forms develop through the homogenization of constituent mediums distributed as electrically small ellipsoids [35, 37].

2.3.5 *Pseudochiral omega mediums*

The Tellegen constitutive relations

$$
\begin{bmatrix} \underline{D}(\underline{r},\omega) \\ \underline{B}(\underline{r},\omega) \end{bmatrix} = \underline{\underline{K}}_\Omega(\omega) \cdot \begin{bmatrix} \underline{E}(\underline{r},\omega) \\ \underline{H}(\underline{r},\omega) \end{bmatrix}, \qquad (2.54)
$$

wherein the 6×6 constitutive dyadic

$$\underline{\underline{\mathbf{K}}}_{\Omega}(\omega) = \begin{bmatrix} \underline{\underline{\epsilon}}_{\Omega}(\omega) & \underline{\underline{\xi}}_{\Omega}(\omega) \\ \underline{\underline{\zeta}}_{\Omega}(\omega) & \underline{\underline{\mu}}_{\Omega}(\omega) \end{bmatrix} \tag{2.55}$$

$$= \begin{bmatrix} \begin{pmatrix} \epsilon_x(\omega) & 0 & 0 \\ 0 & \epsilon_y(\omega) & 0 \\ 0 & 0 & \epsilon_z(\omega) \end{pmatrix} & \begin{pmatrix} 0 & 0 & 0 \\ 0 & 0 & 0 \\ 0 & -\xi(\omega) & 0 \end{pmatrix} \\ \begin{pmatrix} 0 & 0 & 0 \\ 0 & 0 & \xi(\omega) \\ 0 & 0 & 0 \end{pmatrix} & \begin{pmatrix} \mu_x(\omega) & 0 & 0 \\ 0 & \mu_y(\omega) & 0 \\ 0 & 0 & \mu_z(\omega) \end{pmatrix} \end{bmatrix},$$

$$\tag{2.56}$$

describe a *pseudochiral omega medium* [38]. This medium is Lorentz–reciprocal. Such a medium can arise from a microstructure involving arrays of identically oriented Ω–shaped particles, with stems parallel to the Cartesian y axis in the case of eqns. (2.56), embedded in an isotropic dielectric–magnetic host medium [39]. Constitutive relations of this form also describe certain negatively refracting metamaterials which comprise arrays of electrically small split–ring resonators [40].

2.4   Nonhomogeneous mediums

In the preceding sections of this chapter, the descriptions of anisotropy and bianisotropy have been provided in terms of constitutive dyadics which are independent of the position vector $\underline{r}$. However, for many important classes of complex mediums, the macroscopic properties cannot be adequately described without accounting for the dependency of the constitutive properties upon $\underline{r}$. Nonhomogeneity undoubtedly increases the complexity of mathematical analyses; nevertheless, theoretical treatments of electromagnetic properties are well–established for certain nonhomogeneous anisotropic and bianisotropic mediums.

2.4.1   *Periodic nonhomogeneity*

An extensively studied example of anisotropy in nonhomogeneous mediums is furnished by cholesteric liquid crystals (CLCs). The periodic nonhomogeneity of CLCs is a manifestation of their helicoidal arrangement of aciculate molecules. In essence, CLCs are periodically twisted, uniaxial dielectric mediums. Their frequency–domain constitutive relations may be written as[2] [8, 9]

$$
\left.\begin{aligned}
\underline{D}(\underline{r},\omega) &= \underline{\underline{\epsilon}}_{\mathrm{CLC}}(z,\omega) \bullet \underline{E}(\underline{r},\omega) \\
\underline{B}(\underline{r},\omega) &= \mu_0\, \underline{H}(\underline{r},\omega)
\end{aligned}\right\},
\tag{2.57}
$$

where the nonhomogeneous permittivity dyadic

$$
\underline{\underline{\epsilon}}_{\mathrm{CLC}}(z,\omega) = \epsilon_{\mathrm{p}}(\omega)\,\underline{\underline{I}} + [\,\epsilon_{\mathrm{q}}(\omega) - \epsilon_{\mathrm{p}}(\omega)\,]\,\hat{\underline{u}}(z)\,\hat{\underline{u}}(z)
\tag{2.58}
$$

is described in terms of the complex–valued, ω–dependent, scalars ϵ_{p} and ϵ_{q} and the unit vector

$$
\hat{\underline{u}}(z) = \hat{\underline{x}}\,\cos\frac{\pi z}{\Omega_{\mathrm{dp}}} \pm \hat{\underline{y}}\,\sin\frac{\pi z}{\Omega_{\mathrm{dp}}}.
\tag{2.59}
$$

Here $2\Omega_{\mathrm{dp}}$ is the dielectric periodicity along the z direction with which the helicoidal axis is aligned, the '+' sign is chosen to denote structural right–handedness whereas the '−' sign is chosen to denote structural left–handedness.

The helicoidal conformation which is central to CLCs also underpins chiral sculptured thin films (CSTFs) [42]. These comprise parallel helical nanowires deposited on a substrate. The helical shape engenders structural chirality, while the nanowires on the average are characterized by local orthorhombic symmetry. CSTFs are of considerable technological interest because of their optical properties.

The CSTF constitutive relations may expressed as

$$
\left.\begin{aligned}
\underline{D}(\underline{r},\omega) &= \underline{\underline{\epsilon}}_{\mathrm{CSTF}}(z,\omega) \bullet \underline{E}(\underline{r},\omega) \\
\underline{B}(\underline{r},\omega) &= \mu_0\, \underline{H}(\underline{r},\omega)
\end{aligned}\right\},
\tag{2.60}
$$

where the nonhomogeneous permittivity dyadic

$$
\underline{\underline{\epsilon}}_{\mathrm{CSTF}}(z,\omega) = \underline{\underline{S}}_z(z) \bullet \underline{\underline{S}}_y(\psi) \bullet
\begin{pmatrix}
\epsilon_{\mathrm{u}}(\omega) & 0 & 0 \\
0 & \epsilon_{\mathrm{v}}(\omega) & 0 \\
0 & 0 & \epsilon_{\mathrm{w}}(\omega)
\end{pmatrix}
\bullet \underline{\underline{S}}_y^{-1}(\psi) \bullet \underline{\underline{S}}_z^{-1}(z)
\tag{2.61}
$$

---

[2]CLCs are also known as chiral nematic liquid crystals. From an optical point of view, these may be considered as special cases of chiral smectic liquid crystals, which are characterized by constitutive relations of a similar form to Eqs. (2.57) but with the corresponding permittivity dyadic having a biaxial nature [41].

is provided by three complex–valued, ω–dependent parameters ϵ_u, ϵ_v and ϵ_w. The rotation and tilt dyadics

$$\underline{\underline{S}}_y(\psi) = \begin{pmatrix} \cos\psi & 0 & -\sin\psi \\ 0 & 1 & 0 \\ \sin\psi & 0 & \cos\psi \end{pmatrix}, \qquad (2.62)$$

$$\underline{\underline{S}}_z(z) = \begin{pmatrix} \cos\left(\dfrac{\pi z}{\Omega_{hp}}\right) & -h\sin\left(\dfrac{\pi z}{\Omega_{hp}}\right) & 0 \\ h\sin\left(\dfrac{\pi z}{\Omega_{hp}}\right) & \cos\left(\dfrac{\pi z}{\Omega_{hp}}\right) & 0 \\ 0 & 0 & 1 \end{pmatrix}, \qquad (2.63)$$

specify the nanowire morphology wherein Ω_{hp} is the half–pitch of the helical nanowires, $h = \pm 1$ denotes the structural handedness, and $\psi \in [0, \pi/2]$.

The generalization of the CSTF form (2.60)–(2.61) results in

$$\left. \begin{aligned} \underline{D}(\underline{r},\omega) &= \underline{\underline{\epsilon}}_{HBM}(z,\omega) \bullet \underline{E}(\underline{r},\omega) + \underline{\underline{\xi}}_{HBM}(z,\omega) \bullet \underline{H}(\underline{r},\omega) \\ \underline{B}(\underline{r},\omega) &= \underline{\underline{\zeta}}_{HBM}(z,\omega) \bullet \underline{E}(\underline{r},\omega) + \underline{\underline{\mu}}_{HBM}(z,\omega) \bullet \underline{H}(\underline{r},\omega) \end{aligned} \right\}, \qquad (2.64)$$

which are the constitutive relations for a helicoidal bianisotropic medium (HBM) [43]. All four of the 3×3 nonhomogeneous constitutive dyadics herein have the general form

$$\underline{\underline{\beta}}_{HBM}(z,\omega) = \begin{pmatrix} \beta_{11}(z,\omega) & \beta_{12}(z,\omega) & \beta_{13}(z,\omega) \\ \beta_{21}(z,\omega) & \beta_{22}(z,\omega) & \beta_{23}(z,\omega) \\ \beta_{31}(z,\omega) & \beta_{32}(z,\omega) & \beta_{33}(z,\omega) \end{pmatrix}, \qquad (\beta = \epsilon, \xi, \zeta, \mu),$$
$$(2.65)$$

which may be factored as

$$\underline{\underline{\beta}}_{HBM}(z,\omega) = \underline{\underline{S}}_z(z) \bullet \underline{\underline{\beta}}_{HBM}(0,\omega) \bullet \underline{\underline{S}}_z^{-1}(z), \qquad (\beta = \epsilon, \xi, \zeta, \mu). \quad (2.66)$$

For further details on the properties exhibited by CSTFs and HBMs, the reader is referred to the specialist literature [44].

2.4.2 *Gravitationally induced bianisotropy*

Up to this point in this chapter, the implicit backdrop for our description of electromagnetic phenomenons has been flat spacetime. We now digress in order to show that the classical vacuum in curved spacetime is not isotropic — in fact, it may be regarded as a bianisotropic medium.

In the literature on general relativity, the spacetime curvature induced by a gravitational field is conventionally characterized using a symmetric covariant tensor of rank 2, called the *metric* [45]. In keeping with the notational practices implemented throughout this book, we represent the spacetime metric by the symmetric, real–valued, 4×4 matrix $\tilde{\underline{\underline{g}}}(\underline{r}, t)$. The components of $\tilde{\underline{\underline{g}}}(\underline{r}, t)$ are the spacetime metric components. The matrix is both spatially nonhomogeneous and time–dependent, in general.

Upon following a well–established procedure wherein the covariant Maxwell equations are expressed in noncovariant form [46–48], the same spacetime coordinate is fixed as time throughout all spacetime. Then, the electromagnetic response of vacuum in curved spacetime may be represented by the time–domain constitutive relations [22, 49]

$$\left. \begin{array}{l} \tilde{\underline{D}}(\underline{r}, t) = \epsilon_0 \, \tilde{\underline{\underline{\gamma}}}_{\text{GAV}}(\underline{r}, t) \bullet \tilde{\underline{E}}(\underline{r}, t) - \dfrac{1}{c_0} \tilde{\underline{\Gamma}}_{\text{GAV}}(\underline{r}, t) \times \tilde{\underline{H}}(\underline{r}, t) \\[4mm] \tilde{\underline{B}}(\underline{r}, t) = \mu_0 \, \tilde{\underline{\underline{\gamma}}}_{\text{GAV}}(\underline{r}, t) \bullet \tilde{\underline{H}}(\underline{r}, t) + \dfrac{1}{c_0} \tilde{\underline{\Gamma}}_{\text{GAV}}(\underline{r}, t) \times \tilde{\underline{E}}(\underline{r}, t) \end{array} \right\}, \qquad (2.67)$$

in Tellegen form. The 3×3 dyadic $\tilde{\underline{\underline{\gamma}}}_{\text{GAV}}(\underline{r}, t)$ has components

$$\left[\tilde{\underline{\underline{\gamma}}}_{\text{GAV}}(\underline{r}, t) \right]_{\ell m} = - \frac{\left[- \det \tilde{\underline{\underline{g}}}(\underline{r}, t) \right]^{1/2}}{\left[\tilde{\underline{\underline{g}}}(\underline{r}, t) \right]_{1,1}} \left[\tilde{\underline{\underline{g}}}^{-1}(\underline{r}, t) \right]_{\ell+1, m+1},$$

$$(\ell, m \in \{1, 2, 3\}), \quad (2.68)$$

and the 3–vector $\tilde{\underline{\Gamma}}_{\text{GAV}}(\underline{r}, t)$ is given as

$$\tilde{\underline{\Gamma}}_{\text{GAV}}(\underline{r}, t) = \left(\left[\tilde{\underline{\underline{g}}}(\underline{r}, t) \right]_{1,1} \right)^{-1} \begin{pmatrix} \left[\tilde{\underline{\underline{g}}}(\underline{r}, t) \right]_{1,2} \\[2mm] \left[\tilde{\underline{\underline{g}}}(\underline{r}, t) \right]_{1,3} \\[2mm] \left[\tilde{\underline{\underline{g}}}(\underline{r}, t) \right]_{1,4} \end{pmatrix}, \qquad (2.69)$$

wherein $\left[\tilde{\underline{\underline{g}}}(\underline{r}, t) \right]_{\ell,m}$ denotes the (ℓ, m)–th element of $\tilde{\underline{\underline{g}}}(\underline{r}, t)$. Thus, the mathematical description of electromagnetic fields in gravitationally affected vacuum is isomorphic to the description of electromagnetic fields in a fictitious, instantaneously responding medium characterized by the constitutive relations (2.67). Consequently, analytical techniques commonly used to investigate electromagnetic problems (without considering the effects of gravitational fields) may be applied to the study of gravitationally affected vacuum.

The fictitious medium described by Eqs. (2.67) is generally bianisotropic, with the symmetries of its constitutive dyadics being determined by those of the underlying metric. The constitutive dyadics are both spatially nonhomogeneous and time–varying. The medium is neither Lorentz–reciprocal in general (cf. Sec. 1.7.2.1) nor dissipative (cf. Sec. 1.7.2.2), but it does satisfy the Post constraint (1.81) [50]. Since the matrix $\tilde{g}(\underline{r},t)$ is real symmetric, the permittivity and permeability dyadics $\epsilon_o \, \tilde{\underline{\underline{\gamma}}}_{\mathrm{GAV}}(\underline{r},t)$ and $\mu_o \, \tilde{\underline{\underline{\gamma}}}_{\mathrm{GAV}}(\underline{r},t)$, respectively, are of the orthorhombic type and they have the same eigenvectors. In flat spacetime the matrix $\tilde{g}(\underline{r},t)$ simplifies to give the constant diagonal matrix

$$\tilde{g}(\underline{r},t) = \begin{pmatrix} 1 & 0 & 0 & 0 \\ 0 & -1 & 0 & 0 \\ 0 & 0 & -1 & 0 \\ 0 & 0 & 0 & -1 \end{pmatrix}, \tag{2.70}$$

and the constitutive relations (2.67) reduce to the familiar form (2.1).

Finally, the fictitious bianisotropic medium described by the constitutive relations (2.67) has no underlying microstructure, unlike the material mediums described elsewhere in this book. Therefore the distinction between macroscopic and microscopic perspectives, as discussed in Sec. 1.1, is not pertinent to this special case.

References

[1] W.S. Weiglhofer, A flavour of constitutive relations: the linear regime, *Advances in electromagnetics of complex media and metamaterials* (S. Zouhdi, A. Sihvola and M. Arsalane, eds), Kluwer, Dordrecht, The Netherlands, 2003, 61–80.

[2] R.A. Depine and A. Lakhtakia, A new condition to identify isotropic dielectric-magnetic materials displaying negative phase velocity, *Microw Opt Technol Lett* **41** (2004), 315–316.

[3] A. Lakhtakia, M.W. McCall and W.S. Weiglhofer, Negative phase–velocity mediums, *Introduction to complex mediums for optics and electromagnetics* (W.S. Weiglhofer and A. Lakhtakia, eds), SPIE Press, Bellingham, WA, USA, 2003, 347–363.

[4] A. Lakhtakia, *Beltrami fields in chiral media*, World Scientific, Singapore, 1994.

[5] A. Lakhtakia (ed), *Selected papers on natural optical activity*, SPIE Optical Engineering Press, Bellingham, WA, USA, 1990.

[6] M. Born and E. Wolf, *Principles of optics, 7th (expanded) ed*, Cambridge University Press, Cambridge, UK, 1999.

[7] J.F. Nye, *Physical properties of crystals*, Oxford University Press, Oxford, UK, 1985.

[8] S. Chandrasekhar, *Liquid crystals, 2nd ed*, Cambridge University Press, Cambridge, UK, 1992.

[9] P.G. de Gennes and J.A. Prost, *The physics of liquid crystals, 2nd ed*, Clarendon Press, Oxford, UK, 1993.

[10] H.C. Chen, *Theory of electromagnetic waves*, McGraw–Hill, New York, NY, USA, 1983.

[11] W.S. Weiglhofer and A. Lakhtakia, On electromagnetic waves in biaxial bianisotropic media, *Electromagnetics* **19** (1999), 351–362.

[12] T.G. Mackay and W.S. Weiglhofer, Homogenization of biaxial composite materials: dissipative anisotropic properties, *J Opt A: Pure Appl Opt* **2** (2000), 426–432.

[13] N.W. Ashcroft and N.D. Mermin, *Solid state physics*, Saunders, Philadelphia, PA, USA, 1976.

[14] M. Mansuripur, *The physical principles of magneto–optical recording*, Cambridge University Press, Cambridge, UK, 1995.

[15] J.I. Gersten and F.W. Smith, *The physics and chemistry of materials*, Wiley, New York, NY, USA, 2001.

[16] L.B. Felsen and N. Marcuvitz, *Radiation and scattering of waves*, IEEE Press, Piscataway, NJ, USA, 1994.

[17] W.S. Weiglhofer, Constitutive characterization of simple and complex mediums, *Introduction to complex mediums for optics and electromagnetics* (W.S. Weiglhofer and A. Lakhtakia, eds), SPIE Press, Bellingham, WA, USA, 2003, 27–61.

[18] B. Lax and K.J. Button, *Microwave ferrites and ferrimagnetics*, McGraw–Hill, New York, NY, USA, 1962.

[19] T.H. O'Dell, *The electrodynamics of magneto–electric media*, North–Holland, Amsterdam, The Netherlands, 1970.

[20] H. Schmid, Magnetoelectric effects in insulating magnetic materials, *Introduction to complex mediums for optics and electromagnetics* (W.S. Weiglhofer and A. Lakhtakia, eds), SPIE Press, Bellingham, WA, USA, 2003, 167–195.

[21] T.G. Mackay, Linear and nonlinear homogenized composite mediums as metamaterials, *Electromagnetics* **25** (2005), 461–481.

[22] W. Schleich and M.O. Scully, General relativity and modern optics *New trends in atomic physics* (G. Grynberg and R. Stora, eds), Elsevier, Amsterdam, Holland, 1984, 995–1124.

[23] A. Lakhtakia and W.S. Weiglhofer, Lorentz covariance, Occam's razor, and a constraint on linear constitutive relations, *Phys Lett A* **213** (1996), 107–111. Corrections: **222** (1996), 459.

[24] D. Censor, Electrodynamics, topsy–turvy special relativity, and generalized Minkowski constitutive relations for linear and nonlinear systems, *Prog Electromagn Res* **18** (1998), 261–284.

[25] T.G. Mackay and W.S. Weiglhofer, A review of homogenization studies for biaxial bianisotropic materials, *Advances in electromagnetics of complex media and metamaterials* (S. Zouhdi, A. Sihvola and M. Arsalane, eds), Kluwer, Dordrecht, The Netherlands, 2003, 211–228.

[26] G. Gorodetsky, R.M. Hornreich and B.M. Wanklyn, Statistical mechanics and critical behavior of the magnetoelectric effect in $GdVO_4$, *Phys Rev B* **8** (1973), 2263–2267.

[27] A.H. Cooke, S.J. Swithenby and M.R. Wells, Magnetoelectric measurements on holmium phosphate, $HoPO_4$, *Int J Magnetism* **4** (1973), 309–312.

[28] J.C. Bose, On the rotation of plane of polarization of electric waves by a twisted structure, *Proc R Soc Lond* **63** (1898), 146–152.

[29] K.W. Whites and C.Y. Chung, Composite uniaxial bianisotropic chiral materials characterization: comparison of prediction and measured scattering, *J Electromag Waves Applics* **11** (1997), 371–394.

[30] A. Lakhtakia and W.S. Weiglhofer, Electromagnetic waves in a material with simultaneous mirror–conjugated and racemic chirality characteristics, *Electromagnetics* **20** (2000), 481–488. (The first negative sign on the right side of Eq. (26) of this paper should be replaced by a positive sign).

[31] A.A. Sochava, C.R. Simovski and S.A. Tretyakov, Chiral effects and eigenwaves in bi–anisotropic omega structures, *Advances in complex electromagnetic materials* (A. Priou, A. Sihvola, S. Tretyakov and A. Vinogradov, eds), Kluwer, Dordrecht, The Netherlands, 1997, 85–102.

[32] N. Engheta, D.L. Jaggard and M.W. Kowarz, Electromagnetic waves in Faraday chiral media, *IEEE Trans Antennas Propagat* **40** (1992), 367–374.

[33] R.E. Collin, *Foundations for microwave engineering*, McGraw–Hill, New York, NY, USA, 1966.

[34] W.S. Weiglhofer, A. Lakhtakia and B. Michel, On the constitutive parameters of a chiroferrite composite medium, *Microw Opt Technol Lett* **18** (1998), 342–345.

[35] W.S. Weiglhofer and T.G. Mackay, Numerical studies of the constitutive parameters of a chiroplasma composite medium, *Arch Elektron Übertrag* **54** (2000), 259–265.

[36] W.S. Weiglhofer and A. Lakhtakia, The correct constitutive relations of chiroplasmas and chiroferrites, *Microw Opt Technol Lett* **17** (1998), 405–408.

[37] T.G. Mackay, A. Lakhtakia and W.S. Weiglhofer, Ellipsoidal topology, orientation diversity and correlation length in bianisotropic composite mediums, *Arch Elektron Übertrag* **55** (2001), 243–251.

[38] A. Serdyukov, I. Semchenko, S. Tretyakov and A. Sihvola, *Electromagnetics of bi–anisotropic materials*, Gordon and Breach, Amsterdam, The Netherlands, 2001.

[39] M.M.I. Saadoun and N. Engheta, A reciprocal phase shifter using novel pseudochiral or omega medium, *Microw Opt Technol Lett* **5** (1992), 184–188.

[40] X. Chen, B.–I. Wu, J.A. Kong and T.M. Grzegorczyk, Retrieval of the effective constitutive parameters of bianisotropic materials, *Phys Rev E* **71** (2005), 046610. Corrections: **73** (2006), 019905(E).

[41] I. Abdulhalim, Light propagation along the helix of chiral smectics and twisted nematics, *Opt Commun* **64** (1987), 443–448.

[42] V.C. Venugopal and A. Lakhtakia, Sculptured thin films: conception, optical properties and applications, *Electromagnetic fields in unconventional materials and structures* (O.N. Singh and A. Lakhtakia, eds), Wiley, New York, NY, USA, 2000, 151–216.

[43] A. Lakhtakia and W.S. Weiglhofer, On light propagation in helicoidal bianisotropic mediums, *Proc R Soc Lond A* **448** (1995), 419–437. Corrections: **454** (1998), 3275.

[44] A. Lakhtakia and R. Messier, *Sculptured thin films: Nanoengineered morphology and optics*, SPIE Press, Bellingham, WA, USA, 2005.

[45] R. d'Inverno *Introducing Einstein's relativity*, Clarendon Press, Oxford, UK, 1992.

[46] L.D. Landau and E.M. Lifshitz, *The classical theory of fields*, Clarendon Press, Oxford, UK, 1975, §90.

[47] G.V. Skrotskii, The influence of gravitation on the propagation of light, *Soviet Phys–Dokl* **2** (1957) 226–229.

[48] J. Plébanski, Electromagnetic waves in gravitational fields, *Phys Rev* **118** (1960) 1396–1408.

[49] A. Lakhtakia and T.G. Mackay, Towards gravitationally assisted negative refraction of light by vacuum, *J Phys A: Math Gen* **37** (2004), L505–L510. Corrections: **37** (2004), 12093.

[50] A. Lakhtakia, On the genesis of Post constraint in modern electromagnetism, *Optik* **115** (2004), 151–158.

Chapter 3

Spacetime Symmetries and Constitutive Dyadics

The electromagnetic properties of a linear medium are characterized in terms of its constitutive dyadics: the four 3×3 dyadics $\underline{\underline{\epsilon}}_{EH}(\underline{r}, \omega)$, $\underline{\underline{\xi}}_{EH}(\underline{r}, \omega)$, $\underline{\underline{\zeta}}_{EH}(\underline{r}, \omega)$ and $\underline{\underline{\mu}}_{EH}(\underline{r}, \omega)$ in the Tellegen representation (or $\underline{\underline{\epsilon}}_{EB}(\underline{r}, \omega)$, $\underline{\underline{\xi}}_{EB}(\underline{r}, \omega)$, $\underline{\underline{\zeta}}_{EB}(\underline{r}, \omega)$ and $\underline{\underline{\nu}}_{EB}(\underline{r}, \omega)$ in the Boys–Post representation). The symmetries of these constitutive dyadics reflect the underlying spacetime symmetries of the medium. Indeed, the general form of the constitutive dyadics may be deduced from a consideration of macroscopic spacetime symmetries. In this chapter, we largely focus upon those spacetime symmetry operations which bring a medium into self–coincidence, leaving a point within the medium fixed in position. These point symmetry operations — which form the basis of the point group classifications — determine the general forms of the constitutive dyadics.

We begin with a description of the discrete spatial symmetries as characterized within the framework of classical crystallography, and then show how this description may be extended to discrete spacetime symmetries for nonmagnetic and magnetic crystals. Continuous spacetime symmetries are important for artificial anisotropic and bianisotropic mediums, and are also discussed. A brief description of point symmetries combined with translational symmetries, as encountered in the classification of space groups, is provided in the concluding section of this chapter.

3.1 Point groups of classical crystallography

A natural mathematical language for the description of symmetry operations is provided by group theory [1]. A *group* G is defined to be a set (of elements x) which satisfies the following group axioms [2]:

(1) G is closed under multiplication;
(2) multiplication is associative on G;
(3) an identity element e ∈ G exists such that ex = xe = x for all x ∈ G; and
(4) for each x ∈ G, there exists an inverse element x^{-1} ∈ G such that $xx^{-1} = x^{-1}x = $ e.

The order of a group is the number of elements it contains. An important concept is that of *generators* of a group. These are a typically small number of elements which if applied repeatedly — singly or in combination — generate all of the other elements of the group.

The realm of classical crystallography encompasses discrete spatial symmetries. Here we focus on those spatial–symmetry operations which bring a crystal into self–coincidence, but leave a point fixed; i.e., proper and improper rotations, reflections and inversions. These spatial symmetry operations provide the basis for the point group classifications. A discussion of translational symmetry operations is deferred until we consider space groups in the final section of this chapter.

The notations used in the descriptions of point groups require special mention. Two systems are commonly used: the International (or Hermann–Mauguin) and the Schoenflies [3]. In the International system, n signifies an $n$fold proper rotation, whereas $\bar{n}$ denotes rotation followed by reflection in a plane perpendicular to the axis of rotation (i.e., an improper rotation). Reflection in a plane is represented by m. An $n$fold axis with a perpendicular mirror plane is specified by n/m, whereas nm indicates an $n$fold axis contained within a mirror plane. A twofold axis perpendicular to a $n$fold axis is represented as $n2$.

In the Schoenflies system, C_n indicates a proper rotation through $2\pi/n$, while σ (also written as C_s) denotes reflection in a plane. Reflection in a plane perpendicular to a symmetry axis is represented by σ_h, whereas σ_v signifies reflection in a plane containing a symmetry axis. Reflection in a plane containing a symmetry axis, wherein the plane bisects the angle between a pair of twofold rotational axes, is represented by σ_d. The $n$fold improper rotation $S_n \equiv C_{nh}$ arises from combining the proper rotation C_n with the reflection σ_h. The symbol D_n describes groups having an $n$fold axis and twofold axes lying in a perpendicular plane. The combination of D_n with σ_d or σ_h yields D_{nd} or D_{nh}, respectively. The cubic groups have their own notation, with T and O representing 'tetrahedral' and 'octahedral', respectively. While the groups T and O contain only rotations, T_d also has planes of symmetry, and T_h and O_h contain a centre of symmetry.

The symmetry operations of classical crystallography may be represented in terms of real–valued 3×3 matrixes $\underline{\underline{M}}_i$, and the corresponding groups of symmetry operations may be represented as groups of matrixes [3, 4]. For example, invariance under the symmetry operations corresponding to $\underline{\underline{M}}_i$ is represented as

$$\underline{\underline{\epsilon}}(\underline{r},\omega) = \underline{\underline{M}}_i^{-1} \cdot \underline{\underline{\epsilon}}(\underline{r},\omega) \cdot \underline{\underline{M}}_i, \qquad (3.1)$$

for a dielectric medium with permittivity dyadic $\underline{\underline{\epsilon}}(\underline{r},\omega)$. The generating matrixes for the point groups of classical crystallography are provided in Table 3.1, along with descriptions of the symmetry operations themselves. The matrixes presented in Table 3.1 are expressed with respect to the Cartesian coordinate basis vectors $\{\hat{\underline{x}},\hat{\underline{y}},\hat{\underline{z}}\}$. Note that the choice of generating matrixes is not unique.

The generating matrixes listed in Table 3.1, together with the 3×3 identity matrix $\underline{\underline{I}}$, give rise to the 32 classical point groups [5]. The 32 classical point groups are divided into the seven *crystal systems*, namely the triclinic, monoclinic, orthorhombic, tetragonal, trigonal, hexagonal and cubic systems. The symmetries associated with the point groups and their crystal systems are described in Table 3.2.

Within the realm of classical optics, the triclinic, monoclinic and orthorhombic crystal systems characterize mediums which are optically biaxial; the tetragonal, trigonal and hexagonal crystal systems characterize mediums which are optically uniaxial; and the cubic crystal system characterizes optically isotropic mediums. Strictly speaking, these optical classifications only apply within regime of linear electromagnetics [6]. The correspondence between the cubic crystal system and optical isotropy follows because a crystal's optical response may be described by a second–order tensor (i.e., the permittivity dyadic). For physical properties described by higher–order tensors (e.g., the elastic stiffness tensor), the correspondence between the cubic crystal system and isotropy breaks down [7, 8].

3.2 Magnetic point groups

The point groups of classical crystallography pertain to spatial symmetries exclusively. However, magnetic and nonmagnetic crystals cannot be distinguished on this basis. By way of illustration, consider the atoms of ferromagnetic and ferrimagnetic crystals which possess magnetic moments in the absence of an applied magnetic field [9, 10]. Under a spatial symmetry

Table 3.1   Generating matrixes for the point groups of classical crystallography.

Generating matrix	Symmetry operation	Generating matrix	Symmetry operation
$\underline{\underline{M}}_1 = \begin{pmatrix} -1 & 0 & 0 \\ 0 & -1 & 0 \\ 0 & 0 & -1 \end{pmatrix}$	inversion	$\underline{\underline{M}}_2 = \begin{pmatrix} -1 & 0 & 0 \\ 0 & -1 & 0 \\ 0 & 0 & 1 \end{pmatrix}$	twofold rotation about z axis
$\underline{\underline{M}}_3 = \begin{pmatrix} 1 & 0 & 0 \\ 0 & 1 & 0 \\ 0 & 0 & -1 \end{pmatrix}$	reflection in xy plane	$\underline{\underline{M}}_4 = \begin{pmatrix} 1 & 0 & 0 \\ 0 & -1 & 0 \\ 0 & 0 & -1 \end{pmatrix}$	twofold rotation about x axis
$\underline{\underline{M}}_5 = \begin{pmatrix} -1 & 0 & 0 \\ 0 & 1 & 0 \\ 0 & 0 & 1 \end{pmatrix}$	reflection in yz plane	$\underline{\underline{M}}_6 = \begin{pmatrix} 1 & 0 & 0 \\ 0 & -1 & 0 \\ 0 & 0 & 1 \end{pmatrix}$	reflection in xz plane
$\underline{\underline{M}}_7 = \begin{pmatrix} 0 & -1 & 0 \\ 1 & 0 & 0 \\ 0 & 0 & 1 \end{pmatrix}$	fourfold rotation about z axis	$\underline{\underline{M}}_8 = \begin{pmatrix} 0 & -1 & 0 \\ 1 & 0 & 0 \\ 0 & 0 & -1 \end{pmatrix}$	fourfold inversion–rotation about z axis

Table 3.1   continued

Generating matrix	Symmetry operation	Generating matrix	Symmetry operation
$\underline{\underline{M}}_9 = \begin{pmatrix} -1/2 & -\sqrt{3}/2 & 0 \\ \sqrt{3}/2 & -1/2 & 0 \\ 0 & 0 & 1 \end{pmatrix}$	threefold rotation about z axis	$\underline{\underline{M}}_{10} = \begin{pmatrix} 1/2 & -\sqrt{3}/2 & 0 \\ \sqrt{3}/2 & 1/2 & 0 \\ 0 & 0 & -1 \end{pmatrix}$	threefold inversion–rotation about z axis
$\underline{\underline{M}}_{11} = \begin{pmatrix} 1/2 & -\sqrt{3}/2 & 0 \\ \sqrt{3}/2 & 1/2 & 0 \\ 0 & 0 & 1 \end{pmatrix}$	sixfold rotation about z axis	$\underline{\underline{M}}_{12} = \begin{pmatrix} -1/2 & -\sqrt{3}/2 & 0 \\ \sqrt{3}/2 & -1/2 & 0 \\ 0 & 0 & -1 \end{pmatrix}$	sixfold inversion–rotation about z axis
$\underline{\underline{M}}_{13} = \begin{pmatrix} 0 & 0 & 1 \\ 1 & 0 & 0 \\ 0 & 1 & 0 \end{pmatrix}$	threefold rotation about $(1,1,1)$ direction	$\underline{\underline{M}}_{14} = \begin{pmatrix} 0 & -1 & 0 \\ 0 & 0 & -1 \\ -1 & 0 & 0 \end{pmatrix}$	threefold inversion–rotation about $(1,1,1)$ direction

Table 3.2   Thirty two point groups of classical crystallography.

Crystal system [Optical classification]	International symbol (abbrev.)	Schoenflies symbol	No. of symmetry operations	Generating matrixes
Triclinic [biaxial]	1 $\bar{1}$	C_1 $C_i \equiv S_2$	1 2	$\underline{\underline{I}}$ $\underline{\underline{M}}_1$
Monoclinic [biaxial]	2 m $2/m$	C_2 $C_{1h} \equiv C_s$ C_{2h}	2 2 4	$\underline{\underline{M}}_2$ $\underline{\underline{M}}_3$ $\underline{\underline{M}}_2, \underline{\underline{M}}_3$
Orthorhombic [biaxial]	222 $mm2$ mmm	$D_2 \equiv V$ C_{2v} $V_h \equiv D_{2h}$	4 4 8	$\underline{\underline{M}}_2, \underline{\underline{M}}_4$ $\underline{\underline{M}}_2, \underline{\underline{M}}_5$ $\underline{\underline{M}}_3, \underline{\underline{M}}_5, \underline{\underline{M}}_6$
Tetragonal [uniaxial]	4 $\bar{4}$ 422 $4/m$ $4mm$ $\bar{4}2m$ $4/mmm$	C_4 S_4 D_4 C_{4h} C_{4v} $V_d \equiv D_{2d}$ D_{4h}	4 4 8 8 8 8 16	$\underline{\underline{M}}_7$ $\underline{\underline{M}}_8$ $\underline{\underline{M}}_4, \underline{\underline{M}}_7$ $\underline{\underline{M}}_3, \underline{\underline{M}}_7$ $\underline{\underline{M}}_5, \underline{\underline{M}}_7$ $\underline{\underline{M}}_4, \underline{\underline{M}}_8$ $\underline{\underline{M}}_3, \underline{\underline{M}}_5, \underline{\underline{M}}_7$

After [5].

Table 3.2   continued

Crystal system [Optical classification]	International symbol (abbrev.)	Schoenflies symbol	No. of symmetry operations	Generating matrixes
Trigonal [uniaxial]	3	C_3	3	$\underline{\underline{M}}_9$
	$\bar{3}$	$C_{3i} \equiv S_6$	6	$\underline{\underline{M}}_{10}$
	32	D_3	6	$\underline{\underline{M}}_4, \underline{\underline{M}}_9$
	$3m$	C_{3v}	6	$\underline{\underline{M}}_5, \underline{\underline{M}}_9$
	$\bar{3}m$	D_{3d}	12	$\underline{\underline{M}}_5, \underline{\underline{M}}_{10}$
Hexagonal [uniaxial]	6	C_6	6	$\underline{\underline{M}}_{11}$
	$\bar{6}$	C_{3h}	6	$\underline{\underline{M}}_{12}$
	$\bar{6}m2$	D_{3h}	12	$\underline{\underline{M}}_5, \underline{\underline{M}}_{12}$
	622	D_6	12	$\underline{\underline{M}}_4, \underline{\underline{M}}_{11}$
	$6/m$	C_{6h}	12	$\underline{\underline{M}}_3, \underline{\underline{M}}_{11}$
	$6mm$	C_{6v}	12	$\underline{\underline{M}}_5, \underline{\underline{M}}_{11}$
	$6/mmm$	D_{6h}	24	$\underline{\underline{M}}_3, \underline{\underline{M}}_5, \underline{\underline{M}}_{11}$
Cubic [isotropic]	23	T	12	$\underline{\underline{M}}_2, \underline{\underline{M}}_{13}$
	$m3$	T_h	24	$\underline{\underline{M}}_2, \underline{\underline{M}}_{14}$
	432	O	24	$\underline{\underline{M}}_7, \underline{\underline{M}}_{13}$
	$\bar{4}3m$	T_d	24	$\underline{\underline{M}}_8, \underline{\underline{M}}_{13}$
	$m3m$	O_h	48	$\underline{\underline{M}}_7, \underline{\underline{M}}_{14}$

operation which brings similar atoms into coincidence, the atomic magnetic moments may not be brought into coincidence.

In order to characterize the symmetries of such magnetic crystals, we need to consider time–reversal symmetry in addition to spatial symmetries. Let us note that the atomic magnetic moments, otherwise known as spins, are often naively viewed as arising from rotating charges. However, this is merely a convenient analogy. The relationship between spin reversal and time–reversal symmetry is a quantum–mechanical issue.

Magnetic symmetries are explicated using the Wigner time–reversal operator $\mathcal{T}_{\mathrm{w}}$ [11]. Under $\mathcal{T}_{\mathrm{w}}$, the Tellegen constitutive dyadics transform as [12, 13]

$$\left.\begin{aligned}
\mathcal{T}_{\mathrm{w}}\left\{\underline{\underline{\epsilon}}_{\mathrm{EH}}(\underline{r},\omega)\right\} &= \underline{\underline{\epsilon}}^{\mathrm{T}}_{\mathrm{EH}}(\underline{r},\omega) \\
\mathcal{T}_{\mathrm{w}}\left\{\underline{\underline{\xi}}_{\mathrm{EH}}(\underline{r},\omega)\right\} &= -\underline{\underline{\zeta}}^{\mathrm{T}}_{\mathrm{EH}}(\underline{r},\omega) \\
\mathcal{T}_{\mathrm{w}}\left\{\underline{\underline{\zeta}}_{\mathrm{EH}}(\underline{r},\omega)\right\} &= -\underline{\underline{\xi}}^{\mathrm{T}}_{\mathrm{EH}}(\underline{r},\omega) \\
\mathcal{T}_{\mathrm{w}}\left\{\underline{\underline{\mu}}_{\mathrm{EH}}(\underline{r},\omega)\right\} &= \underline{\underline{\mu}}^{\mathrm{T}}_{\mathrm{EH}}(\underline{r},\omega)
\end{aligned}\right\}, \tag{3.2}$$

whereas the Boys–Post constitutive dyadics transform as

$$\left.\begin{aligned}
\mathcal{T}_{\mathrm{w}}\left\{\underline{\underline{\epsilon}}_{\mathrm{EB}}(\underline{r},\omega)\right\} &= \underline{\underline{\epsilon}}^{\mathrm{T}}_{\mathrm{EB}}(\underline{r},\omega) \\
\mathcal{T}_{\mathrm{w}}\left\{\underline{\underline{\xi}}_{\mathrm{EB}}(\underline{r},\omega)\right\} &= \underline{\underline{\zeta}}^{\mathrm{T}}_{\mathrm{EB}}(\underline{r},\omega) \\
\mathcal{T}_{\mathrm{w}}\left\{\underline{\underline{\zeta}}_{\mathrm{EB}}(\underline{r},\omega)\right\} &= \underline{\underline{\xi}}^{\mathrm{T}}_{\mathrm{EB}}(\underline{r},\omega) \\
\mathcal{T}_{\mathrm{w}}\left\{\underline{\underline{\nu}}_{\mathrm{EB}}(\underline{r},\omega)\right\} &= \underline{\underline{\nu}}^{\mathrm{T}}_{\mathrm{EB}}(\underline{r},\omega)
\end{aligned}\right\}. \tag{3.3}$$

Thus, we see that invariance under $\mathcal{T}_{\mathrm{w}}$ corresponds to Lorentz–reciprocity (cf. Sec. 1.7.2.1).

The Wigner time–reversal operator $\mathcal{T}_{\mathrm{w}}$ may be viewed as a *restricted* version of the time–reversal operator $\mathcal{T}$ introduced in Sec. 1.6.1. From this perspective, the actions of $\mathcal{T}_{\mathrm{w}}$ and $\mathcal{T}$ are the same except that $\mathcal{T}_{\mathrm{w}}$ does not alter the signs of the imaginary parts of the constitutive parameters. Thereby, a dissipative medium remains dissipative under the action of $\mathcal{T}_{\mathrm{w}}$, and does not transform into an active medium [14].

The spacetime symmetries of anisotropic and bianisotropic mediums, including magnetic and nonmagnetic mediums, may be classified using the Wigner time–reversal operator $\mathcal{T}_{\mathrm{w}}$ in combination with the point groups of classical crystallography. This is achieved by means of the *magnetic point groups*[1] [3, 4, 13, 15, 16]. The magnetic point groups are divided into three categories:

- The first category describes those mediums which are invariant under $\mathcal{T}_{\mathrm{w}}$. A point group G' in this category is generated by the generators of any classical point group G together with $\mathcal{T}_{\mathrm{w}}$; i.e.,

$$\mathsf{G}' = \mathsf{G} + \mathcal{T}_{\mathrm{w}}\mathsf{G}\,. \qquad (3.4)$$

  Thus, there are 32 discrete point groups G', each containing twice as many symmetry operations as the corresponding classical point group G. Since the symmetry operation $\mathcal{T}_{\mathrm{w}}$ is an element of the point group G', mediums classified in G' are usually nonmagnetic, although some antiferromagnetic materials do belong to this category [4].

- The second category is simply given by the classical point groups. Since $\mathcal{T}_{\mathrm{w}}$ is not a symmetry operation, mediums belonging to this category are generally magnetic and generally not Lorentz–reciprocal. The group symbols coincide for the magnetic point groups of the first and second categories; hence, boldface symbols are used to identify those of the second category of magnetic point groups.

- Magnetic point groups of the third category do not include the Wigner time–reversal operator $\mathcal{T}_{\mathrm{w}}$ explicitly. However, these magnetic point groups include *complementary* symmetry operations in which $\mathcal{T}_{\mathrm{w}}$ features in combination with nonidentity elements of the classical point groups. A magnetic point group of the third category M is constructed from the classical point group G of order $|\mathsf{G}|$ as follows. Choose a subgroup H of G whose order is $|\mathsf{G}|/2$. Then the magnetic point group of the third category is given as

$$\mathsf{M} = \mathsf{H} + \mathcal{T}_{\mathrm{w}}\,(\mathsf{G} - \mathsf{H})\,, \qquad (3.5)$$

  where $\mathsf{G} - \mathsf{H}$ represents those symmetry operations in G which are not present in H. By this process, 58 magnetic point groups M arise from the classical point groups G. Mediums belonging to this category are generally not Lorentz–reciprocal. Examples include gyrotropic mediums and mediums undergoing relative motion [13].

A graphical method of interpretating magnetic point groups, in which spin 'up' and 'down' orientations are labelled as black and white, was introduced by Shubnikov [17]. In this system, magnetic point groups of the first

---

[1]The terminology is rather nonintuitive, but widely used: *magnetic point groups* encompass both magnetic and nonmagnetic mediums.

category are described as singly coloured (either black or white); magnetic point groups of the second category form the gray groups; and magnetic point groups of the third category are identified as the doubly coloured (or black and white) groups.

The form of the constitutive dyadics for the three categories of magnetic point groups may be deduced from the Wigner time–reversal symmetries represented by Eqs. (3.2) and (3.3), together with the spatial symmetries described in Sec. 1.6.2 — as shown by Dmitriev [13]. To begin, consider the magnetic point groups of the second category. These are determined by the classical point groups alone. As previously, we represent the corresponding symmetry operations by the 3×3 real–valued matrixes $\underline{\underline{M}}_i$. Invariance under the symmetry operations of the classical point groups is specified by [13, 18]

$$
\left.
\begin{aligned}
\underline{\underline{\epsilon}}_{\text{EH, EB}}(\underline{r},\omega) &= \underline{\underline{M}}_i^{-1} \bullet \underline{\underline{\epsilon}}_{\text{EH,EB}}(\underline{r},\omega) \bullet \underline{\underline{M}}_i \\
\underline{\underline{\xi}}_{\text{EH,EB}}(\underline{r},\omega) &= \det\left(\underline{\underline{M}}_i\right) \underline{\underline{M}}_i^{-1} \bullet \underline{\underline{\xi}}_{\text{EH,EB}}(\underline{r},\omega) \bullet \underline{\underline{M}}_i \\
\underline{\underline{\zeta}}_{\text{EH,EB}}(\underline{r},\omega) &= \det\left(\underline{\underline{M}}_i\right) \underline{\underline{M}}_i^{-1} \bullet \underline{\underline{\zeta}}_{\text{EH,EB}}(\underline{r},\omega) \bullet \underline{\underline{M}}_i \\
\underline{\underline{\mu}}_{\text{EH}}(\underline{r},\omega) &= \underline{\underline{M}}_i^{-1} \bullet \underline{\underline{\mu}}_{\text{EH}}(\underline{r},\omega) \bullet \underline{\underline{M}}_i \\
\underline{\underline{\nu}}_{\text{EB}}(\underline{r},\omega) &= \underline{\underline{M}}_i^{-1} \bullet \underline{\underline{\nu}}_{\text{EB}}(\underline{r},\omega) \bullet \underline{\underline{M}}_i
\end{aligned}
\right\}. \qquad (3.6)
$$

Implementation of the spatial symmetries represented in Eqs. (3.6) with the generating matrixes given in Table 3.1 yields the forms of the constitutive dyadics for mediums described by the 32 magnetic point groups of the second category. These are listed in Table 3.3 for the Tellegen constitutive dyadics, with the points groups being identified by their Schoenflies symbols.

The forms of the constitutive dyadics corresponding to the magnetic point groups of the first category are straightforwardly deduced from those of the second category, by imposing the Lorentz–reciprocity conditions (1.88) for the Tellegen dyadics and (1.89) for the Boys–Post dyadics. The constitutive dyadic forms for the magnetic point groups of the first category are presented in Table 3.3.

The relation (3.5) is exploited to calculate the constitutive dyadic forms for the magnetic point groups of the third category. The spatial symmetry operations (i.e., those belonging to the subgroup H) impose the conditions (3.6), whereas invariance under the spacetime symmetry operations — i.e., those belonging to $\mathcal{T}_{\mathrm{w}}\,(\mathsf{G}-\mathsf{H})$ — implies [13, 18]

$$\left.\begin{aligned}
\underline{\underline{\epsilon}}_{\mathrm{EH}}(\underline{r},\omega) &= \underline{\underline{M}}_i^{-1} \bullet \underline{\underline{\epsilon}}_{\mathrm{EH}}^{\mathrm{T}}(\underline{r},\omega) \bullet \underline{\underline{M}}_i \\
\underline{\underline{\xi}}_{\mathrm{EH}}(\underline{r},\omega) &= -\det\left(\underline{\underline{M}}_i\right)\underline{\underline{M}}_i^{-1} \bullet \underline{\underline{\zeta}}_{\mathrm{EH}}^{\mathrm{T}}(\underline{r},\omega) \bullet \underline{\underline{M}}_i \\
\underline{\underline{\zeta}}_{\mathrm{EH}}(\underline{r},\omega) &= -\det\left(\underline{\underline{M}}_i\right)\underline{\underline{M}}_i^{-1} \bullet \underline{\underline{\xi}}_{\mathrm{EH}}^{\mathrm{T}}(\underline{r},\omega) \bullet \underline{\underline{M}}_i \\
\underline{\underline{\mu}}_{\mathrm{EH}}(\underline{r},\omega) &= \underline{\underline{M}}_i^{-1} \bullet \underline{\underline{\mu}}_{\mathrm{EH}}^{\mathrm{T}}(\underline{r},\omega) \bullet \underline{\underline{M}}_i
\end{aligned}\right\}, \qquad (3.7)$$

for the Tellegen constitutive dyadics, and

$$\left.\begin{aligned}
\underline{\underline{\epsilon}}_{\mathrm{EB}}(\underline{r},\omega) &= \underline{\underline{M}}_i^{-1} \bullet \underline{\underline{\epsilon}}_{\mathrm{EB}}^{\mathrm{T}}(\underline{r},\omega) \bullet \underline{\underline{M}}_i \\
\underline{\underline{\xi}}_{\mathrm{EB}}(\underline{r},\omega) &= \det\left(\underline{\underline{M}}_i\right)\underline{\underline{M}}_i^{-1} \bullet \underline{\underline{\zeta}}_{\mathrm{EB}}^{\mathrm{T}}(\underline{r},\omega) \bullet \underline{\underline{M}}_i \\
\underline{\underline{\zeta}}_{\mathrm{EB}}(\underline{r},\omega) &= \det\left(\underline{\underline{M}}_i\right)\underline{\underline{M}}_i^{-1} \bullet \underline{\underline{\xi}}_{\mathrm{EB}}^{\mathrm{T}}(\underline{r},\omega) \bullet \underline{\underline{M}}_i \\
\underline{\underline{\nu}}_{\mathrm{EB}}(\underline{r},\omega) &= \underline{\underline{M}}_i^{-1} \bullet \underline{\underline{\nu}}_{\mathrm{EB}}^{\mathrm{T}}(\underline{r},\omega) \bullet \underline{\underline{M}}_i
\end{aligned}\right\}, \qquad (3.8)$$

for the Boys–Post constitutive dyadics. The constitutive dyadic forms for the magnetic point groups of the third category are presented in Table 3.3.

The final entry in Table 3.3, namely the magnetic point group of the second category denoted by $\mathbf{C}_1$, deserves special mention. From the viewpoint of the spacetime symmetries, there are no limitations placed on the 36 parameters of the four constitutive dyadics corresponding to $\mathbf{C}_1$. We remark that the Post constraint (1.80) (or (1.81)) reduces the number of independent complex–valued constitutive parameters to 35.

3.3 Continuous point groups

The magnetic point groups derived from the classical point groups of crystallography categorize those *discrete* symmetry operations which are compatible with a crystal lattice structure. Thus, we see that the permitted rotational symmetries are C_1, C_2, C_3, C_4 and C_6, but that fivefold, sevenfold, and higher–fold rotations are not compatible with the crystal lattice structure. However, noncrystalline mediums, such as may arise through the process of homogenization [19], are not subject to such limitations; instead, these mediums are characterized by *continuous* magnetic point groups.

Continuous magnetic point groups are of infinite order. This may be achieved, for example, by incorporating the infinite–fold rotation C_∞. The highest symmetries are exhibited by isotropic homogeneous mediums. These belong to the continuous magnetic point groups specified by the Schoenflies symbols $\underline{K}_h$, $\underline{K}$, $\mathbf{K_h}$ and $\mathbf{K}$. Here, the underlining indicates

Table 3.3   Constitutive dyadics for the magnetic point groups.

Magnetic point group	$\underline{\underline{\epsilon}}_{EH}$	$\underline{\underline{\xi}}_{EH}$	$\underline{\underline{\zeta}}_{EH}$	$\underline{\underline{\mu}}_{EH}$	No. of params.
1. $T_h, T_d, O_h, \underline{K}_h$ 2. $\mathbf{T_h, T_d, O_h, \underline{K}_h}$ 3. $O_h(T_h), O_h(T_d)$	$\epsilon\underline{\underline{I}}$	$\underline{\underline{0}}$	$\underline{\underline{0}}$	$\mu\underline{\underline{I}}$	2
3. $T_h(T), T_d(T),$ $O_h(O), \underline{K}_h(\underline{K})$	$\epsilon\underline{\underline{I}}$	$\xi\underline{\underline{I}}$	$\xi\underline{\underline{I}}$	$\mu\underline{\underline{I}}$	3
1. $T, O, \underline{K}$ 3. $O(T)$	$\epsilon\underline{\underline{I}}$	$\dot{\xi}\underline{\underline{I}}$	$-\xi\underline{\underline{I}}$	$\mu\underline{\underline{I}}$	3
2. $\mathbf{T, O, \underline{K}}$	$\epsilon\underline{\underline{I}}$	$\xi\underline{\underline{I}}$	$\zeta\underline{\underline{I}}$	$\mu\underline{\underline{I}}$	4

Constitutive dyadics are presented for the 122 discrete magnetic point groups (32 of the first category, 32 of the second category, and 58 of the third category) and the 21 continuous magnetic point groups (7 for each of the first, second and third categories). Point groups are labelled using Schoenflies symbols. After [18].

the continuous nature of these magnetic point groups, and, as previously, boldface is used to distinguish the magnetic point groups of the second category. The groups $\underline{K}_h$ and $\mathbf{K_h}$ contain an infinite number of symmetry axes and symmetry planes passing through the origin of the coordinate system; they also contain a centre of symmetry. The groups $\underline{K}$ and $\mathbf{K}$ similarly contain an infinite number of symmetry axes passing through the origin of the coordinate system, but they do not contain symmetry planes or a centre of symmetry [20]. Isotropic dielectric–magnetic mediums belong to $\underline{K}_h$ and $\mathbf{K_h}$, while isotropic chiral mediums belong to $\underline{K}$. We note that the magnetic point group $\mathbf{K}$ describes the nonreciprocal biisotropic medium with constitutive relations (2.7).

The continuous magnetic point groups of the second category for anisotropic and bianisotropic mediums may be deduced from $\mathbf{K_h}$ by applying appropriate perturbations [21]. The procedure for constructing the continuous magnetic point groups of the first and third categories from

Table 3.3 continued

Magnetic point group	$\underline{\underline{\epsilon}}_{\rm EH}$	$\underline{\underline{\xi}}_{\rm EH}$	$\underline{\underline{\zeta}}_{\rm EH}$	$\underline{\underline{\mu}}_{\rm EH}$	No. of params.
1. $C_{3h}, C_{3i}, C_{4h}, C_{6h},$ $D_{3d}, D_{3h}, D_{4h}, D_{6h},$ $C_{\infty h}, \underline{D_{\infty h}}$ 2. $\mathbf{D_{3d}}, \mathbf{D_{3h}}, \mathbf{D_{4h}}, \mathbf{D_{6h}},$ $\mathbf{D_{\infty h}}$ 3. $C_{6h}(C_{3i}), C_{6h}(C_{3h}),$ $C_{4h}(C_{2h}), D_{6h}(C_{3h}),$ $D_{6h}(D_{3d}), D_{4h}(D_{2h})$	$\begin{pmatrix} \epsilon_{11} & 0 & 0 \\ 0 & \epsilon_{11} & 0 \\ 0 & 0 & \epsilon_{33} \end{pmatrix}$	$\underline{\underline{0}}$	$\underline{\underline{0}}$	$\begin{pmatrix} \mu_{11} & 0 & 0 \\ 0 & \mu_{11} & 0 \\ 0 & 0 & \mu_{33} \end{pmatrix}$	4
1. $C_{3v}, C_{4v}, C_{6v}, \underline{C_{\infty v}}$ 3. $C_{6v}(C_{3v})$	$\begin{pmatrix} \epsilon_{11} & 0 & 0 \\ 0 & \epsilon_{11} & 0 \\ 0 & 0 & \epsilon_{33} \end{pmatrix}$	$\begin{pmatrix} 0 & \xi_{12} & 0 \\ -\xi_{12} & 0 & 0 \\ 0 & 0 & 0 \end{pmatrix}$	$\begin{pmatrix} 0 & \xi_{12} & 0 \\ -\xi_{12} & 0 & 0 \\ 0 & 0 & 0 \end{pmatrix}$	$\begin{pmatrix} \mu_{11} & 0 & 0 \\ 0 & \mu_{11} & 0 \\ 0 & 0 & \mu_{33} \end{pmatrix}$	5
1. D_{2d}	$\begin{pmatrix} \epsilon_{11} & 0 & 0 \\ 0 & \epsilon_{11} & 0 \\ 0 & 0 & \epsilon_{33} \end{pmatrix}$	$\begin{pmatrix} 0 & \xi_{12} & 0 \\ \xi_{12} & 0 & 0 \\ 0 & 0 & 0 \end{pmatrix}$	$\begin{pmatrix} 0 & -\xi_{12} & 0 \\ -\xi_{12} & 0 & 0 \\ 0 & 0 & 0 \end{pmatrix}$	$\begin{pmatrix} \mu_{11} & 0 & 0 \\ 0 & \mu_{11} & 0 \\ 0 & 0 & \mu_{33} \end{pmatrix}$	5
3. $D_{4h}(C_{4v}), D_{3d}(C_{3v}),$ $D_{3h}(C_{3v}), D_{6h}(C_{6v}),$ $\underline{D_{\infty h}(C_{\infty v})}$	$\begin{pmatrix} \epsilon_{11} & 0 & 0 \\ 0 & \epsilon_{11} & 0 \\ 0 & 0 & \epsilon_{33} \end{pmatrix}$	$\begin{pmatrix} 0 & \xi_{12} & 0 \\ -\xi_{12} & 0 & 0 \\ 0 & 0 & 0 \end{pmatrix}$	$\begin{pmatrix} 0 & -\xi_{12} & 0 \\ \xi_{12} & 0 & 0 \\ 0 & 0 & 0 \end{pmatrix}$	$\begin{pmatrix} \mu_{11} & 0 & 0 \\ 0 & \mu_{11} & 0 \\ 0 & 0 & \mu_{33} \end{pmatrix}$	5

Table 3.3   continued

Magnetic point group	$\underline{\underline{\epsilon}}_{EH}$	$\underline{\underline{\xi}}_{EH}$	$\underline{\underline{\zeta}}_{EH}$	$\underline{\underline{\mu}}_{EH}$	No. of params.
3. $D_{4h}(D_{2d})$	$\begin{pmatrix} \epsilon_{11} & 0 & 0 \\ 0 & \epsilon_{11} & 0 \\ 0 & 0 & \epsilon_{33} \end{pmatrix}$	$\begin{pmatrix} 0 & \xi_{12} & 0 \\ \xi_{12} & 0 & 0 \\ 0 & 0 & 0 \end{pmatrix}$	$\begin{pmatrix} 0 & \xi_{12} & 0 \\ \xi_{12} & 0 & 0 \\ 0 & 0 & 0 \end{pmatrix}$	$\begin{pmatrix} \mu_{11} & 0 & 0 \\ 0 & \mu_{11} & 0 \\ 0 & 0 & \mu_{33} \end{pmatrix}$	5
1. 2. D_{2h} $\mathbf{D_{2h}}$	$\begin{pmatrix} \epsilon_{11} & 0 & 0 \\ 0 & \epsilon_{22} & 0 \\ 0 & 0 & \epsilon_{33} \end{pmatrix}$	$\underline{\underline{0}}$	$\underline{\underline{0}}$	$\begin{pmatrix} \mu_{11} & 0 & 0 \\ 0 & \mu_{22} & 0 \\ 0 & 0 & \mu_{33} \end{pmatrix}$	6
1. $D_3, D_4, D_6, \underline{D}_{\infty}$ 3. $D_6(D_3)$	$\begin{pmatrix} \epsilon_{11} & 0 & 0 \\ 0 & \epsilon_{11} & 0 \\ 0 & 0 & \epsilon_{33} \end{pmatrix}$	$\begin{pmatrix} \xi_{11} & 0 & 0 \\ 0 & \xi_{11} & 0 \\ 0 & 0 & \xi_{33} \end{pmatrix}$	$\begin{pmatrix} -\xi_{11} & 0 & 0 \\ 0 & -\xi_{11} & 0 \\ 0 & 0 & -\xi_{33} \end{pmatrix}$	$\begin{pmatrix} \mu_{11} & 0 & 0 \\ 0 & \mu_{11} & 0 \\ 0 & 0 & \mu_{33} \end{pmatrix}$	6
3. $D_{4h}(D_4), D_{3d}(D_3),$ $D_{3h}(D_3), D_{6h}(D_6),$ $D_{2d}(D_2), \underline{D}_{\infty h}(\underline{D}_{\infty})$	$\begin{pmatrix} \epsilon_{11} & 0 & 0 \\ 0 & \epsilon_{11} & 0 \\ 0 & 0 & \epsilon_{33} \end{pmatrix}$	$\begin{pmatrix} \xi_{11} & 0 & 0 \\ 0 & \xi_{11} & 0 \\ 0 & 0 & \xi_{33} \end{pmatrix}$	$\begin{pmatrix} \xi_{11} & 0 & 0 \\ 0 & \xi_{11} & 0 \\ 0 & 0 & \xi_{33} \end{pmatrix}$	$\begin{pmatrix} \mu_{11} & 0 & 0 \\ 0 & \mu_{11} & 0 \\ 0 & 0 & \mu_{33} \end{pmatrix}$	6
1. S_4	$\begin{pmatrix} \epsilon_{11} & 0 & 0 \\ 0 & \epsilon_{11} & 0 \\ 0 & 0 & \epsilon_{33} \end{pmatrix}$	$\begin{pmatrix} \xi_{11} & \xi_{12} & 0 \\ \xi_{12} & -\xi_{11} & 0 \\ 0 & 0 & \xi_{33} \end{pmatrix}$	$\begin{pmatrix} -\xi_{11} & -\xi_{12} & 0 \\ -\xi_{12} & \xi_{11} & 0 \\ 0 & 0 & 0 \end{pmatrix}$	$\begin{pmatrix} \mu_{11} & 0 & 0 \\ 0 & \mu_{11} & 0 \\ 0 & 0 & \mu_{33} \end{pmatrix}$	6

Table 3.3   continued

	Magnetic point group	$\underline{\underline{\epsilon}}_{\rm EH}$	$\underline{\underline{\xi}}_{\rm EH}$	$\underline{\underline{\zeta}}_{\rm EH}$	$\underline{\underline{\mu}}_{\rm EH}$	No. of params.
2. 3.	C_{3h}, C_{3i}, C_{4h}, C_{6h}, $\underline{C_{\infty h}}$ $D_{4h}(C_{4h})$, $D_{3d}(C_{3i})$, $D_{3h}(C_{3h})$, $D_{6h}(C_{6h})$, $\underline{D_{\infty h}}(C_{\infty h})$	$\begin{pmatrix} \epsilon_{11} & \epsilon_{12} & 0 \\ -\epsilon_{12} & \epsilon_{11} & 0 \\ 0 & 0 & \epsilon_{33} \end{pmatrix}$	$\underline{\underline{0}}$	$\underline{\underline{0}}$	$\begin{pmatrix} \mu_{11} & \mu_{12} & 0 \\ -\mu_{12} & \mu_{11} & 0 \\ 0 & 0 & \mu_{33} \end{pmatrix}$	6
2.	C_{3v}, C_{4v}, C_{6v}, $\underline{C_{\infty v}}$	$\begin{pmatrix} \epsilon_{11} & 0 & 0 \\ 0 & \epsilon_{11} & 0 \\ 0 & 0 & \epsilon_{33} \end{pmatrix}$	$\begin{pmatrix} 0 & \xi_{12} & 0 \\ -\xi_{12} & 0 & 0 \\ 0 & 0 & 0 \end{pmatrix}$	$\begin{pmatrix} 0 & \zeta_{12} & 0 \\ -\zeta_{12} & 0 & 0 \\ 0 & 0 & 0 \end{pmatrix}$	$\begin{pmatrix} \mu_{11} & 0 & 0 \\ 0 & \mu_{11} & 0 \\ 0 & 0 & \mu_{33} \end{pmatrix}$	6
2.	D_{2d}	$\begin{pmatrix} \epsilon_{11} & 0 & 0 \\ 0 & \epsilon_{11} & 0 \\ 0 & 0 & \epsilon_{33} \end{pmatrix}$	$\begin{pmatrix} 0 & \xi_{12} & 0 \\ \xi_{12} & 0 & 0 \\ 0 & 0 & 0 \end{pmatrix}$	$\begin{pmatrix} 0 & \zeta_{12} & 0 \\ \zeta_{12} & 0 & 0 \\ 0 & 0 & 0 \end{pmatrix}$	$\begin{pmatrix} \mu_{11} & 0 & 0 \\ 0 & \mu_{11} & 0 \\ 0 & 0 & \mu_{33} \end{pmatrix}$	6
3.	$C_{4h}(S_4)$	$\begin{pmatrix} \epsilon_{11} & 0 & 0 \\ 0 & \epsilon_{11} & 0 \\ 0 & 0 & \epsilon_{33} \end{pmatrix}$	$\begin{pmatrix} \xi_{11} & \xi_{12} & 0 \\ \xi_{12} & -\xi_{11} & 0 \\ 0 & 0 & 0 \end{pmatrix}$	$\begin{pmatrix} \xi_{11} & \xi_{12} & 0 \\ \xi_{12} & -\xi_{11} & 0 \\ 0 & 0 & 0 \end{pmatrix}$	$\begin{pmatrix} \mu_{11} & 0 & 0 \\ 0 & \mu_{11} & 0 \\ 0 & 0 & \mu_{33} \end{pmatrix}$	6

Table 3.3   continued

	Magnetic point group	$\underline{\underline{\epsilon}}_{\mathrm{EH}}$	$\underline{\underline{\xi}}_{\mathrm{EH}}$	$\underline{\underline{\zeta}}_{\mathrm{EH}}$	$\underline{\underline{\mu}}_{\mathrm{EH}}$	No. of params.
3.	$D_{2d}(C_{2v})$	$\begin{pmatrix} \epsilon_{11} & 0 & 0 \\ 0 & \epsilon_{11} & 0 \\ 0 & 0 & \epsilon_{33} \end{pmatrix}$	$\begin{pmatrix} 0 & \xi_{12} & 0 \\ \xi_{21} & 0 & 0 \\ 0 & 0 & 0 \end{pmatrix}$	$\begin{pmatrix} 0 & -\xi_{12} & 0 \\ -\xi_{21} & 0 & 0 \\ 0 & 0 & 0 \end{pmatrix}$	$\begin{pmatrix} \mu_{11} & 0 & 0 \\ 0 & \mu_{11} & 0 \\ 0 & 0 & \mu_{33} \end{pmatrix}$	6
3.	$C_{4v}(C_{2v})$	$\begin{pmatrix} \epsilon_{11} & 0 & 0 \\ 0 & \epsilon_{11} & 0 \\ 0 & 0 & \epsilon_{33} \end{pmatrix}$	$\begin{pmatrix} 0 & \xi_{12} & 0 \\ \xi_{21} & 0 & 0 \\ 0 & 0 & 0 \end{pmatrix}$	$\begin{pmatrix} 0 & \xi_{12} & 0 \\ \xi_{21} & 0 & 0 \\ 0 & 0 & 0 \end{pmatrix}$	$\begin{pmatrix} \mu_{11} & 0 & 0 \\ 0 & \mu_{11} & 0 \\ 0 & 0 & \mu_{33} \end{pmatrix}$	6
1. 3.	$C_3, C_4, C_6, \underline{C}_\infty$ $C_6(C_3)$	$\begin{pmatrix} \epsilon_{11} & 0 & 0 \\ 0 & \epsilon_{11} & 0 \\ 0 & 0 & \epsilon_{33} \end{pmatrix}$	$\begin{pmatrix} \xi_{11} & \xi_{12} & 0 \\ -\xi_{12} & \xi_{11} & 0 \\ 0 & 0 & \xi_{33} \end{pmatrix}$	$\begin{pmatrix} -\xi_{11} & \xi_{12} & 0 \\ -\xi_{12} & -\xi_{11} & 0 \\ 0 & 0 & -\xi_{33} \end{pmatrix}$	$\begin{pmatrix} \mu_{11} & 0 & 0 \\ 0 & \mu_{11} & 0 \\ 0 & 0 & \mu_{33} \end{pmatrix}$	7
3.	$C_{3h}(C_3), C_{3i}(C_3),$ $C_{4h}(C_4), C_{6h}(C_6),$ $\underline{C}_{\infty h}(\underline{C}_\infty)$	$\begin{pmatrix} \epsilon_{11} & 0 & 0 \\ 0 & \epsilon_{11} & 0 \\ 0 & 0 & \epsilon_{33} \end{pmatrix}$	$\begin{pmatrix} \xi_{11} & \xi_{12} & 0 \\ -\xi_{12} & \xi_{11} & 0 \\ 0 & 0 & \xi_{33} \end{pmatrix}$	$\begin{pmatrix} \xi_{11} & -\xi_{12} & 0 \\ \xi_{12} & \xi_{11} & 0 \\ 0 & 0 & \xi_{33} \end{pmatrix}$	$\begin{pmatrix} \mu_{11} & 0 & 0 \\ 0 & \mu_{11} & 0 \\ 0 & 0 & \mu_{33} \end{pmatrix}$	7
3.	$D_4(D_2)$	$\begin{pmatrix} \epsilon_{11} & 0 & 0 \\ 0 & \epsilon_{11} & 0 \\ 0 & 0 & \epsilon_{33} \end{pmatrix}$	$\begin{pmatrix} \xi_{11} & 0 & 0 \\ 0 & \xi_{22} & 0 \\ 0 & 0 & \xi_{33} \end{pmatrix}$	$\begin{pmatrix} -\xi_{22} & 0 & 0 \\ 0 & -\xi_{11} & 0 \\ 0 & 0 & -\xi_{33} \end{pmatrix}$	$\begin{pmatrix} \mu_{11} & 0 & 0 \\ 0 & \mu_{11} & 0 \\ 0 & 0 & \mu_{33} \end{pmatrix}$	7

Table 3.3 continued

	Magnetic point group	$\underline{\underline{\epsilon}}_{EH}$	$\underline{\underline{\xi}}_{EH}$	$\underline{\underline{\zeta}}_{EH}$	$\underline{\underline{\mu}}_{EH}$	No. of params.
1.	C_{2h}	$\begin{pmatrix} \epsilon_{11} & \epsilon_{12} & 0 \\ \epsilon_{12} & \epsilon_{22} & 0 \\ 0 & 0 & \epsilon_{33} \end{pmatrix}$	$\underline{\underline{0}}$	$\underline{\underline{0}}$	$\begin{pmatrix} \mu_{11} & \mu_{12} & 0 \\ \mu_{12} & \mu_{22} & 0 \\ 0 & 0 & \mu_{33} \end{pmatrix}$	8
3.	$D_{2h}(C_{2h})$	$\begin{pmatrix} \epsilon_{11} & \epsilon_{12} & 0 \\ -\epsilon_{12} & \epsilon_{22} & 0 \\ 0 & 0 & \epsilon_{33} \end{pmatrix}$	$\underline{\underline{0}}$	$\underline{\underline{0}}$	$\begin{pmatrix} \mu_{11} & \mu_{12} & 0 \\ -\mu_{12} & \mu_{22} & 0 \\ 0 & 0 & \mu_{33} \end{pmatrix}$	8
1.	C_{2v}	$\begin{pmatrix} \epsilon_{11} & 0 & 0 \\ 0 & \epsilon_{22} & 0 \\ 0 & 0 & \epsilon_{33} \end{pmatrix}$	$\begin{pmatrix} 0 & \xi_{12} & 0 \\ \xi_{21} & 0 & 0 \\ 0 & 0 & 0 \end{pmatrix}$	$\begin{pmatrix} 0 & -\xi_{21} & 0 \\ -\xi_{12} & 0 & 0 \\ 0 & 0 & 0 \end{pmatrix}$	$\begin{pmatrix} \mu_{11} & 0 & 0 \\ 0 & \mu_{22} & 0 \\ 0 & 0 & \mu_{33} \end{pmatrix}$	8
2.	$\mathbf{D_3, D_4, D_6, \underline{D}_\infty}$	$\begin{pmatrix} \epsilon_{11} & 0 & 0 \\ 0 & \epsilon_{11} & 0 \\ 0 & 0 & \epsilon_{33} \end{pmatrix}$	$\begin{pmatrix} \xi_{11} & 0 & 0 \\ 0 & \xi_{11} & 0 \\ 0 & 0 & \xi_{33} \end{pmatrix}$	$\begin{pmatrix} \zeta_{11} & 0 & 0 \\ 0 & \zeta_{11} & 0 \\ 0 & 0 & \zeta_{33} \end{pmatrix}$	$\begin{pmatrix} \mu_{11} & 0 & 0 \\ 0 & \mu_{11} & 0 \\ 0 & 0 & \mu_{33} \end{pmatrix}$	8
3.	$D_{2d}(S_4)$	$\begin{pmatrix} \epsilon_{11} & \epsilon_{12} & 0 \\ -\epsilon_{12} & \epsilon_{11} & 0 \\ 0 & 0 & \epsilon_{33} \end{pmatrix}$	$\begin{pmatrix} \xi_{11} & \xi_{12} & 0 \\ \xi_{12} & -\xi_{11} & 0 \\ 0 & 0 & 0 \end{pmatrix}$	$\begin{pmatrix} \xi_{11} & -\xi_{12} & 0 \\ -\xi_{12} & -\xi_{11} & 0 \\ 0 & 0 & 0 \end{pmatrix}$	$\begin{pmatrix} \mu_{11} & \mu_{12} & 0 \\ -\mu_{12} & \mu_{11} & 0 \\ 0 & 0 & \mu_{33} \end{pmatrix}$	8

Table 3.3 continued

	Magnetic point group	$\underline{\underline{\epsilon}}_{EH}$	$\underline{\underline{\xi}}_{EH}$	$\underline{\underline{\zeta}}_{EH}$	$\underline{\underline{\mu}}_{EH}$	No. of params.
3.	$D_{2h}(C_{2v})$	$\begin{pmatrix} \epsilon_{11} & 0 & 0 \\ 0 & \epsilon_{22} & 0 \\ 0 & 0 & \epsilon_{33} \end{pmatrix}$	$\begin{pmatrix} 0 & \xi_{12} & 0 \\ \xi_{21} & 0 & 0 \\ 0 & 0 & 0 \end{pmatrix}$	$\begin{pmatrix} 0 & \xi_{21} & 0 \\ \xi_{12} & 0 & 0 \\ 0 & 0 & 0 \end{pmatrix}$	$\begin{pmatrix} \mu_{11} & 0 & 0 \\ 0 & \mu_{22} & 0 \\ 0 & 0 & \mu_{33} \end{pmatrix}$	8
1.	D_2	$\begin{pmatrix} \epsilon_{11} & 0 & 0 \\ 0 & \epsilon_{22} & 0 \\ 0 & 0 & \epsilon_{33} \end{pmatrix}$	$\begin{pmatrix} \xi_{11} & 0 & 0 \\ 0 & \xi_{22} & 0 \\ 0 & 0 & \xi_{33} \end{pmatrix}$	$\begin{pmatrix} -\xi_{11} & 0 & 0 \\ 0 & -\xi_{22} & 0 \\ 0 & 0 & -\xi_{33} \end{pmatrix}$	$\begin{pmatrix} \mu_{11} & 0 & 0 \\ 0 & \mu_{22} & 0 \\ 0 & 0 & \mu_{33} \end{pmatrix}$	9
3.	$D_3(C_3), D_4(C_4),$ $D_6(C_6), \underline{D}_\infty(\underline{C}_\infty)$	$\begin{pmatrix} \epsilon_{11} & \epsilon_{12} & 0 \\ -\epsilon_{12} & \epsilon_{11} & 0 \\ 0 & 0 & \epsilon_{33} \end{pmatrix}$	$\begin{pmatrix} \xi_{11} & \xi_{12} & 0 \\ -\xi_{12} & \xi_{11} & 0 \\ 0 & 0 & \xi_{33} \end{pmatrix}$	$\begin{pmatrix} -\xi_{11} & -\xi_{12} & 0 \\ \xi_{12} & -\xi_{11} & 0 \\ 0 & 0 & -\xi_{33} \end{pmatrix}$	$\begin{pmatrix} \mu_{11} & \mu_{12} & 0 \\ -\mu_{12} & \mu_{11} & 0 \\ 0 & 0 & \mu_{33} \end{pmatrix}$	9
3.	$C_{3v}(C_3), C_{4v}(C_4),$ $C_{6v}(C_6), \underline{C}_{\infty v}(\underline{C}_\infty)$	$\begin{pmatrix} \epsilon_{11} & \epsilon_{12} & 0 \\ -\epsilon_{12} & \epsilon_{11} & 0 \\ 0 & 0 & \epsilon_{33} \end{pmatrix}$	$\begin{pmatrix} \xi_{11} & \xi_{12} & 0 \\ -\xi_{12} & \xi_{11} & 0 \\ 0 & 0 & \xi_{33} \end{pmatrix}$	$\begin{pmatrix} \xi_{11} & -\xi_{12} & 0 \\ \xi_{12} & \xi_{11} & 0 \\ 0 & 0 & \xi_{33} \end{pmatrix}$	$\begin{pmatrix} \mu_{11} & \mu_{12} & 0 \\ -\mu_{12} & \mu_{11} & 0 \\ 0 & 0 & \mu_{33} \end{pmatrix}$	9
3.	$D_{2h}(D_2)$	$\begin{pmatrix} \epsilon_{11} & 0 & 0 \\ 0 & \epsilon_{22} & 0 \\ 0 & 0 & \epsilon_{33} \end{pmatrix}$	$\begin{pmatrix} \xi_{11} & 0 & 0 \\ 0 & \xi_{22} & 0 \\ 0 & 0 & \xi_{33} \end{pmatrix}$	$\begin{pmatrix} \xi_{11} & 0 & 0 \\ 0 & \xi_{22} & 0 \\ 0 & 0 & \xi_{33} \end{pmatrix}$	$\begin{pmatrix} \mu_{11} & 0 & 0 \\ 0 & \mu_{22} & 0 \\ 0 & 0 & \mu_{33} \end{pmatrix}$	9

Table 3.3 continued

Magnetic point group	$\underline{\underline{\epsilon}}_{EH}$	$\underline{\underline{\xi}}_{EH}$	$\underline{\underline{\zeta}}_{EH}$	$\underline{\underline{\mu}}_{EH}$	No. of params.
3. $S_4(C_2)$	$\begin{pmatrix} \epsilon_{11} & 0 & 0 \\ 0 & \epsilon_{11} & 0 \\ 0 & 0 & \epsilon_{33} \end{pmatrix}$	$\begin{pmatrix} \xi_{11} & \xi_{12} & 0 \\ \xi_{21} & \xi_{22} & 0 \\ 0 & 0 & \xi_{33} \end{pmatrix}$	$\begin{pmatrix} \xi_{22} & -\xi_{12} & 0 \\ -\xi_{21} & \xi_{11} & 0 \\ 0 & 0 & \xi_{33} \end{pmatrix}$	$\begin{pmatrix} \mu_{11} & 0 & 0 \\ 0 & \mu_{11} & 0 \\ 0 & 0 & \mu_{33} \end{pmatrix}$	9
3. $C_4(C_2)$	$\begin{pmatrix} \epsilon_{11} & 0 & 0 \\ 0 & \epsilon_{11} & 0 \\ 0 & 0 & \epsilon_{33} \end{pmatrix}$	$\begin{pmatrix} \xi_{11} & \xi_{12} & 0 \\ \xi_{21} & \xi_{22} & 0 \\ 0 & 0 & \xi_{33} \end{pmatrix}$	$\begin{pmatrix} -\xi_{22} & \xi_{12} & 0 \\ \xi_{21} & -\xi_{11} & 0 \\ 0 & 0 & -\xi_{33} \end{pmatrix}$	$\begin{pmatrix} \mu_{11} & 0 & 0 \\ 0 & \mu_{11} & 0 \\ 0 & 0 & \mu_{33} \end{pmatrix}$	9
2. $\mathbf{C_{2h}}$	$\begin{pmatrix} \epsilon_{11} & \epsilon_{12} & 0 \\ \epsilon_{21} & \epsilon_{22} & 0 \\ 0 & 0 & \epsilon_{33} \end{pmatrix}$	$\underline{\underline{0}}$	$\underline{\underline{0}}$	$\begin{pmatrix} \mu_{11} & \mu_{12} & 0 \\ \mu_{21} & \mu_{22} & 0 \\ 0 & 0 & \mu_{33} \end{pmatrix}$	10
2. $\mathbf{S_4}$	$\begin{pmatrix} \epsilon_{11} & \epsilon_{12} & 0 \\ -\epsilon_{12} & \epsilon_{11} & 0 \\ 0 & 0 & \epsilon_{33} \end{pmatrix}$	$\begin{pmatrix} \xi_{11} & \xi_{12} & 0 \\ \xi_{12} & -\xi_{11} & 0 \\ 0 & 0 & 0 \end{pmatrix}$	$\begin{pmatrix} \zeta_{11} & \zeta_{12} & 0 \\ \zeta_{12} & -\zeta_{11} & 0 \\ 0 & 0 & 0 \end{pmatrix}$	$\begin{pmatrix} \mu_{11} & \mu_{12} & 0 \\ -\mu_{12} & \mu_{11} & 0 \\ 0 & 0 & \mu_{33} \end{pmatrix}$	10
2. $\mathbf{C_{2v}}$	$\begin{pmatrix} \epsilon_{11} & 0 & 0 \\ 0 & \epsilon_{22} & 0 \\ 0 & 0 & \epsilon_{33} \end{pmatrix}$	$\begin{pmatrix} 0 & \xi_{12} & 0 \\ \xi_{21} & 0 & 0 \\ 0 & 0 & 0 \end{pmatrix}$	$\begin{pmatrix} 0 & \zeta_{12} & 0 \\ \zeta_{21} & 0 & 0 \\ 0 & 0 & 0 \end{pmatrix}$	$\begin{pmatrix} \mu_{11} & 0 & 0 \\ 0 & \mu_{22} & 0 \\ 0 & 0 & \mu_{33} \end{pmatrix}$	10

Table 3.3   continued

Magnetic point group	$\underline{\underline{\epsilon}}_{EH}$	$\underline{\underline{\xi}}_{EH}$	$\underline{\underline{\zeta}}_{EH}$	$\underline{\underline{\mu}}_{EH}$	No. of params.
1. C_i	$\begin{pmatrix} \epsilon_{11} & \epsilon_{12} & \epsilon_{13} \\ \epsilon_{12} & \epsilon_{22} & \epsilon_{23} \\ \epsilon_{13} & \epsilon_{23} & \epsilon_{33} \end{pmatrix}$	$\underline{\underline{0}}$	$\underline{\underline{0}}$	$\begin{pmatrix} \mu_{11} & \mu_{12} & \mu_{13} \\ \mu_{12} & \mu_{22} & \mu_{23} \\ \mu_{13} & \mu_{23} & \mu_{33} \end{pmatrix}$	12
3. $C_{2h}(C_i)$	$\begin{pmatrix} \epsilon_{11} & \epsilon_{12} & \epsilon_{13} \\ \epsilon_{12} & \epsilon_{22} & \epsilon_{23} \\ -\epsilon_{13} & -\epsilon_{23} & \epsilon_{33} \end{pmatrix}$	$\underline{\underline{0}}$	$\underline{\underline{0}}$	$\begin{pmatrix} \mu_{11} & \mu_{12} & \mu_{13} \\ \mu_{12} & \mu_{22} & \mu_{23} \\ -\mu_{13} & -\mu_{23} & \mu_{33} \end{pmatrix}$	12
1. C_s	$\begin{pmatrix} \epsilon_{11} & \epsilon_{12} & 0 \\ \epsilon_{12} & \epsilon_{22} & 0 \\ 0 & 0 & \epsilon_{33} \end{pmatrix}$	$\begin{pmatrix} 0 & 0 & \xi_{13} \\ 0 & 0 & \xi_{23} \\ \xi_{31} & \xi_{32} & 0 \end{pmatrix}$	$\begin{pmatrix} 0 & 0 & -\xi_{31} \\ 0 & 0 & -\xi_{32} \\ -\xi_{13} & -\xi_{23} & 0 \end{pmatrix}$	$\begin{pmatrix} \mu_{11} & \mu_{12} & 0 \\ \mu_{12} & \mu_{22} & 0 \\ 0 & 0 & \mu_{33} \end{pmatrix}$	12
3. $C_{2h}(C_s)$	$\begin{pmatrix} \epsilon_{11} & \epsilon_{12} & 0 \\ \epsilon_{12} & \epsilon_{22} & 0 \\ 0 & 0 & \epsilon_{33} \end{pmatrix}$	$\begin{pmatrix} 0 & 0 & \xi_{13} \\ 0 & 0 & \xi_{23} \\ \xi_{31} & \xi_{32} & 0 \end{pmatrix}$	$\begin{pmatrix} 0 & 0 & \xi_{31} \\ 0 & 0 & \xi_{32} \\ \xi_{13} & \xi_{23} & 0 \end{pmatrix}$	$\begin{pmatrix} \mu_{11} & \mu_{12} & 0 \\ \mu_{12} & \mu_{22} & 0 \\ 0 & 0 & \mu_{33} \end{pmatrix}$	12
3. $C_{2v}(C_s)$	$\begin{pmatrix} \epsilon_{11} & 0 & 0 \\ 0 & \epsilon_{22} & \epsilon_{23} \\ 0 & -\epsilon_{23} & \epsilon_{33} \end{pmatrix}$	$\begin{pmatrix} 0 & \xi_{12} & \xi_{13} \\ \xi_{21} & 0 & 0 \\ \xi_{31} & 0 & 0 \end{pmatrix}$	$\begin{pmatrix} 0 & -\xi_{21} & \xi_{31} \\ -\xi_{12} & 0 & 0 \\ \xi_{13} & 0 & 0 \end{pmatrix}$	$\begin{pmatrix} \mu_{11} & 0 & 0 \\ 0 & \mu_{22} & \mu_{23} \\ 0 & -\mu_{23} & \mu_{33} \end{pmatrix}$	12

Table 3.3 continued

Magnetic point group	$\underline{\underline{\epsilon}}_{EH}$	$\underline{\underline{\xi}}_{EH}$	$\underline{\underline{\zeta}}_{EH}$	$\underline{\underline{\mu}}_{EH}$	No. of params.
2. $\mathbf{C_3, C_4, C_6, \underline{C}_\infty}$	$\begin{pmatrix} \epsilon_{11} & \epsilon_{12} & 0 \\ -\epsilon_{12} & \epsilon_{11} & 0 \\ 0 & 0 & \epsilon_{33} \end{pmatrix}$	$\begin{pmatrix} \xi_{11} & \xi_{12} & 0 \\ -\xi_{12} & \xi_{11} & 0 \\ 0 & 0 & \xi_{33} \end{pmatrix}$	$\begin{pmatrix} \zeta_{11} & \zeta_{12} & 0 \\ -\zeta_{12} & \zeta_{11} & 0 \\ 0 & 0 & \zeta_{33} \end{pmatrix}$	$\begin{pmatrix} \mu_{11} & \mu_{12} & 0 \\ -\mu_{12} & \mu_{11} & 0 \\ 0 & 0 & \mu_{33} \end{pmatrix}$	12
2. $\mathbf{D_2}$	$\begin{pmatrix} \epsilon_{11} & 0 & 0 \\ 0 & \epsilon_{22} & 0 \\ 0 & 0 & \epsilon_{33} \end{pmatrix}$	$\begin{pmatrix} \xi_{11} & 0 & 0 \\ 0 & \xi_{22} & 0 \\ 0 & 0 & \xi_{33} \end{pmatrix}$	$\begin{pmatrix} \zeta_{11} & 0 & 0 \\ 0 & \zeta_{22} & 0 \\ 0 & 0 & \zeta_{33} \end{pmatrix}$	$\begin{pmatrix} \mu_{11} & 0 & 0 \\ 0 & \mu_{22} & 0 \\ 0 & 0 & \mu_{33} \end{pmatrix}$	12
1. C_2	$\begin{pmatrix} \epsilon_{11} & \epsilon_{12} & 0 \\ \epsilon_{21} & \epsilon_{22} & 0 \\ 0 & 0 & \epsilon_{33} \end{pmatrix}$	$\begin{pmatrix} \xi_{11} & \xi_{12} & 0 \\ \xi_{21} & \xi_{22} & 0 \\ 0 & 0 & \xi_{33} \end{pmatrix}$	$\begin{pmatrix} -\xi_{11} & -\xi_{21} & 0 \\ -\xi_{12} & -\xi_{22} & 0 \\ 0 & 0 & -\xi_{33} \end{pmatrix}$	$\begin{pmatrix} \mu_{11} & \mu_{12} & 0 \\ \mu_{12} & \mu_{22} & 0 \\ 0 & 0 & \mu_{33} \end{pmatrix}$	13
3. $D_2(C_2)$	$\begin{pmatrix} \epsilon_{11} & \epsilon_{12} & 0 \\ -\epsilon_{12} & \epsilon_{22} & 0 \\ 0 & 0 & \epsilon_{33} \end{pmatrix}$	$\begin{pmatrix} \xi_{11} & \xi_{12} & 0 \\ \xi_{21} & \xi_{22} & 0 \\ 0 & 0 & \xi_{33} \end{pmatrix}$	$\begin{pmatrix} -\xi_{11} & \xi_{21} & 0 \\ \xi_{12} & -\xi_{22} & 0 \\ 0 & 0 & -\xi_{33} \end{pmatrix}$	$\begin{pmatrix} \mu_{11} & \mu_{12} & 0 \\ -\mu_{12} & \mu_{22} & 0 \\ 0 & 0 & \mu_{33} \end{pmatrix}$	13
3. $C_{2h}(C_2)$	$\begin{pmatrix} \epsilon_{11} & \epsilon_{12} & 0 \\ \epsilon_{12} & \epsilon_{22} & 0 \\ 0 & 0 & \epsilon_{33} \end{pmatrix}$	$\begin{pmatrix} \xi_{11} & \xi_{12} & 0 \\ \xi_{21} & \xi_{22} & 0 \\ 0 & 0 & \xi_{33} \end{pmatrix}$	$\begin{pmatrix} \xi_{11} & \xi_{21} & 0 \\ \xi_{12} & \xi_{22} & 0 \\ 0 & 0 & \xi_{33} \end{pmatrix}$	$\begin{pmatrix} \mu_{11} & \mu_{12} & 0 \\ \mu_{12} & \mu_{22} & 0 \\ 0 & 0 & \mu_{33} \end{pmatrix}$	13

Table 3.3   continued

Magnetic point group	$\underline{\underline{\epsilon}}_{EH}$	$\underline{\underline{\xi}}_{EH}$	$\underline{\underline{\zeta}}_{EH}$	$\underline{\underline{\mu}}_{EH}$	No. of params.
3. $C_{2v}(C_2)$	$\begin{pmatrix} \epsilon_{11} & \epsilon_{12} & 0 \\ -\epsilon_{12} & \epsilon_{22} & 0 \\ 0 & 0 & \epsilon_{33} \end{pmatrix}$	$\begin{pmatrix} \xi_{11} & \xi_{12} & 0 \\ \xi_{21} & \xi_{22} & 0 \\ 0 & 0 & \xi_{33} \end{pmatrix}$	$\begin{pmatrix} \xi_{11} & -\xi_{21} & 0 \\ -\xi_{12} & \xi_{22} & 0 \\ 0 & 0 & \xi_{33} \end{pmatrix}$	$\begin{pmatrix} \mu_{11} & \mu_{12} & 0 \\ -\mu_{12} & \mu_{22} & 0 \\ 0 & 0 & \mu_{33} \end{pmatrix}$	13
2. C_i	$\begin{pmatrix} \epsilon_{11} & \epsilon_{12} & \epsilon_{13} \\ \epsilon_{21} & \epsilon_{22} & \epsilon_{23} \\ \epsilon_{31} & \epsilon_{32} & \epsilon_{33} \end{pmatrix}$	$\underline{\underline{0}}$	$\underline{\underline{0}}$	$\begin{pmatrix} \mu_{11} & \mu_{12} & \mu_{13} \\ \mu_{21} & \mu_{22} & \mu_{23} \\ \mu_{31} & \mu_{32} & \mu_{33} \end{pmatrix}$	18
2. C_s	$\begin{pmatrix} \epsilon_{11} & \epsilon_{12} & 0 \\ \epsilon_{21} & \epsilon_{22} & 0 \\ 0 & 0 & \epsilon_{33} \end{pmatrix}$	$\begin{pmatrix} 0 & 0 & \xi_{13} \\ 0 & 0 & \xi_{23} \\ \xi_{31} & \xi_{32} & 0 \end{pmatrix}$	$\begin{pmatrix} 0 & 0 & \zeta_{13} \\ 0 & 0 & \zeta_{23} \\ \zeta_{31} & \zeta_{32} & 0 \end{pmatrix}$	$\begin{pmatrix} \mu_{11} & \mu_{12} & 0 \\ \mu_{21} & \mu_{22} & 0 \\ 0 & 0 & \mu_{33} \end{pmatrix}$	18
2. C_2	$\begin{pmatrix} \epsilon_{11} & \epsilon_{12} & 0 \\ \epsilon_{21} & \epsilon_{22} & 0 \\ 0 & 0 & \epsilon_{33} \end{pmatrix}$	$\begin{pmatrix} \xi_{11} & \xi_{12} & 0 \\ \xi_{21} & \xi_{22} & 0 \\ 0 & 0 & \xi_{33} \end{pmatrix}$	$\begin{pmatrix} \zeta_{11} & \zeta_{12} & 0 \\ \zeta_{21} & \zeta_{22} & 0 \\ 0 & 0 & \zeta_{33} \end{pmatrix}$	$\begin{pmatrix} \mu_{11} & \mu_{12} & 0 \\ \mu_{21} & \mu_{22} & 0 \\ 0 & 0 & \mu_{33} \end{pmatrix}$	20
1. C_1	$\begin{pmatrix} \epsilon_{11} & \epsilon_{12} & \epsilon_{13} \\ \epsilon_{21} & \epsilon_{22} & \epsilon_{23} \\ \epsilon_{31} & \epsilon_{32} & \epsilon_{33} \end{pmatrix}$	$\begin{pmatrix} \xi_{11} & \xi_{12} & \xi_{13} \\ \xi_{21} & \xi_{22} & \xi_{23} \\ \xi_{31} & \xi_{32} & \xi_{33} \end{pmatrix}$	$\begin{pmatrix} -\xi_{11} & -\xi_{21} & -\xi_{31} \\ -\xi_{12} & -\xi_{22} & -\xi_{32} \\ -\xi_{13} & -\xi_{23} & -\xi_{33} \end{pmatrix}$	$\begin{pmatrix} \mu_{11} & \mu_{12} & \mu_{13} \\ \mu_{12} & \mu_{22} & \mu_{23} \\ \mu_{13} & \mu_{23} & \mu_{33} \end{pmatrix}$	21

Table 3.3 continued

Magnetic point group	$\underline{\underline{\epsilon}}_{EH}$	$\underline{\underline{\xi}}_{EH}$	$\underline{\underline{\zeta}}_{EH}$	$\underline{\underline{\mu}}_{EH}$	No. of params.
3. $C_i(C_1)$	$\begin{pmatrix} \epsilon_{11} & \epsilon_{12} & \epsilon_{13} \\ \epsilon_{12} & \epsilon_{22} & \epsilon_{23} \\ \epsilon_{13} & \epsilon_{23} & \epsilon_{33} \end{pmatrix}$	$\begin{pmatrix} \xi_{11} & \xi_{12} & \xi_{13} \\ \xi_{21} & \xi_{22} & \xi_{23} \\ \xi_{31} & \xi_{32} & \xi_{33} \end{pmatrix}$	$\begin{pmatrix} \xi_{11} & \xi_{21} & \xi_{31} \\ \xi_{12} & \xi_{22} & \xi_{32} \\ \xi_{13} & \xi_{23} & \xi_{33} \end{pmatrix}$	$\begin{pmatrix} \mu_{11} & \mu_{12} & \mu_{13} \\ \mu_{12} & \mu_{22} & \mu_{23} \\ \mu_{13} & \mu_{23} & \mu_{33} \end{pmatrix}$	21
3. $C_2(C_1)$	$\begin{pmatrix} \epsilon_{11} & \epsilon_{12} & \epsilon_{13} \\ \epsilon_{12} & \epsilon_{22} & \epsilon_{23} \\ -\epsilon_{13} & -\epsilon_{23} & \epsilon_{33} \end{pmatrix}$	$\begin{pmatrix} \xi_{11} & \xi_{12} & \xi_{13} \\ \xi_{21} & \xi_{22} & \xi_{23} \\ \xi_{31} & \xi_{32} & \xi_{33} \end{pmatrix}$	$\begin{pmatrix} -\xi_{11} & -\xi_{21} & \xi_{31} \\ -\xi_{12} & -\xi_{22} & \xi_{32} \\ \xi_{13} & \xi_{23} & -\xi_{33} \end{pmatrix}$	$\begin{pmatrix} \mu_{11} & \mu_{12} & \mu_{13} \\ \mu_{12} & \mu_{22} & \mu_{23} \\ -\mu_{13} & -\mu_{23} & \mu_{33} \end{pmatrix}$	21
3. $C_s(C_1)$	$\begin{pmatrix} \epsilon_{11} & \epsilon_{12} & \epsilon_{13} \\ \epsilon_{12} & \epsilon_{22} & \epsilon_{23} \\ -\epsilon_{13} & -\epsilon_{23} & \epsilon_{33} \end{pmatrix}$	$\begin{pmatrix} \xi_{11} & \xi_{12} & \xi_{13} \\ \xi_{21} & \xi_{22} & \xi_{23} \\ \xi_{31} & \xi_{32} & \xi_{33} \end{pmatrix}$	$\begin{pmatrix} \xi_{11} & \xi_{21} & -\xi_{31} \\ \xi_{12} & \xi_{22} & -\xi_{32} \\ -\xi_{13} & -\xi_{23} & \xi_{33} \end{pmatrix}$	$\begin{pmatrix} \mu_{11} & \mu_{12} & \mu_{13} \\ \mu_{12} & \mu_{22} & \mu_{23} \\ -\mu_{13} & -\mu_{23} & \mu_{33} \end{pmatrix}$	21
2. $\mathbf{C_1}$	$\begin{pmatrix} \epsilon_{11} & \epsilon_{12} & \epsilon_{13} \\ \epsilon_{21} & \epsilon_{22} & \epsilon_{23} \\ \epsilon_{31} & \epsilon_{32} & \epsilon_{33} \end{pmatrix}$	$\begin{pmatrix} \xi_{11} & \xi_{12} & \xi_{13} \\ \xi_{21} & \xi_{22} & \xi_{23} \\ \xi_{31} & \xi_{32} & \xi_{33} \end{pmatrix}$	$\begin{pmatrix} \zeta_{11} & \zeta_{12} & \zeta_{13} \\ \zeta_{21} & \zeta_{22} & \zeta_{23} \\ \zeta_{31} & \zeta_{32} & \zeta_{33} \end{pmatrix}$	$\begin{pmatrix} \mu_{11} & \mu_{12} & \mu_{13} \\ \mu_{21} & \mu_{22} & \mu_{23} \\ \mu_{31} & \mu_{32} & \mu_{33} \end{pmatrix}$	36

those of the second category follows in a similar manner to that described in Sec. 3.2 for the discrete magnetic point groups.

In total, there are seven continuous magnetic point groups for each of the three categories of magnetic point groups. The Schoenflies symbols for these are listed in Table 3.4, and the corresponding forms for the constitutive dyadics are provided in Table 3.3.

Table 3.4   Twenty one continuous magnetic point groups.

Category 1	Category 2	Category 3
$\underline{K}_h$	$\mathbf{K_h}$	$\underline{K}_h(\underline{K})$
$\underline{K}$	$\mathbf{K}$	$\underline{D}_{\infty h}(\underline{D}_\infty)$
$\underline{D}_{\infty h}$	$\mathbf{D}_{\infty}\mathbf{h}$	$\underline{D}_{\infty h}(\underline{C}_{\infty v})$
$\underline{D}_\infty$	$\mathbf{D}_\infty$	$\underline{D}_{\infty h}(\underline{C}_{\infty h})$
$\underline{C}_{\infty v}$	$\mathbf{C}_{\infty \mathbf{v}}$	$\underline{D}_\infty(\underline{C}_\infty)$
$\underline{C}_{\infty h}$	$\mathbf{C}_{\infty}\mathbf{h}$	$\underline{C}_{\infty v}(\underline{C}_\infty)$
$\underline{C}_\infty$	$\mathbf{C}_\infty$	$\underline{C}_{\infty h}(\underline{C}_\infty)$

3.4   Space groups

The 32 point groups of classical crystallography classify the spatial arrangements of objects on a three–dimensional lattice in terms of symmetry operations involving rotations, reflections and inversions. Translational symmetries are not taken into consideration. While it is the point group symmetries that determine the forms of the constitutive dyadics, there are important symmetry operations involving translations — e.g., screw–axis rotations and glide–plane reflections — which are not accounted for by the point group classifications. With the inclusion of translational symmetry operations, the number of distinguishable spatial arrangements on a three–dimensional lattice increases to 230. These form the 230 *space groups* of classical crystallography. They are listed in standard references [22].

If the objects arranged on a lattice can be distinguished from each other on the basis of a nongeometric property, e.g., spin, the number of space groups increases greatly. In fact, in the same way that the 32 point groups of classical crystallography give rise to 122 magnetic point groups, the 230 space groups of classical crystallography give rise to 1651 magnetic space groups, by combining the symmetries under the Wigner time–reversal operator with the spatial symmetries including translational symmetries [16, 23]. Detailed descriptions of the magnetic space groups are documented in standard works [17, 24]. Each magnetic space group represents a distinguishable pattern which can be formed on a three–dimensional lattice using a bi–coloured motif.

References

[1] C.E. Baum and H.N. Kritikos, Symmetry in electromagnetics, *Electromagnetic symmetry* (C.E. Baum and H.N. Kritikos, eds), Taylor & Francis, Washington DC, USA, 1995, 1–90.

[2] T.A. Whitelaw, *Introduction to abstract algebra, 3rd ed*, Blackie, Glasgow, UK, 1995.

[3] R. Mirman, *Point groups, space groups, crystals, molecules*, World Scientific, Singapore, 1999.

[4] A.S. Nowick, *Crystal properties via group theory*, Cambridge University Press, Cambridge, UK, 1995.

[5] R.R. Birss, Macroscopic symmetry in space–time, *Rep Prog Phys* **26** (1963), 307–360.

[6] T. Harada, T. Sato and R. Kuroda, Intrinsic birefringence of a chiral sodium chlorate crystal: Is cubic crystal truly optically neutral?, *Chem Phys Lett* **413** (2005), 445–449.

[7] J.F. Nye, *Physical properties of crystals*, Oxford University Press, Oxford, UK, 1985.

[8] D.R. Lovett, *Tensor properties of crystals, 2nd ed*, Institute of Physics, Bristol, UK, 1999.

[9] N.W. Ashcroft and N.D. Mermin, *Solid state physics*, Saunders, Philadelphia, PA, USA, 1976.

[10] B. Lax and K.J. Button, *Microwave ferrites and ferrimagnetics*, McGraw–Hill, New York, NY, USA, 1962.

[11] E.P. Wigner, *Group theory and its application to the quantum mechanics of atomic spectra*, Academic Press, New York, NY, USA, 1959.

[12] K. Suchy, C. Altman and A. Schatzberg, Orthogonal mappings of time–harmonic electromagnetic fields in inhomogeneous (bi)anisotropic media, *Radio Sci* **20** (1985), 149–160.

[13] V. Dmitriev, Group theoretical approach to complex and bianisotropic media description, *Eur Phys J AP* **6** (1999), 49–55.

[14] C. Altman and K. Suchy, *Reciprocity, spatial mapping and time reversal in electromagnetics*, Kluwer, Dordrecht, The Netherlands, 1991.

[15] D.B. Litvin, Point group symmetries, *Introduction to complex mediums for optics and electromagnetics* (W.S. Weiglhofer and A. Lakhtakia, eds), SPIE Press, Bellingham, WA, USA, 2003, 79–102.

[16] R. Lifshitz, Magnetic point groups and space groups, *Encyclopedia of condensed matter physics, Vol 3* (F. Bassani, G.L. Liedl, and P. Wyder, eds), Elsevier, Oxford, UK, 2005, 219–226.

[17] A.V. Schubnikov and V.A. Koptsik, *Symmetry in science and art*, Plenum Press, New York, NY, USA, 1974.

[18] V. Dmitriev, Some general electromagnetic properties of linear homogeneous bianisotropic media following from space and time–reversal symmetry of the second–rank and antisymmetric third–rank constitutive tensors, *Eur Phys J AP* **12** (2000), 3–16.

[19] B. Michel, Recent developments in the homogenization of linear bian-isotropic composite materials, *Electromagnetic fields in unconventional materials and structures* (O.N. Singh and A. Lakhtakia, eds), Wiley, New York, NY, USA, 2000, 39–82.

[20] V.A. Dmitriev, Continuous tensors and general properties of complex and bianisotropic media described by continuous groups of symmetry, *Electron Lett* **34** (1998), 532–534.

[21] V.A. Dmitriev, Symmetry description of continuous homogeneous isotropic media under external perturbation, *Electron Lett* **34** (1998), 745–747.

[22] T. Hahn (ed), *International tables for crystallography. Vol A: Space–group symmetry, 5th ed*, Kluwer, Dordrecht, The Netherlands, 2002.

[23] A.P. Cracknell, *Group theory in solid–state physics*, Taylor & Francis, London, UK, 1975.

[24] A.V. Schubnikov and N.V. Belov, *Colored symmetry*, Pergamon Press, Oxford, UK, 1964.

Chapter 4

Planewave Propagation

The passage of electromagnetic signals through a complex medium is generally a complicated issue, from a theoretical perspective at least. Often it is best broached by considering the propagation of plane waves. Although plane waves themselves are idealizations, being of limitless spatial and temporal extents and possessing infinite energy, their consideration can provide a reasonable understanding of fields far away from sources. Furthermore, realistic signals may be usefully represented as superpositions of plane waves.

Planewave propagation in isotropic dielectric mediums, as well as certain relatively simple, anisotropic dielectric mediums, are widely described in the standard references [1–4]. This is especially true for the scenario wherein the effects of dissipation are neglected, the topic then usually coming under the heading of 'crystal optics'. In this chapter, a comprehensive survey of planewave solutions of the Maxwell curl postulates in anisotropic and bianisotropic mediums, including dissipative and nonhomogeneous mediums, is presented. We begin with a general description of plane waves. As a precursor to the later sections, a brief outline of planewave propagation in isotropic mediums is provided. Thereafter, we discuss the modes of uniform planewave propagation which are supported by various anisotropic and bianisotropic mediums, including nonhomogeneous mediums and some exotic phenomenons which are of current research interest. Our focus throughout is chiefly on nonactive mediums (cf. Sec. 1.7.2.2).

4.1   Uniform and nonuniform plane waves

We recall from Sec. 1.5 that the electric and magnetic field phasors $\underline{E}\,(\underline{r},\omega)$ and $\underline{H}\,(\underline{r},\omega)$, respectively, are conveniently combined in the 6–vector phasor $\underline{\mathbf{F}}(\underline{r},\omega)$, as defined in Eq. (1.40). The corresponding time–domain 6–vector is provided by the inverse temporal Fourier transform

$$\tilde{\underline{\mathbf{F}}}(\underline{r},t) \equiv \begin{pmatrix} \tilde{\underline{E}}\,(\underline{r},t) \\ \tilde{\underline{H}}\,(\underline{r},t) \end{pmatrix} \tag{4.1}$$

$$= \frac{1}{2\pi}\int_{-\infty}^{\infty} \underline{\mathbf{F}}(\underline{r},\omega)\,\exp(-i\omega t)\,d\omega. \tag{4.2}$$

Here we consider the case where $\tilde{\underline{\mathbf{F}}}(\underline{r},t)$ represents a plane wave propagating with wavevector $\underline{k}$ and angular frequency $\omega = \omega_s$; i.e.,

$$\tilde{\underline{\mathbf{F}}}_{\mathrm{pw}}(\underline{r},t) = \mathrm{Re}\left\{\check{\underline{\mathbf{F}}}_{\mathrm{pw}}(\underline{r},\omega_s)\,\exp\left(-i\omega_s t\right)\right\}, \tag{4.3}$$

with

$$\check{\underline{\mathbf{F}}}_{\mathrm{pw}}(\underline{r},\omega_s) \equiv \begin{pmatrix} \check{\underline{E}}_{\mathrm{pw}}(\underline{r},\omega_s) \\ \check{\underline{H}}_{\mathrm{pw}}(\underline{r},\omega_s) \end{pmatrix} \tag{4.4}$$

$$= \check{\underline{\mathbf{F}}}_0(\omega_s)\,\exp\left(i\underline{k}\boldsymbol{\cdot}\underline{r}\right) \tag{4.5}$$

$$\equiv \begin{pmatrix} \check{\underline{E}}_0(\omega_s) \\ \check{\underline{H}}_0(\omega_s) \end{pmatrix}\,\exp\left(i\underline{k}\boldsymbol{\cdot}\underline{r}\right). \tag{4.6}$$

From Eqs. (4.3) and (4.5), we find that the temporal Fourier transform of $\tilde{\underline{\mathbf{F}}}_{\mathrm{pw}}(\underline{r},t)$, namely $\underline{\mathbf{F}}_{\mathrm{pw}}(\underline{r},\omega)$, is related to the complex–valued amplitude $\check{\underline{\mathbf{F}}}_0(\omega_s)$ as (cf. Eq. (1.37))

$$\underline{\mathbf{F}}_{\mathrm{pw}}(\underline{r},\omega) \equiv \begin{pmatrix} \underline{E}_{\mathrm{pw}}(\underline{r},\omega) \\ \underline{H}_{\mathrm{pw}}(\underline{r},\omega) \end{pmatrix} \tag{4.7}$$

$$= \frac{1}{2}\Big[\check{\underline{\mathbf{F}}}_0(\omega_s)\,\exp\left(i\underline{k}\boldsymbol{\cdot}\underline{r}\right)\,\delta(\omega - \omega_s)$$

$$+\,\check{\underline{\mathbf{F}}}_0^*(\omega_s)\,\exp\left(-i\underline{k}\boldsymbol{\cdot}\underline{r}\right)\,\delta(\omega + \omega_s)\Big]. \tag{4.8}$$

Thus the phasor $\underline{\mathbf{F}}_{\mathrm{pw}}(\underline{r},\omega)$ satisfies the symmetry condition (1.34). Let us reiterate the comment made after Eq. (1.37) in Sec. 1.4: the planewave amplitudes $\check{\underline{E}}_{\mathrm{pw}}(\underline{r},\omega)$ and $\check{\underline{H}}_{\mathrm{pw}}(\underline{r},\omega)$ (together with their $\underline{D}$, $\underline{B}$ and $\underline{J}_{\mathrm{e,m}}$ counterparts) satisfy the frequency–domain constitutive relations, Maxwell postulates and boundary conditions in exactly the same way as the frequency–domain phasors $\underline{E}_{\mathrm{pw}}(\underline{r},\omega)$ and $\underline{H}_{\mathrm{pw}}(\underline{r},\omega)$ (together with their $\underline{D}$, $\underline{B}$ and $\underline{J}_{\mathrm{e,m}}$ counterparts).

The wavevector $\underline{k} \in \mathbb{C}^3$, in general, and may be written as

$$\underline{k} = k_{\mathrm{Re}}\,\hat{\underline{k}}_{\mathrm{Re}} + ik_{\mathrm{Im}}\,\hat{\underline{k}}_{\mathrm{Im}}\,, \qquad (4.9)$$

wherein the scalars $k_{\mathrm{Re,Im}} \in \mathbb{R}$ and the unit vectors $\hat{\underline{k}}_{\mathrm{Re,Im}} \in \mathbb{R}^3$. Although $\underline{k}$ varies with ω, for convenience we do not express this dependence explicitly. Furthermore, since the choice $\omega = \omega_{\mathrm{s}}$ is arbitrary, we omit the subscript 's' henceforth. In light of Eq. (4.9), the 6–vector field phasor given by Eq. (4.5) is expressed as

$$\check{\underline{\mathbf{F}}}_{\mathrm{pw}}(\underline{r},\omega) = \check{\underline{\mathbf{F}}}_0(\omega) \exp\left(-k_{\mathrm{Im}}\,\hat{\underline{k}}_{\mathrm{Im}} \bullet \underline{r}\right) \exp\left(ik_{\mathrm{Re}}\,\hat{\underline{k}}_{\mathrm{Re}} \bullet \underline{r}\right). \qquad (4.10)$$

On planes in three–dimensional space specified by

$$\hat{\underline{k}}_{\mathrm{Re}} \bullet \underline{r} = \mathrm{constant}\,, \qquad (4.11)$$

we see from Eq. (4.10) that $\check{\underline{\mathbf{F}}}_{\mathrm{pw}}(\underline{r},\omega)$ has constant phase. The planes of constant phase propagate with velocity

$$\underline{v}_{\mathrm{ph}}(\omega) = \frac{\omega}{k_{\mathrm{Re}}}\,\hat{\underline{k}}_{\mathrm{Re}} \qquad (4.12)$$

in the direction of $\hat{\underline{k}}_{\mathrm{Re}}$. This velocity is called the phase velocity, and its magnitude is the phase speed.

A plane wave may be either *uniform* or *nonuniform*, depending on the relative orientations of $\hat{\underline{k}}_{\mathrm{Re}}$ and $\hat{\underline{k}}_{\mathrm{Im}}$, as follows:

- If $\hat{\underline{k}}_{\mathrm{Re}} = \pm\hat{\underline{k}}_{\mathrm{Im}}$ then the wavevector $\underline{k}$ may be written as

$$\underline{k} = k\,\hat{\underline{k}}\,, \qquad (4.13)$$

  with the complex–valued wavenumber $k = k_{\mathrm{Re}} \pm ik_{\mathrm{Im}}$ and the real-valued unit vector $\hat{\underline{k}} \equiv \hat{\underline{k}}_{\mathrm{Re}}$. In this case, on planes of constant $\hat{\underline{k}} \bullet \underline{r}$, the corresponding phasor $\check{\underline{\mathbf{F}}}_{\mathrm{pw}}(\underline{r},\omega)$ has both constant phase and constant amplitude. Accordingly, these plane waves are called uniform plane waves. Uniform plane waves also crop up in the nondissipative scenario characterized by $k_{\mathrm{Im}} = 0$.

- If $\hat{\underline{k}}_{\mathrm{Re}} \neq \pm\hat{\underline{k}}_{\mathrm{Im}}$ with $k_{\mathrm{Im}} \neq 0$, then the amplitude of $\check{\underline{\mathbf{F}}}_{\mathrm{pw}}(\underline{r},\omega)$ is generally not uniform on planes of constant $\hat{\underline{k}}_{\mathrm{Re}} \bullet \underline{r}$. Accordingly, these plane waves are called nonuniform plane waves.

By definition, the wavevector $\underline{k}$ has complex–valued components (with nonzero imaginary parts) for nonuniform plane waves; in contrast, for uniform plane waves $\underline{k}$ may have either complex– or real–valued components. Most significantly, uniform plane waves with real–valued components of $\underline{k}$ propagate without attenuation. The remainder of this chapter is mostly devoted to uniform plane waves.

4.2   Eigenanalysis

Let us consider the most general linear homogeneous medium — namely, a homogeneous bianisotropic medium, whose Tellegen constitutive relations are expressed as

$$
\left.
\begin{aligned}
\underline{D}(\underline{r},\omega) &= \underline{\underline{\epsilon}}_{\mathrm{EH}}(\omega) \bullet \underline{E}(\underline{r},\omega) + \underline{\underline{\xi}}_{\mathrm{EH}}(\omega) \bullet \underline{H}(\underline{r},\omega) \\
\underline{B}(\underline{r},\omega) &= \underline{\underline{\zeta}}_{\mathrm{EH}}(\omega) \bullet \underline{E}(\underline{r},\omega) + \underline{\underline{\mu}}_{\mathrm{EH}}(\omega) \bullet \underline{H}(\underline{r},\omega)
\end{aligned}
\right\}.
\tag{4.14}
$$

In the absence of sources, the phasors in Eqs. (4.14) are related by the frequency–domain Maxwell curl postulates

$$
\left.
\begin{aligned}
\nabla \times \underline{H}(\underline{r},\omega) + i\omega \underline{D}(\underline{r},\omega) &= \underline{0} \\
\nabla \times \underline{E}(\underline{r},\omega) - i\omega \underline{B}(\underline{r},\omega) &= \underline{0}
\end{aligned}
\right\}.
\tag{4.15}
$$

The combination of Eqs. (4.14) and (4.15) provides us with

$$
\left[\underline{\underline{L}}(\nabla) + i\omega \underline{\underline{K}}_{\mathrm{EH}}(\omega) \right] \bullet \mathbf{F}(\underline{r},\omega) = \underline{0},
\tag{4.16}
$$

wherein the 6–vector/6×6 dyadic notation of Sec. 1.5 has been implemented, with the Tellegen constitutive dyadic

$$
\underline{\underline{K}}_{\mathrm{EH}}(\omega) =
\begin{bmatrix}
\underline{\underline{\epsilon}}_{\mathrm{EH}}(\omega) & \underline{\underline{\xi}}_{\mathrm{EH}}(\omega) \\
\underline{\underline{\zeta}}_{\mathrm{EH}}(\omega) & \underline{\underline{\mu}}_{\mathrm{EH}}(\omega)
\end{bmatrix}.
\tag{4.17}
$$

If we now restrict our attention to planewave solutions, as represented by the phasor given by Eq. (4.5), then the dyadic differential Eq. (4.16) simplifies to

$$
\left[\underline{\underline{L}}(i\underline{k}) + i\omega \underline{\underline{K}}_{\mathrm{EH}}(\omega) \right] \bullet \check{\mathbf{F}}_0(\omega) = \underline{0}.
\tag{4.18}
$$

The algebraic Eq. (4.18) is generally amenable to eigenanalysis, although the process can involve unwieldy expressions, especially for bianisotropic mediums. Symbolic manipulations packages — such as Maple^{TM} and Mathematica^{TM} — can be usefully employed in the analysis.

The existence of nonzero solutions to Eq. (4.18) immediately leads us to the planewave dispersion relation

$$
\det \left[\underline{\underline{L}}(i\underline{k}) + i\omega \underline{\underline{K}}_{\mathrm{EH}}(\omega) \right] = 0.
\tag{4.19}
$$

For uniform plane waves, i.e., $\underline{k} = k\,\hat{\underline{k}}$ with $k \in \mathbb{C}$ and $\hat{\underline{k}} \in \mathbb{R}^3$, the dispersion relation (4.19) can be recast as a polynomial of the fourth degree in k. Accordingly, it has four solutions $k = k_\ell$, $\ell \in [1,4]$.

The dispersion relation (4.19) is quadratic in k^2 for all Lorentz–reciprocal mediums and all Lorentz–nonreciprocal mediums with null–valued magnetoelectric constitutive dyadics (such as dielectric and magnetic gyrotropic mediums). For these mediums two solutions, $k = k_1$ and $k = k_2$, say, hold for propagation co–parallel with $\underline{k}$ and two solutions, $k = k_3$ and $k = k_4$, say, hold for propagation anti–parallel to $\underline{k}$, wherein $k_1 = -k_3$ and $k_2 = -k_4$. If $k_1 \neq k_2$ (or, equivalently, $k_3 \neq k_4$), the medium is said to possess *birefringence*. On the other hand, the term *unirefringence* is used to describe those instances in which $k_1 = k_2$ (or, equivalently, $k_3 = k_4$).

For bianisotropic Lorentz–nonreciprocal mediums (such as Faraday chiral mediums), the dispersion relation generally yields (4.19) four distinct roots — and these do not come in positive/negative pairs.

The solution to Eq. (4.18) may expressed as

$$\breve{\underline{F}}_0(\omega) = \text{adj}\left[\underline{\underline{L}}(i\underline{k}) + i\omega\underline{\underline{K}}_{\text{EH}}(\omega) \right] \cdot \underline{V}, \qquad (4.20)$$

where $\underline{V}$ is an arbitrary 6-vector. Without the boundary conditions being specified, only the relative orientations of $\underline{E}(\underline{r}, \omega)$ and $\underline{H}(\underline{r}, \omega)$ may be deduced in general. From Eq. (4.5) we have that

$$\left.\begin{array}{l} \breve{\underline{E}}_{\text{pw}}(\underline{r}, \omega) = \breve{\underline{E}}_0(\omega) \exp(i\underline{k} \cdot \underline{r}) \\[2mm] \breve{\underline{H}}_{\text{pw}}(\underline{r}, \omega) = \breve{\underline{H}}_0(\omega) \exp(i\underline{k} \cdot \underline{r}) \end{array}\right\}, \qquad (4.21)$$

wherein $\breve{\underline{E}}_0(\omega), \breve{\underline{H}}_0(\omega) \in \mathbb{C}^3$, and these amplitudes may be usefully decomposed as

$$\left.\begin{array}{l} \breve{\underline{E}}_0(\omega) = \breve{\underline{E}}_{0\text{Re}}(\omega) + i\breve{\underline{E}}_{0\text{Im}}(\omega) \\[2mm] \breve{\underline{H}}_0(\omega) = \breve{\underline{H}}_{0\text{Re}}(\omega) + i\breve{\underline{H}}_{0\text{Im}}(\omega) \end{array}\right\}, \qquad (4.22)$$

with $\breve{\underline{E}}_{0\text{Re},0\text{Im}}(\omega), \breve{\underline{H}}_{0\text{Re},0\text{Im}}(\omega) \in \mathbb{R}^3$. The relative orientations and magnitudes of $\breve{\underline{E}}_{0\text{Re}}(\omega)$ and $\breve{\underline{E}}_{0\text{Im}}(\omega)$ determine the polarization state of the electric field, as follows:

- if $\breve{\underline{E}}_{0\text{Re}}(\omega) \times \breve{\underline{E}}_{0\text{Im}}(\omega) = \underline{0}$ then the electric field is *linearly* polarized;
- if $\breve{\underline{E}}_{0\text{Re}}(\omega) \cdot \breve{\underline{E}}_{0\text{Im}}(\omega) = 0$ and $\left|\breve{\underline{E}}_{0\text{Re}}(\omega)\right| = \left|\breve{\underline{E}}_{0\text{Im}}(\omega)\right|$ then the electric field is *circularly* polarized;
- otherwise, the electric field is *elliptically* polarized.

The polarization state of the magnetic field is similarly determined by the relative orientations and magnitudes of $\breve{\underline{H}}_{0\text{Re}}(\omega)$ and $\breve{\underline{H}}_{0\text{Im}}(\omega)$. In optics, the 'polarization state' of a plane wave is conventionally that of its electric field.

For a linearly polarized wave, the plane containing $\breve{\underline{E}}_{0\mathrm{Re}}(\omega)$ (or $\breve{\underline{E}}_{0\mathrm{Im}}(\omega)$ if $\breve{\underline{E}}_{0\mathrm{Re}}(\omega) = \underline{0}$) and $\underline{k}$ is called the *plane of polarization*. A linearly polarized wave may be decomposed into two circularly polarized plane waves of equal amplitude, one being left–circularly polarized and the other right–circularly polarized.

4.3 Reflection and refraction of plane waves

A question of fundamental importance in many applications is the following: What happens when a plane wave is incident upon a boundary separating two distinct half–spaces? As in Sec. 1.2, suppose that the unit vector $\hat{n}$ is normal to the boundary separating half–spaces I and II, directed into half–space II. A plane wave propagating in half–space I, as described by Eq. (4.3) with wavevector $\underline{k}_{\mathrm{i}}$, is incident upon the boundary at point $\underline{r}$. In general, incidence results in:

- two reflected plane waves in half–space I, with wavevectors $\underline{k}_{\mathrm{r}}^{(1)}$ and $\underline{k}_{\mathrm{r}}^{(2)}$, which propagate away from the boundary; and
- two refracted plane waves in half–space II, with wavevectors $\underline{k}_{\mathrm{t}}^{(1)}$ and $\underline{k}_{\mathrm{t}}^{(2)}$, which also propagate away from the boundary[1].

By application of the (frequency–domain) boundary conditions (1.31) and (1.32), we find [4]

$$\underline{k}_{\mathrm{i}} \times \hat{\underline{n}} = \begin{cases} \underline{k}_{\mathrm{r}}^{(1)} \times \hat{\underline{n}} \\ \underline{k}_{\mathrm{r}}^{(2)} \times \hat{\underline{n}} \\ \underline{k}_{\mathrm{t}}^{(1)} \times \hat{\underline{n}} \\ \underline{k}_{\mathrm{t}}^{(2)} \times \hat{\underline{n}} \end{cases} ; \qquad (4.23)$$

that is, the components of the wavevectors of the incident, reflected and refracted plane waves tangential to the boundary must be identical. Furthermore, the wavevectors of the incident, reflected and refracted plane waves all lie in the same plane as $\hat{n}$, this plane being called the *plane of incidence*.

Conventionally, refraction and reflection are treated such that the wavevectors of the refracted and reflected plane waves point away from

---

[1]Note that the *direction* of planewave propagation refers the direction of energy flow density, as provided by the time–averaged Poynting vector defined in Eq. (1.90).

the boundary. Refraction and reflection can then be said to be *positive*. But *negative refraction* and *negative reflection* are also possible, as explained in Sec. 4.8.1 and Sec. 4.8.4, respectively. That is, the uniform refracted plane wave with wavevector $\underline{k}_t^{(\ell)}$ ($\ell = 1, 2$) is

- positively refracted if $\underline{k}_t^{(\ell)} \cdot \hat{\underline{n}} > 0$; and
- negatively refracted if $\underline{k}_t^{(\ell)} \cdot \hat{n} < 0$.

A schematic depiction is provided in Fig. 4.1.

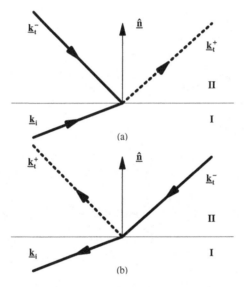

Figure 4.1 Schematic representation of negative and positive refraction for uniform plane waves. In (a), the wavevector of the incident planewave, namely $\underline{k}_i$, is directed towards the boundary; in (b), $\underline{k}_i$ is directed away from the boundary. In both (a) and (b), the wavevector of a negatively refracted plane wave, namely $\underline{k}_t^-$, is directed towards the boundary, whereas the wavevector of a positively refracted plane wave, namely $\underline{k}_t^+$, is directed away from the boundary. The phase velocity of the incident plane wave is positive in (a) and negative in (b); positive and negative phase velocities are described in Sec. 4.8.1.

In a similar vein, the uniform reflected plane wave with wavevector $\underline{k}_r^{(\ell)}$ ($\ell = 1, 2$) is

- positively reflected if $\underline{k}_r^{(\ell)} \cdot \hat{\underline{n}} < 0$; and
- negatively reflected if $\underline{k}_r^{(\ell)} \cdot \hat{\underline{n}} > 0$,

as illustrated in Fig. 4.2.

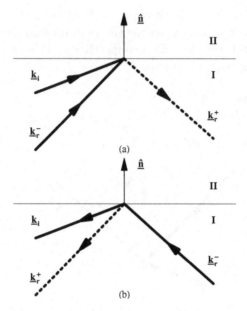

Figure 4.2 Schematic representation of negative and positive reflection for uniform plane waves. In (a), the wavevector of the incident planewave, namely $\underline{k}_i$, is directed towards the boundary; in (b), $\underline{k}_i$ is directed away from the boundary. In both (a) and (b), the wavevector of a negatively reflected plane wave, namely $\underline{k}_r^-$, is directed towards the boundary, whereas the wavevector of a positively reflected plane wave, namely $\underline{k}_r^+$, is directed away from the boundary. The phase velocity of the incident plane wave is positive in (a) and negative in (b); positive and negative phase velocities are described in Sec. 4.8.1.

The phenomenons of negative refraction and negative reflection — which arise within certain constitutive parameter regimes — present many interesting technological possibilities. Materials which exhibit these phenomenons are currently the subject of intense research activity. These issues are taken up in Secs. 4.8.1 and 4.8.4.

The descriptions of positive/negative refraction and reflection for uniform plane waves presented herein can be extended to nonuniform plane waves by replacing the wavevectors with their real parts.

4.4   Uniform plane waves in isotropic mediums

Planewave propagation in isotropic mediums is independent of direction of propagation, which contrasts sharply with planewave propagation in anisotropic and bianisotropic mediums.

4.4.1 *Dielectric–magnetic mediums*

The specialization of the dispersion relation (4.19) appropriate to isotropic dielectric–magnetic mediums, as characterized by the Tellegen constitutive relations (2.3), yields the two roots

$$\left.\begin{array}{l} k_1 = k_2 = \omega\,[\epsilon(\omega)]^{1/2}\,[\mu(\omega)]^{1/2} \\[2mm] k_3 = k_4 = -\omega\,[\epsilon(\omega)]^{1/2}\,[\mu(\omega)]^{1/2} \end{array}\right\}, \qquad (4.24)$$

for nonactive mediums. Hence, isotropic dielectric–magnetic mediums are unirefringent. As regards the orientations of $\breve{\underline{E}}_{\mathrm{pw}}(\underline{r},\omega)$ and $\breve{\underline{H}}_{\mathrm{pw}}(\underline{r},\omega)$, a straightforward manipulation of the Maxwell curl postulates (4.15) delivers the orthogonality relations

$$\left.\begin{array}{l} \underline{k}\cdot\breve{\underline{E}}_{\mathrm{pw}}(\underline{r},\omega) = 0 \\[2mm] \underline{k}\cdot\breve{\underline{H}}_{\mathrm{pw}}(\underline{r},\omega) = 0 \end{array}\right\}. \qquad (4.25)$$

Plane waves in isotropic dielectric–magnetic mediums can be arbitrarily polarized: linearly, circularly, or elliptically. The time–averaged Poynting vector for monochromatic plane waves, as defined in Eq. (1.90), reduces to

$$\langle\breve{\underline{S}}(\underline{r},\omega)\rangle_{\mathrm{t}} = \frac{1}{2}\,\mathrm{Re}\left\{\left[\frac{\epsilon(\omega)}{\mu(\omega)}\right]^{1/2}\right\}\exp\left(-2k_{\mathrm{Im}}\hat{\underline{k}}\cdot\underline{r}\right)\left|\breve{\underline{E}}_0(\omega)\right|^2\hat{\underline{k}} \qquad (4.26)$$

for isotropic dielectric–mediums. Therefore, in this case $\langle\breve{\underline{S}}(\underline{r},\omega)\rangle_{\mathrm{t}}$ is always aligned with the wavevector $\underline{k}$.

4.4.2 *Isotropic chiral mediums*

For isotropic chiral mediums, as characterized by the Tellegen constitutive relations (2.5), the wavenumbers which emerge from the dispersion relation (4.19) are

$$\left.\begin{array}{l} k_1 = -k_3 = \omega\left\{[\epsilon(\omega)\,\mu(\omega)]^{1/2} - i\chi(\omega)\right\} \\[2mm] k_2 = -k_4 = \omega\left\{[\epsilon(\omega)\,\mu(\omega)]^{1/2} + i\chi(\omega)\right\} \end{array}\right\}, \qquad (4.27)$$

for nonactive mediums. Hence, isotropic chiral mediums are birefringent.

In order to determine the associated electromagnetic fields, it is mathematically convenient to introduce the *Beltrami fields* [5]

$$\left.\begin{array}{l} \underline{Q}_1(\underline{r},\omega) = \dfrac{1}{2}\left\{\underline{E}(\underline{r},\omega) + i\left[\dfrac{\mu(\omega)}{\epsilon(\omega)}\right]^{1/2}\underline{H}(\underline{r},\omega)\right\} \\[4mm] \underline{Q}_2(\underline{r},\omega) = \dfrac{1}{2}\left\{\underline{H}(\underline{r},\omega) + i\left[\dfrac{\epsilon(\omega)}{\mu(\omega)}\right]^{1/2}\underline{E}(\underline{r},\omega)\right\} \end{array}\right\}. \qquad (4.28)$$

By doing so, the Maxwell curl postulates (4.15) uncouple, thereby providing the two first–order differential equations

$$\left.\begin{array}{l} \nabla \times \underline{Q}_1(\underline{r}, \omega) - k_1 \underline{Q}_1(\underline{r}, \omega) = \underline{0} \\ \nabla \times \underline{Q}_2(\underline{r}, \omega) + k_2 \underline{Q}_2(\underline{r}, \omega) = \underline{0} \end{array}\right\}. \tag{4.29}$$

Thus, we find that plane waves in an isotropic chiral medium must be circularly polarized. The left–circular polarization state is represented by $\underline{Q}_1$, the right–circular polarization state by $\underline{Q}_2$. Furthermore, since $k_1 \neq k_2$, the right– and left–circularly polarized waves propagate with different phase speeds and with different attenuation rates. This special form of birefringence exhibited by isotropic chiral mediums is known as *circular birefringence*.

Beltrami fields in bianisotropic mediums are considered in Sec. 4.6.4.

4.5   Uniform plane waves in anisotropic mediums

4.5.1   *Uniaxial mediums*

Let us begin our survey of uniform planewave propagation in homogeneous anisotropic mediums with the uniaxial dielectric medium, as characterized by the Tellegen constitutive relations (2.8) with permittivity dyadic given in Eq. (2.9). The distinguished axis $\hat{\underline{u}}$ in Eq. (2.9) is parallel to the sole optic ray axis of the medium.

The wavenumbers extracted from the corresponding specialization of the dispersion relation (4.19) are

$$\left.\begin{array}{l} k_1 = -k_3 = \omega \left[\epsilon(\omega)\mu_0 \right]^{1/2} \\[2mm] k_2 = -k_4 = \omega \left[\dfrac{\epsilon(\omega)\epsilon_u(\omega)\mu_0}{\hat{\underline{k}} \cdot \underline{\underline{\epsilon}}_{\text{uni}}(\omega) \cdot \hat{\underline{k}}} \right]^{1/2} \end{array}\right\}. \tag{4.30}$$

Thus, two distinct modes of planewave propagation are supported: *ordinary* as specified by the wavenumber $k_1 = -k_3$, and *extraordinary* as specified by the wavenumber $k_2 = -k_4$. The phase speed of an extraordinary plane wave depends on the direction of propagation, whereas all ordinary plane waves have the same phase speed regardless of the direction of propagation.

For the unique propagation direction given by $\hat{\underline{k}} = \hat{\underline{u}}$, we observe from Eqs. (4.30) that $k_1 = k_2$ (and, equivalently, $k_3 = k_4$). Hence, all plane waves propagate with only one phase speed in the directions of $\pm\hat{\underline{u}}$. The special direction specified by $\hat{\underline{u}}$ is called the *optic axis*. Notice that the

optic axis for any uniaxial dielectric medium coincides with the sole optic
ray axis of the medium.

The relative orientations of $\underline{k}$, $\underline{\breve{E}}_{pw}(\underline{r},\omega)$ and $\underline{\breve{H}}_{pw}(\underline{r},\omega)$ may be established from Eq. (4.20). For both ordinary and extraordinary plane waves

$$\left.\begin{array}{l} \underline{\breve{H}}_{pw}(\underline{r},\omega) \bullet \underline{k} \\[4pt] \underline{\breve{D}}_{pw}(\underline{r},\omega) \bullet \underline{k} \\[4pt] \underline{\breve{B}}_{pw}(\underline{r},\omega) \bullet \underline{k} \end{array}\right\} = 0, \qquad (4.31)$$

while $\underline{\breve{E}}_{pw}(\underline{r},\omega) \bullet \underline{k} = 0$ for ordinary plane waves, and $\underline{\breve{E}}_{pw}(\underline{r},\omega) \bullet \underline{k} \neq 0$ for
extraordinary plane waves in general.

For the uniaxial dielectric medium with permittivity dyadic specified in
Eq. (2.9), the corresponding time–averaged Poynting vector for monochromatic plane waves is given by [4]

$$\langle \underline{\breve{S}}(\underline{r},\omega) \rangle_t = \frac{\exp\left(-2k_{Im}\,\hat{\underline{k}} \bullet \underline{r}\right)}{2\omega\mu_0} \operatorname{Re}\left\{ |\underline{\breve{E}}_0(\omega)|^2 \underline{k}^* - \left[\underline{\breve{E}}_0(\omega) \bullet \underline{k}^*\right] \underline{\breve{E}}_0^*(\omega) \right\}.$$

$$(4.32)$$

Hence, for ordinary plane waves the time–averaged Poynting vector
is directed along $\hat{\underline{k}}$, whereas it lies in the plane containing $\hat{\underline{k}}$ and
$\operatorname{Re}\left\{ \left[\underline{\breve{E}}_0(\omega) \bullet \underline{k}^*\right] \underline{\breve{E}}_0^*(\omega) \right\}$ for extraordinary plane waves.

Planewave propagation in uniaxial magnetic mediums, as described by
the Tellegen constitutive relations (2.11) with permeability dyadic specified
in Eq. (2.12), is mathematically isomorphic to planewave propagation in
uniaxial dielectric mediums.

Significantly different planewave properties are supported by uniaxial
dielectric–magnetic mediums, as specified by the Tellegen constitutive relations (2.14) with permittivity and permeability dyadics given in Eqs. (2.9)
and (2.12), respectively. The corresponding specialization of the dispersion
relation (4.19) has two independent solutions:

$$\left.\begin{array}{l} k_1 = -k_3 = \omega \left[\dfrac{\epsilon(\omega)\,\epsilon_u(\omega)\,\mu(\omega)}{\hat{\underline{k}} \bullet \underline{\underline{\epsilon}}_{uni}(\omega) \bullet \hat{\underline{k}}}\right]^{1/2} \\[16pt] k_2 = -k_4 = \omega \left[\dfrac{\epsilon(\omega)\,\mu(\omega)\,\mu_u(\omega)}{\hat{\underline{k}} \bullet \underline{\underline{\mu}}_{uni}(\omega) \bullet \hat{\underline{k}}}\right]^{1/2} \end{array}\right\}. \qquad (4.33)$$

All wavenumbers vary according to the direction of propagation — i.e.,
in contrast to the scenario for uniaxial dielectric and uniaxial magnetic
mediums, there is no 'ordinary' planewave mode which propagates with

the same phase speed in all directions. When $\hat{\underline{k}} = \hat{\underline{u}}$, we observe from
Eqs. (4.33) that $k_1 = k_2$ (and, equivalently, $k_3 = k_4$). Thus, *incidental*
unirefringence occurs for propagation parallel to the optic ray axis $\hat{\underline{u}}$, as
is the case for uniaxial dielectric and uniaxial magnetic mediums. Also,
pathological unirefringence, as characterized by $k_1 = k_2$ (and, equivalently,
$k_3 = k_4$) for all $\hat{\underline{k}}$, occurs in the special case identified by [6]

$$\frac{\epsilon(\omega)}{\epsilon_u(\omega)} = \frac{\mu(\omega)}{\mu_u(\omega)}. \qquad (4.34)$$

4.5.2 Biaxial mediums

Next, we focus on a biaxial dielectric medium characterized by the Tellegen
constitutive relations

$$\left. \begin{array}{l} \underline{D}(\underline{r},\omega) = \underline{\underline{\epsilon}}_{\mathrm{bi}}(\omega) \bullet \underline{E}(\underline{r},\omega) \\ \underline{B}(\underline{r},\omega) = \mu_0 \underline{H}(\underline{r},\omega) \end{array} \right\}, \qquad (4.35)$$

where the symmetric permittivity dyadic $\underline{\underline{\epsilon}}_{\mathrm{bi}}(\omega)$ may be of the orthorhombic, monoclinic or triclinic type, as listed in Table 2.1. The corresponding
specialization of the dispersion relation (4.19) is represented by the biquadratic polynomial [7]

$$a_{\mathrm{bi}}(\omega)\,k^4 + b_{\mathrm{bi}}(\omega)\,k^2 + c_{\mathrm{bi}}(\omega) = 0, \qquad (4.36)$$

with coefficients

$$\left. \begin{array}{l} a_{\mathrm{bi}}(\omega) = \hat{\underline{k}} \bullet \underline{\underline{\epsilon}}_{\mathrm{bi}}(\omega) \bullet \hat{\underline{k}} \\ b_{\mathrm{bi}}(\omega) = \omega^2\mu_0\{[\hat{\underline{k}} \bullet \underline{\underline{\epsilon}}_{\mathrm{bi}}(\omega) \bullet \underline{\underline{\epsilon}}_{\mathrm{bi}}(\omega) \bullet \hat{\underline{k}}] - [\hat{\underline{k}} \bullet \underline{\underline{\epsilon}}_{\mathrm{bi}}(\omega) \bullet \hat{\underline{k}}]\,\mathrm{tr}\,\underline{\underline{\epsilon}}_{\mathrm{bi}}(\omega)\,\} \\ c_{\mathrm{bi}}(\omega) = \omega^4\mu_0^2\,\det\underline{\underline{\epsilon}}_{\mathrm{bi}}(\omega) \end{array} \right\}. $$

$$(4.37)$$

The two, generally independent, wavenumbers

$$\left. \begin{array}{l} k_1 = -k_3 = \left\{ \dfrac{-b_{\mathrm{bi}}(\omega) + \left[b_{\mathrm{bi}}^2(\omega) - 4\,a_{\mathrm{bi}}(\omega)\,c_{\mathrm{bi}}(\omega)\right]^{1/2}}{2\,a_{\mathrm{bi}}(\omega)} \right\}^{1/2} \\[4mm] k_2 = -k_4 = \left\{ \dfrac{-b_{\mathrm{bi}}(\omega) - \left[b_{\mathrm{bi}}^2(\omega) - 4\,a_{\mathrm{bi}}(\omega)\,c_{\mathrm{bi}}(\omega)\right]^{1/2}}{2\,a_{\mathrm{bi}}(\omega)} \right\}^{1/2} \end{array} \right. \qquad (4.38)$$

emerge as solutions to Eq. (4.36).

Further insights into planewave propagation in biaxial dielectric mediums may be gained by confining ourselves to the realm of crystal optics,
wherein all components of the permittivity dyadic $\underline{\underline{\epsilon}}_{\mathrm{bi}}(\omega)$ are assumed to

be real–valued. After manipulating Eq. (4.18) and its complex conjugate, we find that [4]

$$\left(\underline{\underline{I}} - \frac{k^4}{\det \underline{\underline{\epsilon}}_{\mathrm{bi}}(\omega)} \, \underline{\underline{\epsilon}}_{\mathrm{bi}}(\omega) \cdot \hat{\underline{k}}\,\hat{\underline{k}} \right) \cdot \left[\underline{\breve{E}}_0(\omega) \times \underline{\breve{E}}_0^*(\omega) \right] = \underline{0}, \qquad (4.39)$$

and

$$\left(\underline{\underline{I}} - \frac{k^4 \, \hat{\underline{k}} \cdot \underline{\underline{\epsilon}}_{\mathrm{bi}}(\omega) \cdot \hat{\underline{k}}}{\det \underline{\underline{\epsilon}}_{\mathrm{bi}}(\omega)} \, \hat{\underline{k}}\,\hat{\underline{k}} \right) \cdot \left[\underline{\breve{H}}_0(\omega) \times \underline{\breve{H}}_0^*(\omega) \right] = \underline{0}. \qquad (4.40)$$

Therefore,

$$\left. \begin{array}{c} \underline{\breve{E}}_0(\omega) \times \underline{\breve{E}}_0^*(\omega) \\[4pt] \underline{\breve{H}}_0(\omega) \times \underline{\breve{H}}_0^*(\omega) \end{array} \right\} = \underline{0}, \qquad (4.41)$$

unless the determinants of the dyadics on the left sides in Eqs. (4.39) and (4.40), respectively, are null–valued. Thus, the following two key attributes of planewave propagation in nondissipative biaxial dielectric mediums may be deduced:

- Plane waves associated with the distinct wavenumbers k_1 and k_2 (or, equivalently, k_3 and k_4) are linearly polarized with mutually orthogonal planes of polarization.
- For two special directions of propagation, given implicitly by

$$b_{\mathrm{bi}}^2(\omega) - 4\,a_{\mathrm{bi}}(\omega)\,c_{\mathrm{bi}}(\omega) = 0, \qquad (4.42)$$

the wavenumbers k_1 and k_2 (and, equivalently, k_3 and k_4) coincide. Along these two directions — known as the optic axes — plane waves can have any polarization state. In contrast to the scenarios for uniaxial dielectric mediums and uniaxial magnetic mediums, the optic axes of a biaxial medium do not coincide with its optic ray axes.

The time–averaged Poynting vector for monochromatic plane waves [4]

$$\langle \underline{\breve{S}}(\underline{r}, \omega) \rangle_{\mathrm{t}} = \frac{\omega \left(\underline{k} \cdot \underline{\breve{E}}_0 \right)^2}{2 \det \left[\omega^2 \mu_0 \underline{\underline{\epsilon}}_{\mathrm{bi}}(\omega) - k^2 \underline{\underline{I}} \right]} \left(\left[\underline{k} \cdot \underline{\underline{\epsilon}}_{\mathrm{bi}}(\omega) \cdot \underline{k} \right] \underline{k} + \omega^2 \mu_0 \underline{\underline{\epsilon}}_{\mathrm{bi}}(\omega) \right.$$
$$\left. \cdot \left\{ \underline{\underline{\epsilon}}_{\mathrm{bi}}(\omega) - \mathrm{tr}\left[\underline{\underline{\epsilon}}_{\mathrm{bi}}(\omega) \right] \underline{\underline{I}} \right\} \cdot \underline{k} + k^2 \left[\underline{\underline{\epsilon}}_{\mathrm{bi}}(\omega) \cdot \underline{k} \right] \right) \qquad (4.43)$$

is not aligned with the wavevector $\underline{k}$ in general.

The description of planewave propagation in biaxial magnetic mediums, as described by the Tellegen constitutive relations (2.22) with permeability dyadic being of the orthorhombic, monoclinic or triclinic type, as specified in Eqs. (2.25), is mathematically isomorphic to that for biaxial dielectric mediums.

The Tellegen constitutive relations (2.26) specify a biaxial dielectric–magnetic medium. The increased parameter space associated with such mediums, as compared with biaxial dielectric mediums or biaxial magnetic mediums, results in a richer palette of planewave attributes. For example, under certain pathological conditions, biaxial dielectric–magnetic mediums can be endowed with only one optic axis [8].

4.5.3   *Gyrotropic mediums*

Let us now turn to an example of a Lorentz–nonreciprocal anisotropic medium. We consider the gyrotropic dielectric medium characterized by the Tellegen constitutive relations (2.36), wherein the components of the antisymmetric permittivity dyadic

$$\underline{\underline{\epsilon}}_{\mathrm{gyro}}(\omega) = \epsilon(\omega)\left(\underline{\underline{I}} - \hat{\underline{u}}\,\hat{\underline{u}}\right) + i\epsilon_g(\omega)\,\hat{\underline{u}} \times \underline{\underline{I}} + \epsilon_u(\omega)\,\hat{\underline{u}}\,\hat{\underline{u}} \qquad (4.44)$$

are complex–valued, in general. The corresponding specialization of the dispersion relation (4.19) yields the biquadratic polynomial [4]

$$a_{\mathrm{gyro}}(\omega)\,k^4 + b_{\mathrm{gyro}}(\omega)\,k^2 + c_{\mathrm{gyro}}(\omega) = 0\,, \qquad (4.45)$$

with coefficients

$$\left.\begin{aligned}
a_{\mathrm{gyro}}(\omega) &= \hat{\underline{k}} \bullet \underline{\underline{\epsilon}}_{\mathrm{gyro}}(\omega) \bullet \hat{\underline{k}} \\
b_{\mathrm{gyro}}(\omega) &= \omega^2\mu_0\hat{\underline{k}} \bullet \left(\mathrm{adj}\left[\underline{\underline{\epsilon}}_{\mathrm{gyro}}(\omega)\right] - \mathrm{tr}\left\{\mathrm{adj}\left[\underline{\underline{\epsilon}}_{\mathrm{gyro}}(\omega)\right]\right\}\underline{\underline{I}}\right) \bullet \hat{\underline{k}} \\
c_{\mathrm{gyro}}(\omega) &= \omega^4\mu_0^2\det\underline{\underline{\epsilon}}_{\mathrm{gyro}}(\omega)
\end{aligned}\right\}. \qquad (4.46)$$

The two, generally independent, wavenumbers

$$\left.\begin{aligned}
k_1 = -k_3 &= \left\{\frac{-b_{\mathrm{gyro}}(\omega) + \left[b_{\mathrm{gyro}}^2(\omega) - 4\,a_{\mathrm{gyro}}(\omega)\,c_{\mathrm{gyro}}(\omega)\right]^{1/2}}{2\,a_{\mathrm{gyro}}(\omega)}\right\}^{1/2} \\
k_2 = -k_4 &= \left[\frac{-b_{\mathrm{gyro}}(\omega) - \left[b_{\mathrm{gyro}}^2(\omega) - 4\,a_{\mathrm{gyro}}(\omega)\,c_{\mathrm{gyro}}(\omega)\right]^{1/2}}{2\,a_{\mathrm{gyro}}(\omega)}\right\}^{1/2}
\end{aligned}\right\} \qquad (4.47)$$

emerge therefrom.

The following two special cases are particularly noteworthy [4]:

- For propagation either co–parallel or antiparallel to $\hat{\underline{u}}$, we have $\hat{\underline{k}} \bullet \hat{\underline{u}} = \pm 1$. In the case of a magnetoplasma, this corresponds to propagation along the direction of the biasing magnetic field (cf. Sec. 2.2.3). The expressions (4.47) then simplify to give

$$\left. \begin{array}{l} k_1 = -k_3 = \omega^2 \mu_0 \left[\epsilon(\omega) - \epsilon_g(\omega) \right]^{1/2} \\ k_2 = -k_4 = \omega^2 \mu_0 \left[\epsilon(\omega) + \epsilon_g(\omega) \right]^{1/2} \end{array} \right\} . \qquad (4.48)$$

When dissipation is negligibly small, we may set $\epsilon(\omega), \epsilon_g(\omega), \epsilon_u(\omega) \in \mathbb{R}$; and then the wavenumbers k_1 and k_2 correspond to left– and right– circularly polarized plane waves, respectively. Let us recall from Sec. 4.2 that a linearly polarized wave may be viewed as the sum of two circularly polarized waves of equal amplitude and opposite sense of rotation. A rotation of the plane of polarization — called *Faraday rotation* — associated with this linearly polarized wave arises from the difference in phase speeds associated with the wavenumbers k_1 and k_2 (or, equivalently, k_3 and k_4).

- For propagation perpendicular to $\hat{u}$ we have $\hat{\underline{k}} \cdot \hat{\underline{u}} = 0$. In the case of a magnetoplasma, this corresponds to propagation perpendicular to the direction of the biasing magnetic field (cf. Sec. 2.2.3). The expressions (4.47) then simplify to give

$$\left. \begin{array}{l} k_1 = -k_3 = \omega^2 \mu_0 \left[\dfrac{\epsilon^2(\omega) - \epsilon_g^2(\omega)}{\epsilon(\omega)} \right]^{1/2} \\ \\ k_2 = -k_4 = \omega^2 \mu_0 \left[\epsilon_u(\omega) \right]^{1/2} \end{array} \right\} . \qquad (4.49)$$

Notice that for a magnetoplasma, the wavenumbers k_1 and k_3 depend upon the strength of the biasing magnetic field, whereas k_2 and k_4 do not. On ignoring dissipation (i.e., setting $\epsilon(\omega), \epsilon_g(\omega), \epsilon_u(\omega) \in \mathbb{R}$), it transpires that the wavenumbers k_1 and k_2 specified by Eqs. (4.49) correspond to elliptically and linearly polarized waves, respectively.

The mathematical description of planewave propagation for magnetic gyrotropic mediums is isomorphic to that for dielectric gyrotropic mediums.

4.6 Uniform plane waves in bianisotropic mediums

Since the parameter space associated with bianisotropic mediums is considerably larger than that associated with anisotropic mediums, a richer and more diverse array of planewave characteristics can be exhibited by bianisotropic mediums. Herein we present four examples of planewave propagation in homogeneous bianisotropic scenarios: The first concerns planewave propagation in a medium which is isotropic dielectric–magnetic when viewed by an observer in a co–moving reference frame, but bianisotropic from the perspective of a non–co–moving observer. As mentioned earlier in Sec. 4.6.1, this bianisotropic medium is in fact unirefringent. The second scenario concerns mediums with simultaneous mirror–conjugated and racemic chirality characteristics, as introduced in Sec. 2.3.3. The third example is that of a Faraday chiral medium, which can support plane waves with four distinct wavenumbers for propagation along an arbitrary axis. Lastly, we consider Beltrami fields in a bianisotropic setting.

The mediums discussed in Secs. 4.6.1 and 4.6.3 are Lorentz–nonreciprocal. We emphasize that the dispersion relation (4.19) reduces to a polynomial that is quadratic in k^2 for all Lorentz–reciprocal mediums, whether bianisotropic or not. Consequently, Lorentz–reciprocal bianisotropic mediums support plane waves with at most two independent wavenumbers for propagation along any direction.

4.6.1   *Mediums moving at constant velocity*

Suppose that a plane wave described by the 6–vector

$$\tilde{\underline{\mathbf{F}}}'_{\mathrm{pw}}(\underline{r}', t') = \mathrm{Re}\left\{ \breve{\underline{\mathbf{F}}}'_0(\omega') \exp\left[i\left(\underline{k}' \cdot \underline{r}' - \omega' t' \right) \right] \right\}, \qquad (4.50)$$

propagates in an inertial reference frame Σ', which moves at a constant velocity $\underline{v} = v\hat{\underline{v}}$ relative to the inertial reference frame Σ. In accordance with the principle of phase invariance [3, 4], the electromagnetic field (4.50) has the planewave form

$$\tilde{\underline{\mathbf{F}}}_{\mathrm{pw}}(\underline{r}, t) = \mathrm{Re}\left\{ \breve{\underline{\mathbf{F}}}_0(\omega) \exp\left[i\left(\underline{k} \cdot \underline{r} - \omega t \right) \right] \right\}, \qquad (4.51)$$

with respect to Σ. Herein $\{t, \underline{r}\}$ and $\{t', \underline{r}'\}$ are related via the Lorentz transformations (1.56), and the 3–vector electric and magnetic field components of $\tilde{\underline{\mathbf{F}}}_{\mathrm{pw}}(\underline{r}, t)$ and $\tilde{\underline{\mathbf{F}}}'_{\mathrm{pw}}(\underline{r}', t')$ (namely, $\tilde{\underline{E}}_{\mathrm{pw}}(\underline{r}, t)$, $\tilde{\underline{H}}_{\mathrm{pw}}(\underline{r}, t)$ and their primed counterparts) are related by the transformations (1.58). The 3–vector electric and magnetic components of $\breve{\underline{\mathbf{F}}}_0(\omega)$ and $\breve{\underline{\mathbf{F}}}'_0(\omega')$ (namely, $\breve{\underline{E}}_0(\omega)$, $\breve{\underline{H}}_0(\omega)$ and their primed counterparts) are related by the transformations (cf. Eqs. (1.58))

$$\left.\begin{aligned}
\breve{\underline{E}}'_0(\omega') &= \gamma \left[\underline{\underline{Y}}^{-1} \cdot \breve{\underline{E}}_0(\omega) + \underline{v} \times \breve{\underline{B}}_0(\omega) \right] \\
\breve{\underline{B}}'_0(\omega') &= \gamma \left[\underline{\underline{Y}}^{-1} \cdot \breve{\underline{B}}_0(\omega) - \frac{1}{c_0^2} \underline{v} \times \breve{\underline{E}}_0(\omega) \right] \\
\breve{\underline{D}}'_0(\omega') &= \gamma \left[\underline{\underline{Y}}^{-1} \cdot \breve{\underline{D}}_0(\omega) + \frac{1}{c_0^2} \underline{v} \times \breve{\underline{H}}_0(\omega) \right] \\
\breve{\underline{H}}'_0(\omega') &= \gamma \left[\underline{\underline{Y}}^{-1} \cdot \breve{\underline{H}}_0(\omega) - \underline{v} \times \breve{\underline{D}}_0(\omega) \right]
\end{aligned}\right\}, \qquad (4.52)$$

where $\breve{\underline{D}}_0(\omega)$ and $\breve{\underline{B}}_0(\omega)$ (and their primed counterparts) are the amplitudes of $\breve{\underline{D}}(\underline{r}, t)$ and $\breve{\underline{B}}(\underline{r}, t)$ (and their primed counterparts), respectively. The wavevector and angular frequency in the two inertial frames are related as [4]

$$\left.\begin{aligned}
\underline{k}' &= \gamma \left(\underline{k} \cdot \underline{\hat{v}} - \frac{\omega v}{c_0^2} \right) \underline{\hat{v}} + \left(\underline{\underline{I}} - \underline{\hat{v}}\,\underline{\hat{v}} \right) \cdot \underline{k} \\
\omega' &= \gamma \left(\omega - \underline{k} \cdot \underline{v} \right)
\end{aligned}\right\}, \qquad (4.53)$$

with γ defined in Eq. $(1.57)_2$.

The following may be inferred from Eqs. (4.53):

- If a medium is unirefringent with respect to Σ', then the medium is also unirefringent with respect to Σ ; i.e., unirefringence is Lorentz–invariant. By the same token, birefringence is also Lorentz–invariant.
- If a plane wave is uniform with respect to Σ' (i.e., $\underline{k}' = k'\underline{\hat{k}}'$ with $\underline{\hat{k}}' \in \mathbb{R}^3$), then the plane wave is generally nonuniform with respect to Σ.

4.6.2 *Mediums with simultaneous mirror–conjugated and racemic chirality characteristics*

Mediums that display simultaneous mirror–conjugated and racemic chirality characteristics may be described by the Tellegen constitutive relations (2.47). Since such mediums are Lorentz–reciprocal, the dispersion relation (4.19) delivers, at most, two independent wavenumbers. For propagation in an arbitrary direction, as represented by the wavevector

$$\underline{k} = k \left(\sin\theta \cos\phi\,\underline{\hat{x}} + \sin\theta \sin\phi\,\underline{\hat{y}} + \cos\theta\,\underline{\hat{z}} \right), \qquad (4.54)$$

the wavenumbers are given by [9, 10]

$$\left.\begin{aligned}
k_1 &= -k_3 = \omega \left\{ \gamma_{\mathrm{MCR}}(\theta, \phi, \omega) \left[\alpha_{\mathrm{MCR}}(\theta, \omega) - \beta_{\mathrm{MCR}}(\theta, \phi, \omega) \right] \right\}^{1/2} \\
k_2 &= -k_4 = \omega \left\{ \gamma_{\mathrm{MCR}}(\theta, \phi, \omega) \left[\alpha_{\mathrm{MCR}}(\theta, \omega) + \beta_{\mathrm{MCR}}(\theta, \phi, \omega) \right] \right\}^{1/2}
\end{aligned}\right\}, \qquad (4.55)$$

where the scalar functions

$$
\begin{aligned}
\alpha_{\mathrm{MCR}}(\theta,\omega) &= 2\epsilon_z(\omega)\mu_z(\omega)\cos^2\theta + [\epsilon_z(\omega)\mu(\omega) + \epsilon(\omega)\mu_z(\omega)]\sin^2\theta \\[4pt]
\beta_{\mathrm{MCR}}(\theta,\phi,\omega) &= \Big\{ [\epsilon_z(\omega)\mu(\omega) - \epsilon(\omega)\mu_z(\omega)]^2 \\
&\quad -4\epsilon_z(\omega)\mu_z(\omega)\xi^2(\omega)\cos^2 2\phi \Big\}^{1/2}\sin^2\theta \\[4pt]
\gamma_{\mathrm{MCR}}(\theta,\phi,\omega) &= \frac{1}{2}\left[\epsilon(\omega)\mu(\omega) + \xi^2(\omega)\right]\Big\{ [\epsilon(\omega)\sin^2\theta + \epsilon_z(\omega)\cos^2\theta] \\
&\quad \times [\mu(\omega)\sin^2\theta + \mu_z(\omega)\cos^2\theta] \\
&\quad +\xi^2(\omega)\sin^4\theta\cos^2 2\phi \Big\}^{-1}
\end{aligned}
\tag{4.56}
$$

are introduced. Thus, the medium which exhibits simultaneous mirror–conjugated and racemic chirality characteristics is generally birefringent. However, there are exceptions: the medium is unirefringent with respect to propagation along the racemic (i.e., z) axis. That is, for $\theta = 0$ only one independent wavenumber emerges from the dispersion relation (4.19). Furthermore, pathological unirefringence [6] arises when

$$
\epsilon_z(\omega)\mu(\omega) - \epsilon(\omega)\mu_z(\omega) = 2\left[\epsilon_z(\omega)\mu_z(\omega)\right]^{1/2}\xi(\omega)\cos 2\phi. \tag{4.57}
$$

4.6.3 *Faraday chiral mediums*

The Faraday chiral medium (FCM) was introduced in Sec. 2.3.4. The frequency–domain constitutive relations of a FCM are given in the Tellegen representation by Eqs. (2.50), with the 3×3 constitutive dyadics given in Eqs. (2.51).

The corresponding specialization of the dispersion relation (4.19) delivers a quartic polynomial in k with four distinct roots. For definiteness — and without loss of generality — let us suppose that the biasing magnetic field is directed along the z axis; i.e, $\hat{u} = \hat{z}$. Thus, the FCM is specified by the 6×6 constitutive dyadic given in Eq. (2.52). In addition, we focus on the propagation of uniform plane waves in the xz plane, as characterized by the wavevector

$$
\underline{k} = k\left(\sin\theta\,\hat{\underline{x}} + \cos\theta\,\hat{\underline{z}}\right). \tag{4.58}
$$

The four wavenumbers then emerge as the roots of the quartic polynomial [11]

$$
a_4(\theta,\omega)\left(\frac{k}{\omega}\right)^4 + a_3(\theta,\omega)\left(\frac{k}{\omega}\right)^3 + a_2(\theta,\omega)\left(\frac{k}{\omega}\right)^2 + a_1(\theta,\omega)\frac{k}{\omega} + a_0(\omega) = 0, \tag{4.59}
$$

with coefficients

$$a_4(\theta,\omega) = \left[\epsilon(\omega)\sin^2\theta + \epsilon_u(\omega)\cos^2\theta\right]\left[\mu(\omega)\sin^2\theta + \mu_u(\omega)\cos^2\theta\right]$$
$$- \left[\xi(\omega)\sin^2\theta + \xi_z(\omega)\cos^2\theta\right]^2$$

$$a_3(\theta,\omega) = 2\cos\theta\Bigg(\sin^2\theta\Big\{\mu_g(\omega)\left[\epsilon(\omega)\xi_u(\omega) - \epsilon_u(\omega)\xi(\omega)\right]$$
$$+\epsilon_g(\omega)\left[\mu(\omega)\xi_u(\omega) - \mu_u(\omega)\xi(\omega)\right]$$
$$+\xi_g(\omega)\left[\mu(\omega)\epsilon_u(\omega) + \epsilon(\omega)\mu_u(\omega) - 2\xi(\omega)\xi_g(\omega)\right]\Big\}$$
$$+2\cos^2\theta\xi_g(\omega)\left[\epsilon_u(\omega)\mu_u(\omega) - \xi_u^2(\omega)\right]\Bigg)$$

$$a_2(\theta,\omega) = \sin^2\theta\Big(\mu(\omega)\mu_u(\omega)\left[\epsilon_g^2(\omega) - \epsilon^2(\omega)\right]$$
$$+ \left[\xi^2(\omega) + \xi_g^2(\omega)\right]\left[\mu(\omega)\epsilon_u(\omega) + \epsilon(\omega)\mu_u(\omega)\right]$$
$$-2\xi(\omega)\left\{\xi_u(\omega)\left[\xi_g^2(\omega) - \xi^2(\omega)\right] + \mu_g(\omega)\epsilon_u(\omega)\xi_g(\omega)\right\}$$
$$-2\epsilon_g(\omega)\left\{\xi_u(\omega)\left[\mu_g(\omega)\xi(\omega) - \mu(\omega)\xi_g(\omega)\right] + \mu_u(\omega)\xi(\omega)\xi_g(\omega)\right\}$$
$$-\epsilon(\omega)\big\{\epsilon_u(\omega)\left[\mu^2(\omega) - \mu_g^2(\omega)\right] + 2\xi_u(\omega)\left[\mu(\omega)\xi(\omega)\right.$$
$$\left.-\mu_g(\omega)\xi_g(\omega)\right]\big\}\Big) + 2\cos^2\theta\left[\epsilon_u(\omega)\mu_u(\omega) - \xi_u^2(\omega)\right]$$
$$\times\left[3\xi_g^2(\omega) - \xi^2(\omega) - \epsilon_g(\omega)\mu_g(\omega) - \epsilon(\omega)\mu(\omega)\right]$$

$$a_1(\theta,\omega) = 4\cos\theta\left[\epsilon_u(\omega)\mu_u(\omega) - \xi_u^2(\omega)\right]\Big\{\xi(\omega)\left[\epsilon_g(\omega)\mu(\omega) + \epsilon(\omega)\mu_g(\omega)\right]$$
$$+\xi_g(\omega)\left[\xi_g^2(\omega) - \xi^2(\omega) - \epsilon(\omega)\mu(\omega) - \epsilon_g(\omega)\mu_g(\omega)\right]\Big\}$$

$$a_0(\omega) = \left[\epsilon_u(\omega)\mu_u(\omega) - \xi_u^2(\omega)\right]\Big\{\left[\epsilon^2(\omega) - \epsilon_g^2(\omega)\right]\left[\mu^2(\omega) - \mu_g^2(\omega)\right]$$
$$+ \left[\xi_g^2(\omega) - \xi^2(\omega)\right]^2 - 2\left[\epsilon(\omega)\mu(\omega) + \epsilon_g(\omega)\mu_g(\omega)\right]$$
$$\times \left[\xi_g^2(\omega) + \xi^2(\omega)\right] + 4\xi(\omega)\xi_g(\omega)\left[\epsilon(\omega)\mu_g(\omega) + \mu(\omega)\epsilon_g(\omega)\right]\Big\}$$

$$\tag{4.60}$$

The wavenumbers may be extracted by standard algebraic or numerical methods [12].

The mathematical description of planewave propagation in FCMs simplifies considerably in the particular case $\hat{\underline{k}} = \hat{\underline{u}}$; i.e., for propagation along the direction of the biasing magnetic field. The four wavenumbers

$$k_1 = \omega\left\{\left[\epsilon(\omega) + \epsilon_g(\omega)\right]^{1/2}\left[\mu(\omega) + \mu_g(\omega)\right]^{1/2} - \xi(\omega) - \xi_g(\mu)\right\}$$
$$k_2 = \omega\left\{-\left[\epsilon(\omega) + \epsilon_g(\omega)\right]^{1/2}\left[\mu(\omega) + \mu_g(\omega)\right]^{1/2} - \xi(\omega) - \xi_g(\mu)\right\}$$
$$k_3 = \omega\left\{\left[\epsilon(\omega) - \epsilon_g(\omega)\right]^{1/2}\left[\mu(\omega) - \mu_g(\omega)\right]^{1/2} + \xi(\omega) - \xi_g(\mu)\right\}$$
$$k_4 = \omega\left\{-\left[\epsilon(\omega) - \epsilon_g(\omega)\right]^{1/2}\left[\mu(\omega) - \mu_g(\omega)\right]^{1/2} + \xi(\omega) - \xi_g(\mu)\right\}$$

$$\tag{4.61}$$

are delivered by the dispersion relation (4.19). Since $k_1 + k_3 \neq 0$ and $k_2 + k_4 \neq 0$, the FCM is not Lorentz–reciprocal. The orientations of the electromagnetic field phasors are determined by analyzing Eq. (4.20): thereby, it is found that there is no component of the electric field in the direction of $\hat{u}$, whereas the time–averaged Poynting vector is aligned parallel to $\hat{u}$ [11].

4.6.4   *Beltrami fields in a bianisotropic medium*

By definition, a Beltrami field $\underline{Q}(\underline{r}, \omega)$ is either parallel or antiparallel to its own circulation [13]; i.e.,

$$\nabla \times \underline{Q}(\underline{r}, \omega) = \varkappa_{\mathrm{Bel}} \, \underline{Q}(\underline{r}, \omega), \tag{4.62}$$

where $\varkappa_{\mathrm{Bel}} \neq 0$ is akin to a wavenumber. In the context of electromagnetics, such Beltrami fields represent circularly polarized plane waves [5]. Let us consider the propagation of these plane waves in a general bianisotropic medium, as specified by the Tellegen constitutive dyadic given in Eq. (4.17). Notice that by a simple rearrangement, the combination of the source–free Maxwell curl postulates (4.15) and the constitutive relations (4.14) as represented by Eq. (4.16) may be rewritten as

$$\underline{\underline{L}}'(\nabla) \bullet \underline{F}(\underline{r}, \omega) = i\omega \, \underline{\underline{K}}'_{\mathrm{EH}}(\omega) \bullet \underline{F}(\underline{r}, \omega), \tag{4.63}$$

where

$$\left.\begin{aligned} \underline{\underline{L}}'(\nabla) &= \begin{pmatrix} \nabla \times \underline{I} & \underline{0} \\ \underline{0} & \nabla \times \underline{I} \end{pmatrix} \\[2ex] \underline{\underline{K}}'_{\mathrm{EH}}(\omega) &= \begin{pmatrix} \underline{\underline{\zeta}}_{\mathrm{EH}}(\omega) & \underline{\underline{\mu}}_{\mathrm{EH}}(\omega) \\ -\underline{\underline{\epsilon}}_{\mathrm{EH}}(\omega) & -\underline{\underline{\xi}}_{\mathrm{EH}}(\omega) \end{pmatrix} \end{aligned}\right\}. \tag{4.64}$$

Now, in order for the 6–vector $\underline{F}(\underline{r}, \omega)$ to represent a Beltrami solution, it must satisfy the differential equation

$$\underline{\underline{L}}'(\nabla) \bullet \underline{F}(\underline{r}, \omega) = \varkappa_{\mathrm{Bel}} \, \underline{F}(\underline{r}, \omega) \equiv i\omega\kappa_{\mathrm{Bel}} \, \underline{F}(\underline{r}, \omega), \tag{4.65}$$

where κ_{Bel} is an inverse speed. Hence, comparing Eqs. (4.63) and (4.65), we see that

$$\underline{\underline{K}}'_{\mathrm{EH}}(\omega) \bullet \underline{F}(\underline{r}, \omega) = \kappa_{\mathrm{Bel}} \, \underline{F}(\underline{r}, \omega). \tag{4.66}$$

Thus, Beltrami fields arise as eigenvectors of $\underline{\underline{K}}'_{\mathrm{EH}}(\omega)$ [14].

The question arises: can Beltrami solutions, corresponding to circularly polarized plane waves propagating along the $\hat{z}$ axis, say, be found for a generally bianisotropic medium? The answer is 'yes', provided that the following four conditions are satisfied [15]:

$$\begin{aligned} &\left(\left[\underline{\underline{\zeta}}_{\mathrm{EH}}(\omega) \right]_{31} \pm i \left[\underline{\underline{\zeta}}_{\mathrm{EH}}(\omega) \right]_{32} \right) \left(\left[\underline{\underline{\xi}}_{\mathrm{EH}}(\omega) \right]_{31} \pm i \left[\underline{\underline{\xi}}_{\mathrm{EH}}(\omega) \right]_{32} \right) \\ &= \left(\left[\underline{\underline{\epsilon}}_{\mathrm{EH}}(\omega) \right]_{31} \pm i \left[\underline{\underline{\epsilon}}_{\mathrm{EH}}(\omega) \right]_{32} \right) \left(\left[\underline{\underline{\mu}}_{\mathrm{EH}}(\omega) \right]_{31} \pm i \left[\underline{\underline{\mu}}_{\mathrm{EH}}(\omega) \right]_{32} \right), \tag{4.67} \end{aligned}$$

$$\left[\left(\left[\underline{\underline{\zeta}}_{\mathrm{EH}}(\omega)\right]_{11} - \left[\underline{\underline{\zeta}}_{\mathrm{EH}}(\omega)\right]_{22}\right) \pm i \left(\left[\underline{\underline{\zeta}}_{\mathrm{EH}}(\omega)\right]_{12} + \left[\underline{\underline{\zeta}}_{\mathrm{EH}}(\omega)\right]_{21}\right)\right]$$
$$\times \left[\left(\left[\underline{\underline{\xi}}_{\mathrm{EH}}(\omega)\right]_{11} - \left[\underline{\underline{\xi}}_{\mathrm{EH}}(\omega)\right]_{22}\right) \pm i \left(\left[\underline{\underline{\xi}}_{\mathrm{EH}}(\omega)\right]_{12} + \left[\underline{\underline{\xi}}_{\mathrm{EH}}(\omega)\right]_{21}\right)\right]$$
$$= \left[\left(\left[\underline{\underline{\epsilon}}_{\mathrm{EH}}(\omega)\right]_{11} - \left[\underline{\underline{\epsilon}}_{\mathrm{EH}}(\omega)\right]_{22}\right) \pm i \left(\left[\underline{\underline{\epsilon}}_{\mathrm{EH}}(\omega)\right]_{12} + \left[\underline{\underline{\epsilon}}_{\mathrm{EH}}(\omega)\right]_{21}\right)\right]$$
$$\times \left[\left(\left[\underline{\underline{\mu}}_{\mathrm{EH}}(\omega)\right]_{11} - \left[\underline{\underline{\mu}}_{\mathrm{EH}}(\omega)\right]_{22}\right) \pm i \left(\left[\underline{\underline{\mu}}_{\mathrm{EH}}(\omega)\right]_{12} + \left[\underline{\underline{\mu}}_{\mathrm{EH}}(\omega)\right]_{21}\right)\right], \tag{4.68}$$

$$-\alpha_j(\omega)\left[\left(\left[\underline{\underline{\mu}}_{\mathrm{EH}}(\omega)\right]_{11} - \left[\underline{\underline{\mu}}_{\mathrm{EH}}(\omega)\right]_{22}\right) \pm i \left(\left[\underline{\underline{\mu}}_{\mathrm{EH}}(\omega)\right]_{12} + \left[\underline{\underline{\mu}}_{\mathrm{EH}}(\omega)\right]_{21}\right)\right]$$
$$= \left(\left[\underline{\underline{\zeta}}_{\mathrm{EH}}(\omega)\right]_{11} - \left[\underline{\underline{\zeta}}_{\mathrm{EH}}(\omega)\right]_{22}\right) \pm i \left(\left[\underline{\underline{\zeta}}_{\mathrm{EH}}(\omega)\right]_{12} + \left[\underline{\underline{\zeta}}_{\mathrm{EH}}(\omega)\right]_{21}\right), \tag{4.69}$$

$$-\alpha_j(\omega)\left(\left[\underline{\underline{\mu}}_{\mathrm{EH}}(\omega)\right]_{31} \pm i \left[\underline{\underline{\mu}}_{\mathrm{EH}}(\omega)\right]_{32}\right) = \left(\left[\underline{\underline{\zeta}}_{\mathrm{EH}}(\omega)\right]_{31} \pm i \left[\underline{\underline{\zeta}}_{\mathrm{EH}}(\omega)\right]_{32}\right), \tag{4.70}$$

wherein the index $j = p, q$ labels the two admittances

$$\left.\begin{aligned} \alpha_p(\omega) &= \frac{k_p - \omega \left(\left[\underline{\underline{\zeta}}_{\mathrm{EH}}(\omega)\right]_{21} \pm i \left[\underline{\underline{\zeta}}_{\mathrm{EH}}(\omega)\right]_{22}\right)}{\omega \left(\left[\underline{\underline{\mu}}_{\mathrm{EH}}(\omega)\right]_{21} \pm i \left[\underline{\underline{\mu}}_{\mathrm{EH}}(\omega)\right]_{22}\right)} \\ \alpha_q(\omega) &= \frac{k_q - \omega \left(\left[\underline{\underline{\zeta}}_{\mathrm{EH}}(\omega)\right]_{21} \pm i \left[\underline{\underline{\zeta}}_{\mathrm{EH}}(\omega)\right]_{22}\right)}{\omega \left(\left[\underline{\underline{\mu}}_{\mathrm{EH}}(\omega)\right]_{21} \pm i \left[\underline{\underline{\mu}}_{\mathrm{EH}}(\omega)\right]_{22}\right)} \end{aligned}\right\}, \tag{4.71}$$

with the corresponding wavenumbers being

$$\left.\begin{aligned} k_p &= \frac{\omega}{2}\left\{\left[\underline{\underline{\zeta}}_{\mathrm{EH}}(\omega)\right]_{21} \pm i \left[\underline{\underline{\zeta}}_{\mathrm{EH}}(\omega)\right]_{22} - \left(\left[\underline{\underline{\xi}}_{\mathrm{EH}}(\omega)\right]_{21} \pm i \left[\underline{\underline{\xi}}_{\mathrm{EH}}(\omega)\right]_{22}\right)\right. \\ &\quad + \left[\left(\left[\underline{\underline{\zeta}}_{\mathrm{EH}}(\omega)\right]_{21} \pm i \left[\underline{\underline{\zeta}}_{\mathrm{EH}}(\omega)\right]_{22} + \left[\underline{\underline{\xi}}_{\mathrm{EH}}(\omega)\right]_{21} \pm i \left[\underline{\underline{\xi}}_{\mathrm{EH}}(\omega)\right]_{22}\right)^2\right. \\ &\quad -4\left(\left[\underline{\underline{\epsilon}}_{\mathrm{EH}}(\omega)\right]_{21} \pm i \left[\underline{\underline{\epsilon}}_{\mathrm{EH}}(\omega)\right]_{22}\right) \\ &\quad \left.\left.\times \left(\left[\underline{\underline{\mu}}_{\mathrm{EH}}(\omega)\right]_{21} \pm i \left[\underline{\underline{\mu}}_{\mathrm{EH}}(\omega)\right]_{22}\right)\right]^{1/2}\right\} \\ k_q &= \frac{\omega}{2}\left\{\left[\underline{\underline{\zeta}}_{\mathrm{EH}}(\omega)\right]_{21} \pm i \left[\underline{\underline{\zeta}}_{\mathrm{EH}}(\omega)\right]_{22} - \left(\left[\underline{\underline{\xi}}_{\mathrm{EH}}(\omega)\right]_{21} \pm i \left[\underline{\underline{\xi}}_{\mathrm{EH}}(\omega)\right]_{22}\right)\right. \\ &\quad - \left[\left(\left[\underline{\underline{\zeta}}_{\mathrm{EH}}(\omega)\right]_{21} \pm i \left[\underline{\underline{\zeta}}_{\mathrm{EH}}(\omega)\right]_{22} + \left[\underline{\underline{\xi}}_{\mathrm{EH}}(\omega)\right]_{21} \pm i \left[\underline{\underline{\xi}}_{\mathrm{EH}}(\omega)\right]_{22}\right)^2\right. \\ &\quad -4\left(\left[\underline{\underline{\epsilon}}_{\mathrm{EH}}(\omega)\right]_{21} \pm i \left[\underline{\underline{\epsilon}}_{\mathrm{EH}}(\omega)\right]_{22}\right) \\ &\quad \left.\left.\times \left(\left[\underline{\underline{\mu}}_{\mathrm{EH}}(\omega)\right]_{21} \pm i \left[\underline{\underline{\mu}}_{\mathrm{EH}}(\omega)\right]_{22}\right)\right]^{1/2}\right\} \end{aligned}\right\}. \tag{4.72}$$

As the conditions (4.67)–(4.70) are easily satisfied by isotropic dielectric–magnetic mediums and isotropic chiral mediums, circularly polarized plane waves can propagate in these mediums in any direction. The conditions (4.67)–(4.70) are also satisfied by Faraday chiral mediums, as characterized by Eqs. (2.50) and (2.51), for propagation parallel (or antiparallel) to the distinguished axis $\hat{u}$.

4.7 Plane waves in nonhomogeneous mediums

We now turn to planewave propagation in two types of nonhomogeneous mediums. The first is the general class of helicoidal bianisotropic mediums (HBMs). These represent a wide range of technologically important materials such as cholesteric liquid crystals (CLCs) and chiral sculptured thin films (CSTFs). The second is gravitationally affected vacuum, which is of particular interest in view of recent debate on whether certain spacetime metrics support propagation with negative phase velocity.

The mediums discussed in Secs. 4.7.1 and 4.7.2 are atypical of nonhomogeneous mediums insofar as analytical (or semi–analytical) methods may be fruitfully deployed to explore planewave propagation therein. Generally, wave propagation studies for nonhomogeneous mediums pose formidable challenges to theorists, particularly if closed–form solutions are sought.

4.7.1 *Periodic nonhomogeneity*

We recall from Sec. 2.4.1 that the constitutive relations for a HBM with its helicoidal axis aligned with the z axis may be stated in the Tellegen representation as Eqs. (2.64), with constitutive dyadics given by Eqs. (2.65) and (2.66). On combining the constitutive relations (2.64) and the source–free Maxwell curl postulates (4.15), the system of partial differential equations

$$\left. \begin{array}{l} \nabla \times \breve{\underline{E}}(\underline{r},\omega) = i\omega\left[\underline{\underline{\zeta}}_{\mathrm{HBM}}(z,\omega) \bullet \breve{\underline{E}}(\underline{r},\omega) + \underline{\underline{\mu}}_{\mathrm{HBM}}(z,\omega) \bullet \breve{\underline{H}}(\underline{r},\omega)\right] \\[2mm] \nabla \times \breve{\underline{H}}(\underline{r},\omega) = -i\omega\left[\underline{\underline{\epsilon}}_{\mathrm{HBM}}(z,\omega) \bullet \breve{\underline{E}}(\underline{r},\omega) + \underline{\underline{\xi}}_{\mathrm{HBM}}(z,\omega) \bullet \breve{\underline{H}}(\underline{r},\omega)\right] \end{array} \right\} \quad (4.73)$$

emerges. Herein, $\breve{\underline{E}}(\underline{r},\omega)$ and $\breve{\underline{H}}(\underline{r},\omega)$ are the complex–valued amplitudes of the corresponding monochromatic fields, as specified in Eq. (1.36); and the fact that monochromatic field amplitudes given in Eq. (1.36) satisfy the Maxwell postulates and constitutive relations in the same way as do their temporal Fourier transform counterparts specified in Eq. (1.25) has been exploited.

As a key step towards eliminating the z–dependence on the right side of Eq. (4.73) for propagation along the helicoidal axis, the Oseen transformation [16]

$$\left. \begin{aligned} \breve{\underline{E}}'(\underline{r},\omega) &= \underline{\underline{S}}_z^{-1}(z) \bullet \breve{\underline{E}}(\underline{r},\omega) \\ \breve{\underline{H}}'(\underline{r},\omega) &= \underline{\underline{S}}_z^{-1}(z) \bullet \breve{\underline{H}}(\underline{r},\omega) \end{aligned} \right\} \tag{4.74}$$

is implemented. Thereby, Eqs. (4.73) may be recast as

$$\left. \begin{aligned} \underline{\underline{S}}_z^{-1}(z) \bullet \left(\nabla \times \underline{\underline{I}} \right) \bullet \underline{\underline{S}}_z(z) \bullet \breve{\underline{E}}'(\underline{r},\omega) &= \\ i\omega \left[\underline{\underline{\zeta}}_{\mathrm{HBM}}(z,\omega) \bullet \breve{\underline{E}}'(\underline{r},\omega) + \underline{\underline{\mu}}_{\mathrm{HBM}}(z,\omega) \bullet \breve{\underline{H}}'(\underline{r},\omega) \right] \\ \underline{\underline{S}}_z^{-1}(z) \bullet \left(\nabla \times \underline{\underline{I}} \right) \bullet \underline{\underline{S}}_z(z) \bullet \breve{\underline{H}}'(\underline{r},\omega) &= \\ -i\omega \left[\underline{\underline{\epsilon}}_{\mathrm{HBM}}(z,\omega) \bullet \breve{\underline{E}}'(\underline{r},\omega) + \underline{\underline{\xi}}_{\mathrm{HBM}}(z,\omega) \bullet \breve{\underline{H}}'(\underline{r},\omega) \right] \end{aligned} \right\} . \tag{4.75}$$

Upon introducing the planewave representations

$$\left. \begin{aligned} \breve{\underline{E}}'(\underline{r},\omega) &= \breve{\underline{E}}'_0(z,\omega) \exp\left(i\kappa_{\mathrm{HBM}} x \right) \\ \breve{\underline{H}}'(\underline{r},\omega) &= \breve{\underline{H}}'_0(z,\omega) \exp\left(i\kappa_{\mathrm{HBM}} x \right) \end{aligned} \right\} , \tag{4.76}$$

wherein κ_{HBM} may be interpreted as a wavenumber in the xy plane, Eqs. (4.75) generally reduce to the 4×4 matrix differential equation

$$\frac{\partial}{\partial z} \breve{\underline{F}}'_0(z,\omega) = \underline{\underline{M}}'(z,\omega) \bullet \breve{\underline{F}}'_0(z,\omega) . \tag{4.77}$$

Herein, the column 4–vector

$$\breve{\underline{F}}'_0(z,\omega) = \begin{bmatrix} \hat{\underline{x}} \bullet \breve{\underline{E}}'_0(z,\omega) \\ \hat{\underline{y}} \bullet \breve{\underline{E}}'_0(z,\omega) \\ \hat{\underline{x}} \bullet \breve{\underline{H}}'_0(z,\omega) \\ \hat{\underline{y}} \bullet \breve{\underline{H}}'_0(z,\omega) \end{bmatrix} , \tag{4.78}$$

and the 4×4 matrix function $\underline{\underline{M}}'(z,\omega)$ — which is too cumbersome to explicitly reproduce here (see [17], for example, for the full details) — may be conveniently expressed as

$$\begin{aligned} \underline{\underline{M}}'(z) = \underline{\underline{A}}' + \kappa_{\mathrm{HBM}} &\left[\underline{\underline{C}}'_{1,1} \exp\left(\frac{i\pi z}{\Omega_{\mathrm{hp}}} \right) + \underline{\underline{C}}'_{1,-1} \exp\left(-\frac{i\pi z}{\Omega_{\mathrm{hp}}} \right) \right] \\ + \kappa_{\mathrm{HBM}}^2 &\left[\underline{\underline{C}}'_{2,2} \exp\left(\frac{i2\pi z}{\Omega_{\mathrm{hp}}} \right) + \underline{\underline{C}}'_{2,0} + \underline{\underline{C}}'_{2,-2} \exp\left(-\frac{i2\pi z}{\Omega_{\mathrm{hp}}} \right) \right] , \end{aligned} \tag{4.79}$$

with the 4×4 matrixes $\underline{\underline{A}}'$, $\underline{\underline{C}}'_{1,\pm1}$, $\underline{\underline{C}}'_{2,0}$ and $\underline{\underline{C}}'_{2,\pm2}$ being independent of z and κ_{HBM} but not of the half–pitch Ω_{hp} and ω.

For axial propagation, $\kappa_{\mathrm{HBM}} = 0$ and a closed–form solution to Eq. (4.77) is provided as [18]

$$\check{\underline{F}}'_0(z,\omega) = \exp\left(i\underline{\underline{A}}'z\right)\check{\underline{F}}'_0(0,\omega). \tag{4.80}$$

For nonaxial propagation, $\kappa_{\mathrm{HBM}} \neq 0$ and the solution to Eq. (4.77) may be expressed in terms of a power series in z through exploiting the representation (4.79) [19]. Alternatively, a piecewise uniform approximation may be implemented [17].

4.7.2 *Gravitationally affected vacuum*

The electromagnetic properties of vacuum in generally curved spacetime are equivalent to those of a fictitious, instantaneously responding, bianisotropic medium, as outlined in Sec. 2.4.2. The same spacetime coordinate is designated as time throughout all spacetime. Then, the time–domain constitutive relations (2.67) for the fictitious medium are expressed in terms of the 3×3 dyadic $\tilde{\underline{\underline{\gamma}}}_{\mathrm{GAV}}(\underline{r},t)$ and 3–vector $\tilde{\underline{\Gamma}}_{\mathrm{GAV}}(\underline{r},t)$, which are derived from the spacetime metric and specified in Eqs. (2.68) and (2.69), respectively.

The nonhomogeneous nature of the equivalent medium represented by Eqs. (2.67) may be effectively dealt with by implementing a piecewise uniform approximation [20]. This technique is widely used in solving differential equations with nonhomogeneous coefficients [21] and also for nonaxial propagation in CLCs and CSTFs [17]. We focus our attention on a spacetime–neighbourhood which is sufficiently small so that the nonuniform quantities $\tilde{\underline{\underline{\gamma}}}_{\mathrm{GAV}}(\underline{r},t)$ and $\tilde{\underline{\Gamma}}_{\mathrm{GAV}}(\underline{r},t)$ may be replaced by their uniform equivalents $\tilde{\underline{\underline{\gamma}}}_{\mathrm{GAV}}$ and $\tilde{\underline{\Gamma}}_{\mathrm{GAV}}$, respectively. This approximation does not amount to the implementation of locally flat spacetime, being simply a partitioning of all spacetime with the time coordinate having been chosen globally prior to the partitioning. Thereby, the propagation of planewaves within the neighbourhood may be analyzed using the methods described in Sec. 4.2. Global solutions may then be developed by stitching together the planewave solutions from adjacent spacetime–neighbourhoods.

Now let us proceed to establish planewave solutions

$$\left.\begin{array}{l}\tilde{\underline{E}}(\underline{r},t) = \mathrm{Re}\left\{\check{\underline{E}}_0(\omega)\exp\left[i\left(\underline{k}\cdot\underline{r}-\omega t\right)\right]\right\} \\[2mm] \tilde{\underline{H}}(\underline{r},t) = \mathrm{Re}\left\{\check{\underline{H}}_0(\omega)\exp\left[i\left(\underline{k}\cdot\underline{r}-\omega t\right)\right]\right\}\end{array}\right\}, \tag{4.81}$$

within a neighbourhood which is sufficiently small that the nonuniform constitutive relations (2.67) may be replaced by their uniform counterparts [22]

$$\left.\begin{aligned}
\underline{\tilde{D}}(\underline{r},t) &= \epsilon_0\, \bar{\underline{\underline{\gamma}}}_{\text{GAV}} \bullet \underline{\tilde{E}}(\underline{r},t) - \frac{1}{c_0}\bar{\underline{\Gamma}}_{\text{GAV}} \times \underline{\tilde{H}}(\underline{r},t) \\[2mm]
\underline{\tilde{B}}(\underline{r},t) &= \mu_0\, \bar{\underline{\underline{\gamma}}}_{\text{GAV}} \bullet \underline{\tilde{H}}(\underline{r},t) + \frac{1}{c_0}\bar{\underline{\Gamma}}_{\text{GAV}} \times \underline{\tilde{E}}(\underline{r},t)
\end{aligned}\right\}. \tag{4.82}$$

The amplitudes $\underline{\breve{E}}_0(\omega)$ and $\underline{\breve{H}}_0(\omega)$ in Eqs. (4.81) are determined by the boundary/initial conditions. Combining the source–free Maxwell curl postulates

$$\left.\begin{aligned}
\nabla \times \underline{\tilde{H}}(\underline{r},t) - \frac{\partial}{\partial t}\underline{\tilde{D}}(\underline{r},t) &= \underline{0} \\[2mm]
\nabla \times \underline{\tilde{E}}(\underline{r},t) + \frac{\partial}{\partial t}\underline{\tilde{B}}(\underline{r},t) &= \underline{0}
\end{aligned}\right\} \tag{4.83}$$

with the constitutive relations (4.82), we arrive at the eigenvector equation

$$\left\{\left[\left(\frac{\omega}{c_0}\right)^2 \det\left(\bar{\underline{\underline{\gamma}}}_{\text{GAV}}\right) - \underline{p}_{\text{GAV}} \bullet \bar{\underline{\underline{\gamma}}}_{\text{GAV}} \bullet \underline{p}_{\text{GAV}}\right]\underline{\underline{I}} \right.$$

$$\left. + \underline{p}_{\text{GAV}}\underline{p}_{\text{GAV}} \bullet \bar{\underline{\underline{\gamma}}}_{\text{GAV}}\right\} \bullet \underline{\breve{E}}_0(\omega) = \underline{0}, \tag{4.84}$$

where the vector

$$\underline{p}_{\text{GAV}}(\omega) = \underline{k} - \frac{\omega}{c_0}\bar{\underline{\Gamma}}_{\text{GAV}}. \tag{4.85}$$

The dispersion relation

$$\left[\underline{p}_{\text{GAV}}(\omega) \bullet \bar{\underline{\underline{\gamma}}}_{\text{GAV}} \bullet \underline{p}_{\text{GAV}}(\omega) - \left(\frac{\omega}{c_0}\right)^2 \det\left(\bar{\underline{\underline{\gamma}}}_{\text{GAV}}\right)\right]^2 = 0 \tag{4.86}$$

thereby arises, from which the four roots

$$\left.\begin{aligned}
k_1 = -k_3 &= \frac{\omega}{c_0\, \hat{\underline{k}} \bullet \bar{\underline{\underline{\gamma}}}_{\text{GAV}} \bullet \hat{\underline{k}}}\left\{\hat{\underline{k}} \bullet \bar{\underline{\underline{\gamma}}}_{\text{GAV}} \bullet \bar{\underline{\Gamma}}_{\text{GAV}} + \left[\left(\hat{\underline{k}} \bullet \bar{\underline{\underline{\gamma}}}_{\text{GAV}} \bullet \bar{\underline{\Gamma}}_{\text{GAV}}\right)^2 \right.\right. \\
&\left.\left. -\hat{\underline{k}} \bullet \bar{\underline{\underline{\gamma}}}_{\text{GAV}} \bullet \hat{\underline{k}}\left(\bar{\underline{\Gamma}}_{\text{GAV}} \bullet \bar{\underline{\underline{\gamma}}}_{\text{GAV}} \bullet \bar{\underline{\Gamma}}_{\text{GAV}} - \det\bar{\underline{\underline{\gamma}}}_{\text{GAV}}\right)\right]^{1/2}\right\} \\[3mm]
k_2 = -k_4 &= \frac{\omega}{c_0\, \hat{\underline{k}} \bullet \bar{\underline{\underline{\gamma}}}_{\text{GAV}} \bullet \hat{\underline{k}}}\left\{\hat{\underline{k}} \bullet \bar{\underline{\underline{\gamma}}}_{\text{GAV}} \bullet \bar{\underline{\Gamma}}_{\text{GAV}} - \left[\left(\hat{\underline{k}} \bullet \bar{\underline{\underline{\gamma}}}_{\text{GAV}} \bullet \bar{\underline{\Gamma}}_{\text{GAV}}\right)^2 \right.\right. \\
&\left.\left. -\hat{\underline{k}} \bullet \bar{\underline{\underline{\gamma}}}_{\text{GAV}} \bullet \hat{\underline{k}}\left(\bar{\underline{\Gamma}}_{\text{GAV}} \bullet \bar{\underline{\underline{\gamma}}}_{\text{GAV}} \bullet \bar{\underline{\Gamma}}_{\text{GAV}} - \det\bar{\underline{\underline{\gamma}}}_{\text{GAV}}\right)\right]^{1/2}\right\}
\end{aligned}\right\}, \tag{4.87}$$

emerge. Just as described in Sec. 4.6.1, the wavenumbers k_1 and k_2 (or, equivalently, k_3 and k_4) given by Eqs. (4.87) are not independent, and the medium represented by Eqs. (2.67) is unirefringent accordingly.

The general solution to Eq. (4.84) may be expressed as the sum

$$\breve{\underline{E}}_0(\omega) = \breve{A}_p(\omega)\,\hat{\underline{e}}_p(\omega) + \breve{A}_q(\omega)\,\hat{\underline{e}}_q(\omega), \tag{4.88}$$

wherein the complex–valued amplitude scalars $\breve{A}_{p,q}(\omega)$ are determined from boundary/initial conditions, and the unit vectors $\hat{\underline{e}}_{p,q}(\omega)$ are taken as

$$\left. \begin{aligned} \hat{\underline{e}}_p(\omega) &= \frac{\bar{\bar{\gamma}}^{-1} \cdot \hat{\underline{w}}_{GAV}(\omega)}{|\bar{\bar{\gamma}}^{-1}_{GAV} \cdot \hat{\underline{w}}_{GAV}(\omega)|} \\[2ex] \hat{\underline{e}}_q(\omega) &= \frac{\bar{\bar{\gamma}}^{-1}_{GAV} \cdot \left[\underline{p}_{GAV}(\omega) \times \hat{\underline{e}}_p(\omega) \right]}{|\bar{\bar{\gamma}}^{-1}_{GAV} \cdot \left[\underline{p}_{GAV}(\omega) \times \hat{\underline{e}}_p(\omega) \right]|} \end{aligned} \right\}. \tag{4.89}$$

The unit vector $\hat{\underline{w}}_{GAV}(\omega)$ in Eq. (4.89)$_1$ is orthogonal to $\underline{p}_{GAV}(\omega)$, i.e., $\hat{\underline{w}}_{GAV}(\omega) \cdot \underline{p}_{GAV}(\omega) = 0$, but is otherwise arbitrary. In a similar manner,

$$\breve{\underline{H}}_0(\omega) = \breve{A}_p(\omega)\,\hat{\underline{h}}_p(\omega) + \breve{A}_q(\omega)\,\hat{\underline{h}}_q(\omega), \tag{4.90}$$

where the unit vectors

$$\hat{\underline{h}}_{p,q}(\omega) = \frac{1}{\omega\mu_0}\bar{\bar{\gamma}}^{-1}_{GAV} \cdot \left[\underline{p}_{GAV}(\omega) \times \hat{\underline{e}}_{p,q}(\omega) \right]. \tag{4.91}$$

From Eqs. (4.88) and (4.90), we find that the corresponding time–averaged Poynting vector for monochromatic plane waves

$$\langle \breve{\underline{S}}(\underline{r},\omega) \rangle_t = \frac{1}{2\omega\mu_0 \det \bar{\bar{\gamma}}_{GAV}} \Big[|\breve{A}_p(\omega)|^2\,\hat{\underline{e}}_p(\omega) \cdot \bar{\bar{\gamma}}_{GAV} \cdot \hat{\underline{e}}_p(\omega) $$
$$+ |\breve{A}_q(\omega)|^2\,\hat{\underline{e}}_q(\omega) \cdot \bar{\bar{\gamma}}_{GAV} \cdot \hat{\underline{e}}_q(\omega) \Big]\,\bar{\bar{\gamma}} \cdot \underline{p}_{GAV}(\omega) \tag{4.92}$$

is aligned with $\bar{\bar{\gamma}}_{GAV} \cdot \underline{p}_{GAV}(\omega)$.

Let us close this section by emphasizing that a local observer is cognizant only of a flat spacetime, and will therefore choose the time coordinate such that $\bar{\bar{\gamma}}_{GAV} = \underline{\underline{I}}$ and $\bar{\underline{\Gamma}}_{GAV} = \underline{0}$ [20]. The spacetime metric being real symmetric, this choice is in accord with the equivalence principle [23].

4.8 Exotic planewave phenomenons

We now survey some examples of exotic planewave phenomenons, which are currently active topics of research. The relatively large parameter space associated with anisotropic mediums — and even more so with bianisotropic mediums — increases the propensity for exotic behaviour. In recent years, the prospects of such behaviour being exhibited by negatively refracting materials — sometimes called *metamaterials* — has attracted a considerable degree of attention, from both theorists and experimentalists alike. As yet, there is no universally accepted agreement on what constitutes a metamaterial; a plausible working definition is: an artificial composite material which exhibits properties that are either not exhibited at all by its constituent materials or not exhibited to the same extent by its constituent materials [24]. We describe the manifestation of metamaterials as homogenized composite mediums in Sec. 6.7.

4.8.1 *Plane waves with negative phase velocity*

Let us consider the orientation of the phase velocity $\underline{v}_{\mathrm{ph}}(\omega)$, as defined in Eq. (4.12), relative to the time–averaged Poynting vector for monochromatic plane waves $\langle \underline{\breve{S}}(\underline{r},\omega)\rangle_{\mathrm{t}}$, as defined in Eq. (1.90). The phase velocity may be classified as being

- *positive* if $\underline{v}_{\mathrm{ph}}(\omega) \cdot \langle \underline{\breve{S}}(\underline{r},\omega)\rangle_{\mathrm{t}} > 0$;
- *orthogonal* if $\underline{v}_{p} \cdot \langle \underline{S}(\underline{r},\omega)\rangle_{\mathrm{t}} = 0$; or
- *negative* if $\underline{v}_{\mathrm{ph}}(\omega) \cdot \langle \underline{\breve{S}}(\underline{r},\omega)\rangle_{\mathrm{t}} < 0$.

Conventional materials support planewave propagation with positive phase velocity (PPV). However, due to recent progress in the development of novel materials, attention is turning increasingly towards parameter regimes giving rise to negative phase velocity (NPV). Furthermore, the concept of *infinite* phase velocity — which may arise at the boundary between NPV and PPV propagation in isotropic dielectric–magnetic mediums — has come up [25]. In the case of nonuniform planewave propagation, orthogonal phase velocity arises at the transition from PPV to NPV (or vice versa) [26, 27].

For isotropic dielectric–magnetic mediums, the phenomenon of NPV propagation is closely associated with negative refraction [28, 29]. An explosion of interest in this topic took place at the beginning of this century, following an experimental report of negative refraction in the microwave regime exhibited by a material which comprised a periodic array of metallic wire and split–ring inclusions embossed on circuit boards [30]. Subsequent efforts by experimentalists and theorists have been directed towards higher–frequency regimes, with negative refraction in the visible regime being achieved most recently, albeit with considerable associated dissipation [31, 32]. The prospect of fabricating lenses with extremely high resolving power from negatively refracting materials has fuelled much of this work [33]. However, dissipation continues to be a significant roadblock [34, 35], which has prompted the consideration of negatively refracting metamaterials with active constituents [36, 37].

A simple construction method for isotropic dielectric–magnetic materials which support NPV propagation has recently been proposed. The method is based on the homogenization of a random assembly of two different types of electrically small spherical particles. The two types of constituent particles, type 'a' and type 'b', can each be made of an isotropic, homogeneous, dielectric–magnetic material, with permittivities ϵ_a and ϵ_b, and permeabilities μ_a and μ_b, respectively. Provided that $\epsilon_{a,b}$ and $\mu_{a,b}$ lie within certain parameter ranges, with the real parts of $\epsilon_{a,b}$ being negative–valued and the real parts of $\mu_{a,b}$ being positive–valued (or vice versa), the bulk constituent materials do not support NPV propagation whereas the corresponding homogenized composite medium (HCM) does. Whether or not NPV propagation is supported by the HCM depends upon the relative proportions of the constituents, as well as the size [38] and spatial distribution [39] of the constituent particles.

As compared with isotropic dielectric–magnetic mediums, anisotropic mediums offer greater scope for achieving NPV propagation on account of their larger parameter spaces [11, 40]. Bianisotropic HCMs which support NPV propagation have been conceptualized — as Faraday chiral mediums [11], for example — which arise from component materials that do not themselves support NPV propagation. Also, chiral mediums can support NPV propagation provided that the magnetoelectric coupling is sufficiently strong [10, 41]. Furthermore, isotropic chiral mediums can support NPV and PPV propagation simultaneously, depending upon the polarization state [42].

The phenomenons of NPV propagation and negative refraction are not exclusive to the realm of metamaterials. Negative refraction has been observed in metallic ferromagnetic materials [43], as well as possibly in the eyes of certain lobsters and moths [44]. Furthermore, there are noteworthy manifestations of NPV in relativistic scenarios, with potentially important consequences for observational astronomy, for example. The property of supporting NPV propagation is not Lorentz covariant (cf. Sec. 1.6.3) and, indeed, neither is the phenomenon of negative refraction [45]. Thus, a medium may support PPV propagation from the perspective of one inertial observer, but support NPV propagation from the perspective of another inertial observer [27, 46]. Also, we note that a Gaussian beam, launched from vacuum into a moving medium, may be deflected in a negative or positive sense relative to the interface normal depending on the moving medium's component of velocity parallel to the interface [27, 47].

The covariant analogue of the NPV condition $\underline{v}_{ph} \cdot \langle \underline{\breve{S}}(\underline{r}, \omega) \rangle_t < 0$ for an inertial observer is available [48]. However, there is no *a priori* reason for the orientation of the phase velocity relative to the time–averaged Poynting vector to be covariant (unlike the electromagnetic fields which are necessarily covariant entities).

Classical vacuum itself — which does not support NPV propagation in any inertial reference frame in flat spacetime — can support NPV propagation for certain curved spacetimes, such as the Kerr spacetime [49] and Schwarzschild–anti–de Sitter spacetime [50], according to the noncovariant approach to gravitational electromagnetics pioneered by Tamm [51–54]. In this noncovariant approach, a particular spacetime coordinate has to be chosen as time globally, not locally.

Finally in this section, we note that NPV has commonly been understood to betoken negative refraction [55, 56], and indeed the two phenomenons are effectively synonymous for uniform planewave propagation in isotropic dielectric–magnetic mediums. However, certain uniaxial dielectric mediums described by indefinite permittivity dyadics [57] and pseudochiral omega mediums [26] can support NPV propagation in conjunction with positive refraction (and vice versa). Furthermore, nonuniform plane waves with NPV can propagate in an isotropic dielectric medium [26], whereas negative refraction in this medium is impossible. The independence of NPV and negative refraction has also been reported for certain active materials [58]. Therefore, the correspondence between NPV and negative refraction is only strictly appropriate in the case of uniform planewave propagation in passive, isotropic dielectric–magnetic mediums.

4.8.2 *Hyperbolic dispersion relations*

In Sec. 4.5.1, we described the propagation of ordinary and extraordinary plane waves, with wavenumbers given in Eqs. (4.30), in a uniaxial dielectric medium characterized by the Tellegen constitutive relations (2.8) with permittivity dyadic given in Eq. (2.9). Let us now consider the constitutive parameter regimes delineated by the real–valued quantity

$$\gamma_\epsilon(\omega) = \begin{cases} \dfrac{\epsilon_u(\omega)}{\epsilon(\omega)} & \text{for } \epsilon_u(\omega), \epsilon(\omega) \in \mathbb{R} \\[2ex] \dfrac{\text{Re}\left\{\epsilon_u(\omega)\right\}}{\text{Re}\left\{\epsilon(\omega)\right\}} & \text{for } \epsilon_u(\omega), \epsilon(\omega) \in \mathbb{C} \end{cases}. \tag{4.93}$$

The upper expression is appropriate to nondissipative mediums whereas the lower expression is appropriate to dissipative mediums. The electromagnetic/optical properties of uniaxial mediums with $\gamma_\epsilon(\omega) > 0$, which category includes naturally occurring uniaxial crystals, have long been established [1, 4].

Uniaxial mediums characterized by $\gamma_\epsilon(\omega) < 0$ are exotic. Such mediums are sometimes called indefinite mediums since the corresponding permittivity dyadic $\underline{\underline{\epsilon}}_{\text{uni}}(\omega)$ (for nondissipative mediums) or $\text{Re}\left\{\underline{\underline{\epsilon}}_{\text{uni}}(\omega)\right\}$ (for dissipative mediums) is indefinite [59].

Let us suppose that the distinguished axis of the dielectric medium coincides with the x axis, i.e., $\hat{u} = \hat{x}$; and the ordinary and extraordinary wavevectors, namely $\underline{k}^{\text{or}}$ and $\underline{k}^{\text{ex}}$ respectively, lie in the xz plane, i.e.,

$$\left. \begin{aligned} \underline{k}^{\text{or}} &= k_x\hat{x} + k_z^{\text{or}}\hat{z} \\ \underline{k}^{\text{ex}} &= k_x\hat{x} + k_z^{\text{ex}}\hat{z} \end{aligned} \right\}. \tag{4.94}$$

In conformity with planar boundary value problems [4] and potential optical devices [60], we choose $k_x \in \mathbb{R}$ and $k_z^{\text{or,ex}} \in \mathbb{C}$.

As provided by Eq. (4.30)$_1$, the components of the ordinary wavevector are required to satisfy

$$k_x^2 + \left(k_z^{\text{or}}\right)^2 = \omega^2\epsilon(\omega)\mu_0. \tag{4.95}$$

Therefore, ordinary plane waves propagate in a nondissipative medium only when $\epsilon(\omega) > 0$ and $\omega^2\epsilon(\omega)\mu_0 > k_x^2$. In geometric terms, the wavevector components then have a circular representation in (k_x, k_z^{or}) space.

In contrast, we deduce from Eq. (4.30)$_2$ that the components of the extraordinary wavevector satisfy

$$k_x^2\epsilon_u(\omega) + \left(k_z^{\text{ex}}\right)^2\epsilon(\omega) = \omega^2\epsilon(\omega)\epsilon_u(\omega)\mu_0. \tag{4.96}$$

The conditions to be satisfied for the propagation of extraordinary plane waves in a nondissipative medium depend upon the sign of $\gamma_\epsilon(\omega)$:

- If $\gamma_\epsilon(\omega) > 0$ then we require $\omega^2\epsilon(\omega)\mu_0 > k_x^2$. This implies that $\epsilon(\omega) > 0$ and $\epsilon_u(\omega) > 0$. In geometric terms, the wavevector components have an elliptical representation in (k_x, k_z^{ex}) space.

- If $\gamma_\epsilon(\omega) < 0$ then we require (a) $\omega^2 \epsilon(\omega) \mu_0 < k_x^2$ when $\epsilon(\omega) > 0$; or (b) $-\infty < k_x < \infty$ when $\epsilon(\omega) < 0$. In geometric terms, the wavevector components have a hyperbolic representation in (k_x, k_z^{ex}) space.

Numerical illustrations of hyperbolic dispersion relations — and near–hyperbolic dispersion relations in the case of weakly dissipative mediums — can be found elsewhere [61].

Interest in mediums with $\gamma_\epsilon(\omega) < 0$ stems from their potential applications in negatively refracting scenarios [62] and in diffraction gratings [60], for example. In connection with negative refraction, it is noteworthy that nondissipative uniaxial dielectric mediums with $\gamma_\epsilon(\omega) < 0$ do not support NPV propagation [61], but negative refraction may be achieved via a two–slab assembly of complementary mediums each with $\gamma_\epsilon(\omega) < 0$ [62]. Although uniaxial dielectric mediums with $\gamma_\epsilon(\omega) < 0$ do not occur in nature (as far as we are aware), they can be conceptualized as metamaterials by means of homogenization [61].

Greater scope for exotic behaviour — and negative refraction, in particular [63, 64] — is offered by anisotropic dielectric–magnetic mediums, characterized by a permittivity dyadic $\underline{\underline{\epsilon}}_{\mathrm{aniso}}(\omega)$ and a permeability dyadic $\underline{\underline{\mu}}_{\mathrm{aniso}}(\omega)$ such that

- both $\underline{\underline{\epsilon}}_{\mathrm{aniso}}(\omega)$ and $\underline{\underline{\mu}}_{\mathrm{aniso}}(\omega)$ are indefinite, in the case of nondissipative mediums; or
- both $\mathrm{Re}\left\{\underline{\underline{\epsilon}}_{\mathrm{aniso}}(\omega)\right\}$ and $\mathrm{Re}\left\{\underline{\underline{\mu}}_{\mathrm{aniso}}(\omega)\right\}$ are indefinite, in the case of dissipative mediums.

Such mediums may also be usefully employed as linear polarizers [65] and waveguides [66]. Their reflection and transmission properties are comprehensively described in the recent literature [67–69].

4.8.3 *Voigt waves*

As described in Sec. 4.5.1–4.5.3, two distinct plane waves generally propagate in each direction in anisotropic mediums. However, under certain special circumstances, the two plane waves may coalesce to form a *Voigt wave*. Experimental observations of these waves were reported by Woldemar Voigt more than 100 years ago [70]. While the theoretical basis for Voigt waves was developed in the 1950s [71], this topic has attracted renewed attention lately as a consequence of advances in metamaterial technologies. Recent work has demonstrated the possibility of constructing complex composite materials which support the propagation of Voigt waves, using component materials which do not themselves support Voigt–wave propagation [72]. In connection with Sec. 4.7, we note that the emergence of Voigt waves has also been investigated in certain periodically nonhomogeneous mediums [73].

Let us now describe the circumstances under which Voigt waves can arise in anisotropic dielectric mediums [74], as well as in bianisotropic mediums [75].

4.8.3.1   *Anisotropic dielectric mediums*

Let us take a general anisotropic dielectric medium, specified by the Tellegen constitutive relations

$$
\left.
\begin{aligned}
\underline{D}(\underline{r}, \omega) &= \underline{\underline{\epsilon}}_{\text{aniso}}(\omega) \bullet \underline{E}(\underline{r}, \omega) \\
\underline{B}(\underline{r}, \omega) &= \mu_0 \underline{H}(\underline{r}, \omega)
\end{aligned}
\right\} ,
\tag{4.97}
$$

with complex–valued permittivity dyadic

$$
\underline{\underline{\epsilon}}_{\text{aniso}}(\omega) =
\begin{pmatrix}
\epsilon_{11}(\omega) & \epsilon_{12}(\omega) & \epsilon_{13}(\omega) \\
\epsilon_{21}(\omega) & \epsilon_{22}(\omega) & \epsilon_{23}(\omega) \\
\epsilon_{31}(\omega) & \epsilon_{32}(\omega) & \epsilon_{33}(\omega)
\end{pmatrix} ,
\tag{4.98}
$$

as the setting for our analysis. For definiteness — and without loss of generality — we examine uniform planewave propagation parallel to the z axis; i.e., $\hat{\underline{k}} = \hat{\underline{z}}$. Accordingly, we consider plane waves of the form

$$
\tilde{\mathbf{F}}(\underline{r}, t) = \text{Re}
\left\{
\begin{bmatrix}
\breve{E}_x(z, \omega) \\
\breve{E}_y(z, \omega) \\
\breve{E}_z(z, \omega) \\
\breve{H}_x(z, \omega) \\
\breve{H}_y(z, \omega) \\
\breve{H}_z(z, \omega)
\end{bmatrix}
\exp\left(-i\omega t\right)
\right\}
\tag{4.99}
$$

$$
= \text{Re}
\left\{
\begin{bmatrix}
\breve{E}_{0x}(\omega) \\
\breve{E}_{0y}(\omega) \\
\breve{E}_{0z}(\omega) \\
\breve{H}_{0x}(\omega) \\
\breve{H}_{0y}(\omega) \\
\breve{H}_{0z}(\omega)
\end{bmatrix}
\exp\left[i\left(kz - \omega t\right)\right]
\right\} .
\tag{4.100}
$$

Combining Eqs. (4.97) and (4.99) with the anisotropic dielectric specialization of Eq. (4.18), we first eliminate $\breve{E}_{0z}(\omega)$ and $\breve{H}_{0z}(\omega)$, and then eliminate

$\check{H}_{0x}(\omega)$ and $\check{H}_{0y}(\omega)$, to obtain

$$\frac{\omega^2 \mu_0}{\epsilon_{33}(\omega)} \begin{pmatrix} \delta_{11}(\omega) & \delta_{12}(\omega) \\ \delta_{12}(\omega) & \delta_{22}(\omega) \end{pmatrix} \begin{bmatrix} \check{E}_{0x}(\omega) \\ \check{E}_{0y}(\omega) \end{bmatrix} = k^2 \begin{bmatrix} \check{E}_{0x}(\omega) \\ \check{E}_{0y}(\omega) \end{bmatrix}, \qquad (4.101)$$

where

$$\left. \begin{aligned} \delta_{11}(\omega) &= \epsilon_{11}(\omega)\,\epsilon_{33}(\omega) - \epsilon_{13}(\omega)\,\epsilon_{31}(\omega) \\ \delta_{12}(\omega) &= \epsilon_{12}(\omega)\,\epsilon_{33}(\omega) - \epsilon_{13}(\omega)\,\epsilon_{32}(\omega) \\ \delta_{21}(\omega) &= \epsilon_{21}(\omega)\,\epsilon_{33}(\omega) - \epsilon_{23}(\omega)\,\epsilon_{31}(\omega) \\ \delta_{22}(\omega) &= \epsilon_{22}(\omega)\,\epsilon_{33}(\omega) - \epsilon_{23}(\omega)\,\epsilon_{32}(\omega) \end{aligned} \right\}. \qquad (4.102)$$

In the case of isotropic dielectric mediums, the matrix on the left side of Eq. (4.101) has only one eigenvalue but two linearly independent eigenvectors. The normal scenario for anisotropic dielectric mediums is that of birefringence, wherein the matrix in Eq. (4.101) has two distinct eigenvalues and two independent eigenvectors. The general solution for the x and y components of the electric field amplitude may then be expressed as

$$\begin{bmatrix} \check{E}_x(z,\omega) \\ \check{E}_y(z,\omega) \end{bmatrix} = C_1 \begin{bmatrix} \check{E}_{x1}(\omega) \\ \check{E}_{y1}(\omega) \end{bmatrix} \exp\left(ik_1 z\right) + C_2 \begin{bmatrix} \check{E}_{x2}(\omega) \\ \check{E}_{y2}(\omega) \end{bmatrix} \exp\left(ik_2 z\right).$$
$$(4.103)$$

Herein, the eigenvectors $\left[\check{E}_{x1}(\omega),\, \check{E}_{y1}(\omega) \right]^{\mathrm{T}}$ and $\left[\check{E}_{x2}(\omega),\, \check{E}_{y2}(\omega) \right]^{\mathrm{T}}$ correspond to the two wavenumbers

$$\left. \begin{aligned} k_1 &= \omega \left[\frac{\mu_0}{2\epsilon_{33}(\omega)} \left(\left[\delta_{11}(\omega) + \delta_{22}(\omega) \right] \right. \right. \\ &\qquad \left. \left. + \left\{ \left[\delta_{11}(\omega) - \delta_{22}(\omega) \right]^2 + 4\delta_{12}(\omega)\delta_{21}(\omega) \right\}^{1/2} \right) \right]^{1/2} \\ k_2 &= \omega \left[\frac{\mu_0}{2\epsilon_{33}(\omega)} \left(\left[\delta_{11}(\omega) + \delta_{22}(\omega) \right] \right. \right. \\ &\qquad \left. \left. - \left\{ \left[\delta_{11}(\omega) - \delta_{22}(\omega) \right]^2 + 4\delta_{12}(\omega)\delta_{21}(\omega) \right\}^{1/2} \right) \right]^{1/2} \end{aligned} \right\}, \qquad (4.104)$$

respectively, whereas $C_{1,2}$ are amplitude coefficients which may be determined from boundary conditions.

Anomalously, it is possible that the matrix in Eq. (4.101) has only one independent eigenvector — and therefore only one eigenvalue k. Accordingly, the matrix cannot be a scalar multiple of the identity matrix. In this case, the general solution for the x and y components of the electric field amplitude may be expressed as

$$\begin{bmatrix} \breve{E}_x(z,\omega) \\ \breve{E}_y(z,\omega) \end{bmatrix} = \left(C_1 \begin{bmatrix} \breve{E}_{x1}(\omega) \\ \breve{E}_{y1}(\omega) \end{bmatrix} + ikz\, C_2 \begin{bmatrix} \breve{E}_{x2}(\omega) \\ \breve{E}_{y2}(\omega) \end{bmatrix} \right) \exp(ikz), \quad (4.105)$$

which signifies a Voigt wave. A prominent distinguishing feature of the Voigt wave solution represented by Eq. (4.105) is that the plane wave amplitude has a linear dependence on propagation distance.

Sufficient conditions for Voigt–wave propagation are [74]

- $[\delta_{11}(\omega) - \delta_{22}(\omega)]^2 + 4\delta_{12}(\omega)\delta_{21}(\omega) = 0$, and
- $|\delta_{12}(\omega)| + |\delta_{12}(\omega)| \neq 0$.

These conditions cannot be satisfied by uniaxial dielectric mediums, but can be satisfied by certain non–orthorhombic biaxial dielectric mediums and gyrotropic mediums [76].

4.8.3.2 *Bianisotropic mediums*

Now let us turn to the most general linear scenario in which plane waves, as specified by Eq. (4.99), propagate in the bianisotropic medium characterized by the Tellegen constitutive relations (4.14) and 6×6 constitutive dyadic $\underline{\underline{\mathbf{K}}}_{EH}(\omega)$ given in Eq. (4.18). The components of the corresponding 3×3 constitutive dyadics are written as

$$\underline{\underline{\eta}}_{EH}(\omega) = \begin{pmatrix} \eta_{11}(\omega) & \eta_{12}(\omega) & \eta_{13}(\omega) \\ \eta_{21}(\omega) & \eta_{22}(\omega) & \eta_{23}(\omega) \\ \eta_{31}(\omega) & \eta_{32}(\omega) & \eta_{33}(\omega) \end{pmatrix}, \quad (\eta = \epsilon, \zeta, \xi, \mu). \quad (4.106)$$

The algebraic Eq. (4.18) — which represents the combination of the source–free Maxwell curl postulates for planewave solutions with the constitutive relations — directly delivers the z components of the field amplitudes per

$$\begin{bmatrix} \breve{E}_{0z}(\omega) \\ \breve{H}_{0z}(\omega) \end{bmatrix} = \underline{\underline{F}}(\omega) \bullet \begin{bmatrix} \breve{E}_{0x}(\omega) \\ \breve{E}_{0y}(\omega) \\ H_{0x}(\omega) \\ H_{0y}(\omega) \end{bmatrix}, \quad (4.107)$$

with the 2×4 matrix

$$\underline{\underline{F}}(\omega) = -\begin{pmatrix} \epsilon_{33}(\omega) & \xi_{33}(\omega) \\ \zeta_{33}(\omega) & \mu_{33}(\omega) \end{pmatrix}^{-1} \begin{pmatrix} \epsilon_{31}(\omega) & \epsilon_{32}(\omega) & \xi_{31}(\omega) & \xi_{32}(\omega) \\ \zeta_{31}(\omega) & \zeta_{32}(\omega) & \mu_{31}(\omega) & \mu_{32}(\omega) \end{pmatrix},$$
$$(4.108)$$

provided that $\epsilon_{33}(\omega)\mu_{33}(\omega) - \xi_{33}(\omega)\zeta_{33}(\omega) \neq 0$. Thus, Eq. (4.18) reduces to the eigenvector equation

$$\omega \underline{\underline{R}}(\omega) \cdot \breve{\underline{F}}_0(\omega) = k \breve{\underline{F}}_0(\omega). \tag{4.109}$$

Herein the 4×4 matrix

$$\underline{\underline{R}}(\omega) = \big(\, \underline{R}_1(\omega) \quad \underline{R}_2(\omega) \quad \underline{R}_3(\omega) \quad \underline{R}_4(\omega)\,\big) \tag{4.110}$$

consists of the four column 4–vectors

$$\underline{R}_1(\omega) = \begin{pmatrix} \zeta_{21}(\omega) + \zeta_{23}(\omega)\left[\underline{\underline{F}}(\omega)\right]_{11} + \mu_{23}(\omega)\left[\underline{\underline{F}}(\omega)\right]_{21} \\ -\zeta_{11}(\omega) - \zeta_{13}(\omega)\left[\underline{\underline{F}}(\omega)\right]_{11} - \mu_{13}(\omega)\left[\underline{\underline{F}}(\omega)\right]_{21} \\ -\epsilon_{21}(\omega) - \epsilon_{23}(\omega)\left[\underline{\underline{F}}(\omega)\right]_{11} - \xi_{23}(\omega)\left[\underline{\underline{F}}(\omega)\right]_{21} \\ \epsilon_{11}(\omega) + \epsilon_{13}(\omega)\left[\underline{\underline{F}}(\omega)\right]_{11} + \xi_{13}(\omega)\left[\underline{\underline{F}}(\omega)\right]_{21} \end{pmatrix}$$

$$\underline{R}_2(\omega) = \begin{pmatrix} \zeta_{22}(\omega) + \zeta_{23}(\omega)\left[\underline{\underline{F}}(\omega)\right]_{12} + \mu_{23}(\omega)\left[\underline{\underline{F}}(\omega)\right]_{22} \\ -\zeta_{12}(\omega) - \zeta_{13}(\omega)\left[\underline{\underline{F}}(\omega)\right]_{12} - \mu_{13}(\omega)\left[\underline{\underline{F}}(\omega)\right]_{22} \\ -\epsilon_{22}(\omega) - \epsilon_{23}(\omega)\left[\underline{\underline{F}}(\omega)\right]_{12} - \xi_{23}(\omega)\left[\underline{\underline{F}}(\omega)\right]_{22} \\ \epsilon_{12}(\omega) + \epsilon_{13}(\omega)\left[\underline{\underline{F}}(\omega)\right]_{12} + \xi_{13}(\omega)\left[\underline{\underline{F}}(\omega)\right]_{22} \end{pmatrix}$$

$$\underline{R}_3(\omega) = \begin{pmatrix} \mu_{21}(\omega) + \zeta_{23}(\omega)\left[\underline{\underline{F}}(\omega)\right]_{13} + \mu_{23}(\omega)\left[\underline{\underline{F}}(\omega)\right]_{23} \\ -\mu_{11}(\omega) - \zeta_{13}(\omega)\left[\underline{\underline{F}}(\omega)\right]_{13} - \mu_{13}(\omega)\left[\underline{\underline{F}}(\omega)\right]_{23} \\ -\xi_{21}(\omega) - \epsilon_{23}(\omega)\left[\underline{\underline{F}}(\omega)\right]_{13} - \xi_{23}(\omega)\left[\underline{\underline{F}}(\omega)\right]_{23} \\ \xi_{11}(\omega) + \epsilon_{13}(\omega)\left[\underline{\underline{F}}(\omega)\right]_{13} + \xi_{13}(\omega)\left[\underline{\underline{F}}(\omega)\right]_{23} \end{pmatrix}$$

$$\underline{R}_4(\omega) = \begin{pmatrix} \mu_{22}(\omega) + \zeta_{23}(\omega)\left[\underline{\underline{F}}(\omega)\right]_{14} + \mu_{23}(\omega)\left[\underline{\underline{F}}(\omega)\right]_{24} \\ -\mu_{12}(\omega) - \zeta_{13}(\omega)\left[\underline{\underline{F}}(\omega)\right]_{14} - \mu_{13}(\omega)\left[\underline{\underline{F}}(\omega)\right]_{24} \\ -\xi_{22}(\omega) - \epsilon_{23}(\omega)\left[\underline{\underline{F}}(\omega)\right]_{14} - \xi_{23}(\omega)\left[\underline{\underline{F}}(\omega)\right]_{24} \\ \xi_{12}(\omega) + \epsilon_{13}(\omega)\left[\underline{\underline{F}}(\omega)\right]_{14} + \xi_{13}(\omega)\left[\underline{\underline{F}}(\omega)\right]_{24} \end{pmatrix}$$

$$\left. \right\} , \tag{4.111}$$

with $\left[\underline{\underline{F}}(\omega)\right]_{mn}$ denoting the entry in the mth row and nth column for the
matrix $\underline{\underline{F}}(\omega)$; and the column 4–vector

$$\underline{\breve{F}}_0(\omega) = \begin{bmatrix} \breve{E}_{0x}(\omega) \\ \breve{E}_{0y}(\omega) \\ \breve{H}_{0x}(\omega) \\ \breve{H}_{0y}(\omega) \end{bmatrix}. \tag{4.112}$$

Voigt wave solutions can arise whenever the algebraic multiplicity a_k
of an eigenvalue of $\underline{\underline{R}}(\omega)$ exceeds its geometric multiplicity g_k [77]. Since,
in general, $\underline{\underline{R}}(\omega)$ can have four distinct eigenvalues, there is the possibility
of more general Voigt wave solutions than represented in Eq. (4.105). For
example, if $\underline{\underline{R}}(\omega)$ had an eigenvalue with $a_k = 3$ or 4 and $g_k = 1$, then the
corresponding planewave solution would contain terms proportional to the
second or third power of the propagation distance. However, it is mathematically cumbersome to establish explicit conditions for the emergence of
such Voigt waves in terms of the components of $\underline{\underline{K}}_{EH}(\omega)$.

4.8.4 *Negative reflection*

As set out in Sec. 4.3, there are parallels between negative reflection and
negative refraction. Both of these phenomenons involve the emergence of a
wavevector on the opposite side of the normal vector to a planar boundary,
as compared to conventional descriptions of reflection and refraction (i.e.,
as compared to descriptions of positive reflection and positive refraction).
Furthermore, from a technological viewpoint, it is highly significant that
both phenomenons may be exploited to construct planar lenses [78].

Negative reflection can arise in certain isotropic chiral mediums [78, 79],
anisotropic mediums [80] and bianisotropic mediums [81]. In a manner akin
to NPV propagation, negative reflection occurs in isotropic chiral mediums
[78, 79] and Faraday chiral mediums [81] provided that the magnetoelectric
constitutive parameters are sufficiently large. Furthermore, from numerical
studies involving a Faraday chiral medium with sufficiently large magnetoelectric coupling, we see that a PPV plane wave incident on a boundary
gives rise to a negatively reflected plane wave with NPV and a positively
reflected plane wave with PPV. Conversely, when the incident plane wave
has NPV, a positively reflected plane wave with NPV and a negatively
reflected plane wave with PPV emerge [81].

4.8.5 *Counterposition of wavevector and time–averaged Poynting vector*

When a uniform plane wave is launched from a planar boundary — either by refraction or reflection, for example — into a homogeneous medium, it is possible for the wave vector $\underline{k}$ and the time–averaged Poynting vector for monochromatic plane waves $\langle\,\underline{\breve{S}}(\underline{r},\omega)\rangle_t$ to be counterposed; i.e.,

$$\left.\begin{array}{l}\dfrac{\underline{k}}{|\underline{k}|}\times\hat{\underline{n}}=-\dfrac{\langle\,\underline{\breve{S}}(\underline{r},\omega)\rangle_t}{\left|\langle\,\underline{\breve{S}}(\underline{r},\omega)\rangle_t\right|}\times\hat{\underline{n}}\ \text{for}\quad \underline{k}\bullet\hat{\underline{n}}>0\\[1.5em] \dfrac{\underline{k}}{|\underline{k}|}\times\hat{\underline{n}}=\dfrac{\langle\,\underline{\breve{S}}(\underline{r},\omega)\rangle_t}{\left|\langle\,\underline{\breve{S}}(\underline{r},\omega)\rangle_t\right|}\times\hat{\underline{n}}\quad\text{for}\quad \underline{k}\bullet\hat{\underline{n}}<0\end{array}\right\},\qquad(4.113)$$

where $\hat{\underline{n}}$ is the unit vector normal to the boundary, and directed into the medium. See Fig. 4.3 for a schematic representation.

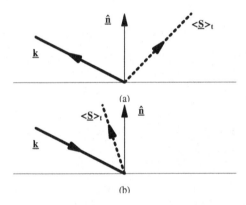

Figure 4.3   Schematic representation of counterposition of the wavevector and the time–averaged Poynting vector for uniform plane waves. In (a), the wavevector $\underline{k}$ is directed away from the boundary; in (b), $\underline{k}$ is directed towards the boundary. In both (a) and (b), the plane wave propagates away from the boundary, since $\langle\,\underline{\breve{S}}(\underline{r},\omega)\rangle_t\bullet\hat{\underline{n}}(\underline{r})>0$.

It is clear from Eq. (4.26) that the counterposition conditions (4.113) cannot be satisfied for propagation in an isotropic dielectric–magnetic medium (at rest). However, if the wavevector and time–averaged Poynting vector are not constrained to be parallel or anti–parallel, then the possibility of counterposition arises. For certain constitutive parameter regimes, this counterposition phenomenon occurs in anisotropic dielectric–magnetic mediums [82], as has been confirmed experimentally [83]. Also, counterposition can be induced by uniform motion in a medium which does not support counterposition at rest. The counterposition induced by relative motion occurs whether the medium supports negative or positive refraction when it is at rest [27, 84]. Thus, counterposition — like NPV and negative refraction — is not Lorentz covariant [27, 45].

The wavevector/time–averaged Poynting vector counterposition, as described by the conditions (4.113), is quite distinct from negative refraction. The two phenomenons have been confused in the literature [85, 86], with counterposition sometimes being referred to as amphoteric refraction [83]. Refraction — whether negative or positive — concerns the orientation of the refracted wavevector relative to the normal to the interface, as described in Sec. 4.3. The orientation of the refracted time–averaged Poynting vector is irrelevant to whether or not the refraction is negative or positive.

In the case of nonuniform planewave propagation, counterposition — as specified by the conditions (4.113) with $\underline{k}$ replaced by Re $\{\underline{k}\}$ therein — can arise even in isotropic dielectric mediums (at rest) [26, 27].

References

[1] M. Born and E. Wolf, *Principles of optics, 7th (expanded) ed*, Cambridge University Press, Cambridge, UK, 1999.

[2] J.F. Nye, *Physical properties of crystals*, Oxford University Press, Oxford, UK, 1985.

[3] J.A. Kong, *Electromagnetic wave theory*, Wiley, New York, NY, USA, 1986.

[4] H.C. Chen, *Theory of electromagnetic waves*, McGraw–Hill, New York, NY, USA, 1983.

[5] A. Lakhtakia, *Beltrami fields in chiral media*, World Scientific, Singapore, 1994.

[6] A. Lakhtakia, V.K. Varadan and V.V. Varadan, Plane waves and canonical sources in a gyroelectromagnetic uniaxial medium, *Int J Electron* **71** (1991), 853–861.

[7] B. Michel, A Fourier space approach to the pointwise singularity of an anisotropic dielectric medium, *Int J Appl Electromagn Mech* **8** (1997), 219–227.

[8] N.–H. Shen, Q. Win, J. Chen, Y.–X. Fan, J. Ding, H.–T. Wang, Y. Tian and N.–B. Ming, Optically uniaxial left–handed materials, *Phys Rev B* **72** (2005), 153104.

[9] A. Lakhtakia and W.S. Weiglhofer, Electromagnetic waves in a material with simultaneous mirror–conjugated and racemic chirality characteristics, *Electromagnetics* **20** (2000), 481–488. (The first negative sign on the right side of Eq. (26) of this paper should be replaced by a positive sign).

[10] T.G. Mackay and A. Lakhtakia, Negative phase velocity in a material with simultaneous mirror–conjugated and racemic chirality characteristics, *New J Phys* **7** (2005), 165.

[11] T.G. Mackay and A. Lakhtakia, Plane wave with negative phase velocity in Faraday chiral mediums, *Phys Rev E* **69** (2004), 026602.

[12] M. Abramowitz and I.A. Stegun (eds), *Handbook of mathematical functions*, Dover, New York, NY, USA, 1965.

[13] O. Bjørgum and T. Godal, On Beltrami vector fields and flows ($\nabla \times \mathbf{v} = \Omega \mathbf{v}$). Part II. The case when Ω is constant in space, *Naturvitenskapelig rekke no. 13*, Universitet i Bergen, Bergen, Norway, 1952.

[14] A. Lakhtakia, Beltrami field phasors are eigenvectors of 6×6 linear constitutive dyadics, *Microw Opt Technol Lett* **30** (2001), 127–128.

[15] A. Lakhtakia, Conditions for circularly polarized plane wave propagation in a linear bianisotropic medium, *Electromagnetics* **22** (2002), 123–127.

[16] C.W. Oseen, The theory of liquid crystals, *J Chem Soc Faraday Trans II* **29** (1933), 883–899.

[17] A. Lakhtakia and R. Messier, *Sculptured thin films: Nanoengineered morphology and optics*, SPIE Press, Bellingham, WA, USA, 2005.

[18] A. Lakhtakia and W.S. Weiglhofer, On light propagation in helicoidal bianisotropic mediums, *Proc R Soc Lond A* **448** (1995), 419–437. Corrections: **454** (1998), 3275.

[19] A. Lakhtakia and W.S. Weiglhofer, Further results on light propagation in helicoidal bianisotropic mediums: oblique propagation, *Proc R Soc Lond A* **453** (1997), 93–105. Corrections: **454** (1998), 3275.

[20] T.G. Mackay, A. Lakhtakia and S. Setiawan, Global and local perspectives of gravitationally assisted negative–phase–velocity propagation of electromagnetic waves in vacuum, *Phys Lett A* **336** (2005), 89–96.

[21] J.D. Hoffman, *Numerical methods for engineers and scientists*, McGraw–Hill, New York, USA, 1992.

[22] A. Lakhtakia and T.G. Mackay, Towards gravitationally assisted negative refraction of light by vacuum, *J Phys A: Math Gen* **37** (2004), L505–L510. Corrections: **37** (2004), 12093.

[23] B.F. Schutz, *A first course in general relativity*, Cambridge University Press, Cambridge, UK, 1985.

[24] R.M. Walser, Metamaterials: an introduction, in *Introduction to complex mediums for optics and electromagnetics* (W.S. Weiglhofer and A. Lakhtakia, eds), SPIE Press, Bellingham, WA, USA, 2003, 295–316.

[25] A. Lakhtakia and T.G. Mackay, Infinite phase velocity as the boundary between positive and negative phase velocities, *Microw Opt Technol Lett* **41** (2004), 165–166.

[26] T.G. Mackay and A. Lakhtakia, Negative refraction, negative phase velocity and counterposition in bianisotropic materials and metamaterials, *Phys Rev B* **79** (2009), 235121.

[27] T.G. Mackay and A. Lakhtakia, Counterposition and negative phase velocity in uniformly moving dissipative materials, *J Phys A: Math Theor* **42** (2009), 415401.

[28] A. Lakhtakia, M.W. McCall and W.S. Weiglhofer, Negative phase–velocity mediums, *Introduction to complex mediums for optics and electromagnetics*

(W.S. Weiglhofer and A. Lakhtakia, eds), SPIE Press, Bellingham, WA, USA, 2003, 347–363.

[29] S.A. Ramakrishna, Physics of negative refractive index materials, *Rep Prog Phys* **68** (2005), 449–521.

[30] R.A. Shelby, D.R. Smith and S. Schultz, Experimental verification of a negative index of refraction, *Science* **292** (2001), 77–79.

[31] G. Dolling, M. Wegener, C.M. Soukoulis and S. Linden, Negative–index metamaterial at 780 nm wavelength, *Opt Lett* **32** (2007), 53–55.

[32] V.M. Shalaev, Optical negative–index metamaterials, *Nature Photon* **1** (2007), 41–48.

[33] J.B. Pendry, Negative refraction, *Contemp Phys* **45** (2004), 191–202.

[34] J. Yao, Z. Liu, Y. Liu, Y. Wang, C. Sun, G. Bartal, A.M. Stacy and X. Zhang, Optical negative refraction in bulk metamaterials of nanowires, *Science* **321** (2008), 930.

[35] Y.-J. Jen, A. Lakhtakia, C.-W. Yu and C.-T. Lin, Vapor–deposited thin films with negative real refractive index in the visible regime, *Opt Express* **17** (2009), 7784–7789.

[36] M.A. Noginov, Compensation of surface plasmon loss by gain in dielectric medium, *J Nanophoton* **2** (2008), 021855.

[37] A. Bratkovsky, E. Ponizovskaya, S.-Y. Wang, P. Holmström, L. Thylén, Y. Fu and H. Ågren, A metal–wire/quantum–dot composite metamaterial with negative ε and compensated optical loss, *Appl Phys Lett* **93** (2008), 193106.

[38] A. Lakhtakia and T.G. Mackay, Negative phase velocity in isotropic dielectric–magnetic media via homogenization: Part II, *Microw Opt Technol Lett* **48** (2006), 709–712.

[39] T.G. Mackay and A. Lakhtakia, Correlation length and negative phase velocity in isotropic dielectric–magnetic materials, *J Appl Phys* **100** (2006), 063533.

[40] L. Hu and Z. Lin, Imaging properties of uniaxially anisotropic negative refractive index materials, *Phys Lett A* **313** (2003), 316–324.

[41] T.G. Mackay, Plane waves with negative phase velocity in isotropic chiral mediums, *Microw Opt Technol Lett* **45** (2005), 120–121, 2005. Corrections: **47** (2005), 406.

[42] T.G. Mackay and A. Lakhtakia, Simultaneous negative– and positive–phase-velocity propagation in an isotropic chiral medium, *Microw Opt Technol Lett* **49** (2007), 1245–1246.

[43] A. Pimenov, A. Loidl, K. Gehrke, V. Moshnyaga and K. Samwer, Negative refraction observed in a metallic ferromagnet in the gigahertz frequency range, *Phys Rev Lett* **98** (2007), 197401.

[44] D.G. Stavenga, Invertebrate superposition eyes — structures that behave like metamaterial with negative refractive index, *J Eur Opt Soc – Rapid Pub* **1** (2006), 06010.

[45] T.G. Mackay and A. Lakhtakia, Negative and positive refraction are not Lorentz covariant, http://arxiv.org/abs/0907.5278

[46] T.G. Mackay and A. Lakhtakia, Negative phase velocity in a uniformly moving, homogeneous, isotropic, dielectric–magnetic medium, *J Phys A: Math Gen* **37** (2004), 5697–5711.

[47] T.G. Mackay and A. Lakhtakia, Concealment by uniform motion, *J Eur Opt Soc – Rapid Pub* **2** (2007), 07003.

[48] M.W. McCall, A covariant theory of negative phase velocity propagation, *Metamaterials* **2** (2008), 92–100.

[49] T.G. Mackay, A. Lakhtakia and S. Setiawan, Electromagnetic negative-phase–velocity propagation in the ergosphere of a rotating black hole, *New J Phys* **7** (2005), 171.

[50] T.G. Mackay, A. Lakhtakia and S. Setiawan, Electromagnetic waves with negative phase velocity in Schwarzschild–de Sitter spacetime, *Europhys Lett* **71** (2005), 925–931.

[51] G.V. Skrotskii, The influence of gravitation on the propagation of light, *Soviet Phys–Dokl* **2** (1957), 226–229.

[52] J. Plébanski, Electromagnetic waves in gravitational fields, *Phys Rev* **118** (1960), 1396–1408.

[53] L.D. Landau and E.M. Lifshitz, *The classical theory of fields*, Claredon Press, Oxford, UK, 1975, §90.

[54] W. Schleich and M.O. Scully, General relativity and modern optics, *New trends in atomic physics* (G. Grynberg and R Stora, eds), Elsevier, Amsterdam, Holland, 1984, 995–1124.

[55] A. Lakhtakia, M.W. McCall and W.S. Weiglhofer, Brief overview of recent developments on negative phase-velocity mediums (alias left-handed materials), *Arch Elektron Übertrag* **56** (2002), 407–410.

[56] M.W. McCall, A. Lakhtakia and W.S. Weiglhofer, The negative index of refraction demystified, *Eur J Phys* **23** (2002), 353–359.

[57] P.A. Belov, Backward waves and negative refraction in uniaxial dielectrics with negative dielectric permittivity along the anisotropy axis, *Microw Opt Technol Lett* **37** (2003), 259–263.

[58] A. Lakhtakia, T.G. Mackay and J.B. Geddes III, On the inapplicability of a negative–phase–velocity condition as a negative–refraction condition for active materials, *Microw Opt Technol Lett* **51** (2009), 1230.

[59] T.A. Whitelaw, *Introduction to linear algebra, 2nd ed*, Blackie, Glasgow, UK, 1991.

[60] R.A. Depine and A. Lakhtakia, Diffraction by a grating made of a uniaxial dielectric–magnetic medium exhibiting negative refraction, *New J Physics* **7** 2005, 158.

[61] T.G. Mackay, A. Lakhtakia and R.A. Depine, Uniaxial dielectric media with hyperbolic dispersion relations, *Microw Opt Technol Lett* **48** (2006), 363–367.

[62] G.X. Li, H.L. Tam, F.Y. Wang and K.W. Cheah, Superlens from complementary anisotropic metamaterials, *J Appl Phys* **102** (2007), 116101.

[63] D.R. Smith and D. Schurig, Electromagnetic wave propagation in media with indefinite permittivity and permeability tensors, *Phys Rev Lett* **90** (2003), 077405.

122   *Electromagnetic Anisotropy and Bianisotropy: A Field Guide*

[64] D.R. Smith, P. Kolinko and D. Schurig, Negative refraction in indefinite media, *J Opt Soc Am B* **21** (2004), 1032–1043.
[65] Z. Ma, P. Wang, Y. Cao, H. Tang and H. Ming, Linear polarizer made of indefinite media, *Appl Phys B* **84** (2006), 261–264.
[66] G.–d. Xu, T. Pan, T.–c. Zang and J. Sun, characteristics of guided waves in indefinite–medium waveguides, *Opt Commun* **281** (2008), 2819–2825.
[67] Y. Xiang, X. Dai and S. Wen, Total reflection of electromagnetic waves propagating from an isotropic medium to an indefinite metamaterial, *Opt Commun* **274** (2007), 248–253.
[68] N.–H. Shen, Q. Wang, J. Chen, Y.–X. Fan, J. Ding and H.–T. Wang, Total transmission of electromagnetic waves at interfaces asociated with an indefinite medium, *J Opt Soc Am B* **23** (2006), 904–912.
[69] W. Yan, L. Shen and L. Ran, Surface modes at the interfaces between isotropic media and indefinite media, *J Opt Soc Am A* **24** (2007), 530–535.
[70] W. Voigt, On the behaviour of pleochroitic crystals along directions in the neighbourhood of an optic axis, *Phil Mag* **4** (1902), 90–97.
[71] S. Pancharatnum, Light propagation in absorbing crystals possessing optical activity — Electromagnetic theory, *Proc Ind Acad Sci A* **48** (1958), 227–244.
[72] T.G. Mackay and A. Lakhtakia, Voigt wave propagation in biaxial composite materials, *J Opt A: Pure Appl Opt* **5** (2003), 91–95.
[73] A. Lakhtakia, Anomalous axial propagation in helicoidal bianisotropic media, *Opt Commun* **157** (1998), 193–201.
[74] J. Gerardin and A. Lakhtakia, Conditions for Voigt wave propagation in linear, homogeneous, dielectric mediums, *Optik* **112** (2001), 493–495.
[75] M.V. Berry, The optical singularities of bianisotropic crystals, *Proc R Soc Lond A* **461** (2005), 2071–2098.
[76] V.M. Agranovich and V.L. Ginzburg, *Crystal optics with spatial dispersion, and excitons*, Springer, Berlin, Germany, 1984.
[77] W.E. Boyce and R.C. DiPrima, *Elementary differential equations and boundary value problems, 6th ed*, Wiley, New York, NY, USA, 1997.
[78] C. Zhang and T.J. Cui, Negative reflections of electromagnetic waves in strong chiral medium, *Appl Phys Lett* **91** (2007), 194101.
[79] M. Faryad and Q.A. Naqvi, Cylindrical reflector in chiral medium supporting simultaneously positive phase velocity and negative phase velocity, *J Electromag Waves Applics* **22** (2008), 563–572.
[80] Y. Wang, X. Zha and J. Yan, Reflection and refraction of light at the interface of a uniaxial bicrystal, *Europhys Lett* **72** (2005), 830–836.
[81] T.G. Mackay and A. Lakhtakia, Negative reflection in a Faraday chiral medium, *Microw Opt Technol Lett* **50** (2008), 1368–1371.
[82] A. Lakhtakia and M.W. McCall, Counterposed phase velocity and energy-transport velocity vectors in a dielectric–magnetic uniaxial medium, *Optik* **115** (2004), 28–30.
[83] Y. Zhang, B. Fluegel and A. Mascarenhas, Total negative refraction in real crystals for ballistic electrons and light, *Phys Rev Lett* **91** (2003), 157404.

[84] T.G. Mackay and A. Lakhtakia, Counterposition and negative refraction due to uniform motion, *Microw Opt Technol Lett* **49** (2007), 874–876.

[85] A.J. Hoffman, L. Alekseyev, S.S. Howard, K.J. Franz, D. Wasserman, V.A. Podolskiy, E.E. Narimanov, D.L. Sivco and C. Gmachl, Negative refraction in semiconductor metamaterials, *Nature Mater* **6** (2007), 946–950.

[86] T.M. Grzegorczyk and J.A. Kong, Electrodynamics of moving media inducing positive and negative refraction, *Phys Rev B* **74** (2006), 033102.

Chapter 5

Dyadic Green Functions

Often in practical situations one wishes to find the (frequency–domain) electromagnetic field generated by a given distribution of sources immersed a specific linear, homogeneous medium. By virtue of the linearity of the Maxwell postulates, this fundamental issue may be tackled by means of dyadic Green functions (DGFs) [1, 2]. However, the explicit delineation of DGFs for complex mediums poses formidable challenges to theorists. Closed–form representations of DGFs are available for isotropic mediums and for some classes of relatively simple anisotropic and bianisotropic medium, but not for general anisotropic and bianisotropic mediums. Notwithstanding that, integral representations of DGFs can always be found. Furthermore, for certain applications — such as in the determination of the scattering response of electrically small particles, used in homogenization studies — suitable approximative solutions may be constructed using only the singular part of the DGF, which gives rise to the depolarization dyadic. A survey of DGFs and depolarization dyadics for unbounded anisotropic and bianisotropic mediums is presented in this chapter. All of the mediums considered are homogeneous, with the exception of HBMs in Sec. 5.4.2.

5.1 Definition and properties

As defined in Eq. (1.43), $\mathbf{Q}(\underline{r}, \omega)$ denotes the source 6–vector immersed in the homogeneous bianisotropic medium characterized by the Tellegen 6×6 constitutive dyadic $\underline{\underline{\mathbf{K}}}_{\mathrm{EH}}(\omega)$. The relationship between $\mathbf{Q}(\underline{r}, \omega)$ and $\mathbf{F}(\underline{r}, \omega)$ is dictated by the frequency–domain Maxwell curl postulates (1.42). The linear nature of the differential Eq. (1.42) implies that its solution may be expressed in terms of a 6×6 DGF $\underline{\underline{\mathbf{G}}}(\underline{r}, \underline{r}', \omega)$. Thus, the field at position $\underline{r}$

is given as

$$\underline{\mathbf{F}}(\underline{r},\omega) = \underline{\mathbf{F}}^{\text{cf}}(\underline{r},\omega) + \int_{V'} \underline{\underline{\mathbf{G}}}(\underline{r},\underline{r}',\omega) \cdot \underline{\mathbf{Q}}(\underline{r}',\omega)\, d^3\underline{r}', \qquad (5.1)$$

wherein the source points $\underline{r}'$ are confined to the region of integration V'. The 6–vector $\underline{\mathbf{F}}^{\text{cf}}(\underline{r},\omega)$ denotes the corresponding complementary function satisfying

$$\left[\underline{\underline{\mathbf{L}}}(\nabla) + i\omega\underline{\underline{\mathbf{K}}}_{\text{EH}}(\omega)\right] \cdot \underline{\mathbf{F}}^{\text{cf}}(\underline{r},\omega) = \underline{\mathbf{0}} \qquad (5.2)$$

everywhere.

By construction, the DGF is the solution of the differential equation

$$\left[\underline{\underline{\mathbf{L}}}(\nabla) + i\omega\underline{\underline{\mathbf{K}}}_{\text{EH}}(\omega)\right] \cdot \underline{\underline{\mathbf{G}}}(\underline{r},\underline{r}',\omega) = \delta(\underline{r} - \underline{r}')\,\underline{\underline{\mathbf{I}}}, \qquad (5.3)$$

where

$$\delta(\underline{r}) = \frac{1}{2\pi^3} \int_{-\infty}^{\infty} \exp\left(i\underline{q} \cdot \underline{r}\right)\, d^3\underline{q} \qquad (5.4)$$

is the Dirac delta function. Informally, the DGF may be viewed as representing the 'response' of the medium at $\underline{r}$ to a point 'source' located at $\underline{r}'$. Let us note that the DGF is required to satisfy the Sommerfeld radiation condition [3].

At this point, we look ahead to anticipate the spectral representation of the DGF, which follows from taking the spatial Fourier transform of both sides of Eq. (5.3), as presented in Sec. 5.4.1. It follows therefrom that the DGF is translationally invariant. Consequently, in the following we write $\underline{\underline{\mathbf{G}}}(\underline{r} - \underline{r}',\omega)$ in lieu of $\underline{\underline{\mathbf{G}}}(\underline{r},\underline{r}',\omega)$.

The 6×6 DGF $\underline{\underline{\mathbf{G}}}(\underline{r} - \underline{r}',\omega)$ may be expressed in terms of its four component 3×3 DGFs per

$$\underline{\underline{\mathbf{G}}}(\underline{r} - \underline{r}',\omega) = \begin{bmatrix} \underline{\underline{\mathbf{G}}}^{\text{ee}}(\underline{r} - \underline{r}',\omega) & \underline{\underline{\mathbf{G}}}^{\text{em}}(\underline{r} - \underline{r}',\omega) \\ \underline{\underline{\mathbf{G}}}^{\text{me}}(\underline{r} - \underline{r}',\omega) & \underline{\underline{\mathbf{G}}}^{\text{mm}}(\underline{r} - \underline{r}',\omega) \end{bmatrix}. \qquad (5.5)$$

These 3×3 DGFs satisfy the differential equations [2]

$$\left.\begin{aligned} \underline{\underline{L}}_{\text{e}}(\nabla,\omega) \cdot \underline{\underline{G}}^{\text{ee}}(\underline{r} - \underline{r}',\omega) &= i\omega\underline{\underline{I}}\,\delta\left(\underline{r} - \underline{r}'\right) \\ \underline{\underline{L}}_{\text{e}}(\nabla,\omega) \cdot \underline{\underline{G}}^{\text{em}}(\underline{r} - \underline{r}',\omega) &= -\left[\nabla \times \underline{\underline{I}} + i\omega\underline{\underline{\xi}}_{\text{EH}}(\omega)\right] \cdot \underline{\underline{\mu}}_{\text{EH}}^{-1}(\omega)\,\delta\left(\underline{r} - \underline{r}'\right) \\ \underline{\underline{L}}_{\text{m}}(\nabla,\omega) \cdot \underline{\underline{G}}^{\text{me}}(\underline{r} - \underline{r}',\omega) &= \left[\nabla \times \underline{\underline{I}} - i\omega\underline{\underline{\zeta}}_{\text{EH}}(\omega)\right] \cdot \underline{\underline{\varepsilon}}_{\text{EH}}^{-1}(\omega)\,\delta\left(\underline{r} - \underline{r}'\right) \\ \underline{\underline{L}}_{\text{m}}(\nabla,\omega) \cdot \underline{\underline{G}}^{\text{mm}}(\underline{r} - \underline{r}',\omega) &= i\omega\underline{\underline{I}}\,\delta\left(\underline{r} - \underline{r}'\right) \end{aligned}\right\},$$

$$(5.6)$$

with the linear differential operators

$$
\left.
\begin{aligned}
\underline{\underline{L}}_{\mathrm{e}}(\nabla,\omega) &= \left[\nabla \times \underline{\underline{I}} + i\omega\,\underline{\underline{\xi}}_{\mathrm{EH}}(\omega)\right] \bullet \underline{\underline{\mu}}_{\mathrm{EH}}^{-1}(\omega) \bullet \left[\nabla \times \underline{\underline{I}} - i\omega\,\underline{\underline{\zeta}}_{\mathrm{EH}}(\omega)\right] \\
&\quad - \omega^2\,\underline{\underline{\epsilon}}_{\mathrm{EH}}(\omega) \\
\underline{\underline{L}}_{\mathrm{m}}(\nabla,\omega) &= \left[\nabla \times \underline{\underline{I}} - i\omega\,\underline{\underline{\zeta}}_{\mathrm{EH}}(\omega)\right] \bullet \underline{\underline{\epsilon}}_{\mathrm{EH}}^{-1}(\omega) \bullet \left[\nabla \times \underline{\underline{I}} + i\omega\,\underline{\underline{\xi}}_{\mathrm{EH}}(\omega)\right] \\
&\quad - \omega^2\,\underline{\underline{\mu}}_{\mathrm{EH}}(\omega)
\end{aligned}
\right\} .
$$

$$(5.7)$$

Thus, if $\underline{\underline{G}}^{\mathrm{ee}}(\underline{r} - \underline{r}',\omega)$ is known, then $\underline{\underline{G}}^{\mathrm{me}}(\underline{r} - \underline{r}',\omega)$ may be calculated directly as

$$
\underline{\underline{G}}^{\mathrm{me}}(\underline{r}-\underline{r}',\omega) = \frac{1}{i\omega}\,\underline{\underline{\mu}}_{\mathrm{EH}}^{-1}(\omega) \bullet \left[\nabla \times \underline{\underline{I}} - i\omega\,\underline{\underline{\zeta}}_{\mathrm{EH}}(\omega)\right] \bullet \underline{\underline{G}}^{\mathrm{ee}}(\underline{r}-\underline{r}',\omega), \quad (5.8)
$$

in lieu of solving the differential equation for $\underline{\underline{G}}^{\mathrm{me}}(\underline{r} - \underline{r}',\omega)$ in Eq. (5.6). Similarly, $\underline{\underline{G}}^{\mathrm{em}}(\underline{r} - \underline{r}',\omega)$ may be determined as

$$
\underline{\underline{G}}^{\mathrm{em}}(\underline{r}-\underline{r}',\omega) = -\frac{1}{i\omega}\,\underline{\underline{\epsilon}}_{\mathrm{EH}}^{-1}(\omega) \bullet \left[\nabla \times \underline{\underline{I}} + i\omega\,\underline{\underline{\xi}}_{\mathrm{EH}}(\omega)\right] \bullet \underline{\underline{G}}^{\mathrm{mm}}(\underline{r}-\underline{r}',\omega),
$$

$$(5.9)$$

provided that $\underline{\underline{G}}^{\mathrm{mm}}(\underline{r} - \underline{r}',\omega)$ is known. In addition, once one of the four 3×3 DGFs in Eq. (5.5) is known, the others can often be deduced from symmetry considerations.

5.2 Closed–form representations

Various ingenious methods, frequently relying upon manipulations of dyadic operators and/or field transformations, have been devised in the pursuit of closed–form representations for DGFs [4, 5]. Such methods have often been developed on an *ad hoc* basis. In contrast, the following dyadic scalarization and factorization technique is generally applicable, at least in principle: Formally, the 6×6 DGF may be expressed as [4]

$$
\underline{\underline{G}}(\underline{r}-\underline{r}',\omega) = \left[\underline{\underline{L}}(\nabla) + i\omega\underline{\underline{K}}(\omega)\right]^{\circledast} g(\underline{r}-\underline{r}',\omega). \quad (5.10)
$$

Herein, the scalar Green function $g(\underline{r} - \underline{r}',\omega)$ is the solution of

$$
\mathcal{H}(\nabla,\underline{r},\omega)g(\underline{r}-\underline{r}',\omega) = \delta(\underline{r}-\underline{r}'), \quad (5.11)
$$

with $\mathcal{H}(\nabla,\underline{r},\omega)$ being a scalar fourth–order differential operator, while the adjoint operation indicated by the superscript $\circledast$ is defined implicitly via

$$
\left[\underline{\underline{L}}(\nabla) + i\omega\underline{\underline{K}}(\omega)\right] \bullet \left[\underline{\underline{L}}(\nabla) + i\omega\underline{\underline{K}}(\omega)\right]^{\circledast} = \mathcal{H}(\nabla,\underline{r},\omega)\,\underline{\underline{I}}. \quad (5.12)
$$

The adjoint operator $\left[\underline{\underline{\mathbf{L}}}(\nabla) + i\omega \underline{\underline{\mathbf{K}}}(\omega)\right]^{\circledast}$ for anisotropic and bianisotropic mediums may be constructed from straightforward, but lengthy, matrix–algebraic operations. However, solutions to the fourth–order partial differential Eq. (5.11) are generally elusive.

The existence of closed–form representations for DGFs is closely allied to the factorization properties of $\mathcal{H}(\nabla, \underline{r}, \omega)$. In many instances where $\mathcal{H}(\nabla, \underline{r}, \omega)$ is expressible as a product of two second–order differential operators, closed–form representations for DGFs have been established [5, 6].

An alternative approach for the determination of DGFs is provided by spatial Fourier transformations. Upon transforming the spatial coordinates in Eq. (5.3), the differential equation is converted into a soluble algebraic equation. However, the process of inverse Fourier transformation, which is required to extract an explicit DGF representation, is generally problematic.

5.2.1 *Isotropic mediums*

5.2.1.1 *Dielectric–magnetic mediums*

The 3×3 DGF $\underline{\underline{G}}^{ee}_{iso}(\underline{r} - \underline{r}', \omega)$ for the isotropic dielectric–magnetic medium described by the Tellegen constitutive relations (2.3) emerges as a result of applying $\nabla \bullet$ from the left to both sides of Eq. (5.6)$_1$. Thereby, we find that

$$\underline{\underline{G}}^{ee}_{iso}(\underline{r} - \underline{r}', \omega) = i\omega\,\mu(\omega)\left[\underline{\underline{I}} + \frac{1}{\omega^2\,\epsilon(\omega)\,\mu(\omega)}\,\nabla\nabla\right]g_{iso}(\underline{r} - \underline{r}', \omega), \quad (5.13)$$

with the scalar Green function $g_{iso}(\underline{r} - \underline{r}', \omega)$ satisfying the scalar differential equation

$$\left[\nabla^2 + \omega^2\,\epsilon(\omega)\,\mu(\omega)\right]g_{iso}(\underline{r} - \underline{r}', \omega) = -\delta(\underline{r} - \underline{r}'). \quad (5.14)$$

The solution of Eq. (5.14) is known very well as [7]

$$g_{iso}(\underline{r} - \underline{r}', \omega) = \frac{1}{4\pi\,|\underline{r} - \underline{r}'|}\,\exp\left\{i\omega\,\left[\epsilon(\omega)\,\mu(\omega)\right]^{1/2}\,|\underline{r} - \underline{r}'|\right\}. \quad (5.15)$$

At locations outside the source region (i.e., $\underline{r} \neq \underline{r}'$), the explicit representation [7, 8]

$$\underline{\underline{G}}^{ee}_{iso}(\underline{R}, \omega) = i\omega\,\mu(\omega)\left[\left(\underline{\underline{I}} - \hat{\underline{R}}\,\hat{\underline{R}}\right)g_{iso}(\underline{R}, \omega)\right.$$

$$+ \frac{i}{\omega\,\left[\epsilon(\omega)\,\mu(\omega)\right]^{1/2}\,R}\left(\underline{\underline{I}} - 3\hat{\underline{R}}\,\hat{\underline{R}}\right)g_{iso}(\underline{R}, \omega)$$

$$\left. - \frac{1}{\omega^2\,\epsilon(\omega)\,\mu(\omega)\,R^2}\left(\underline{\underline{I}} - 3\hat{\underline{R}}\,\hat{\underline{R}}\right)g_{iso}(\underline{R}, \omega)\right] \quad (5.16)$$

follows from combining Eqs. (5.13) and (5.15), with $\underline{R} = R\,\hat{\underline{R}} = \underline{r} - \underline{r}'$. At locations inside the source region the DGF is singular, which issue is taken up in Sec. 5.5.

In light of the expression (5.13) for $\underline{\underline{G}}^{\mathrm{ee}}_{\mathrm{iso}}(\underline{r} - \underline{r}', \omega)$, the dual DGF

$$\underline{\underline{G}}^{\mathrm{mm}}_{\mathrm{iso}}(\underline{r} - \underline{r}', \omega) = \frac{\epsilon(\omega)}{\mu(\omega)}\, \underline{\underline{G}}^{\mathrm{ee}}_{\mathrm{iso}}(\underline{r} - \underline{r}', \omega) \qquad (5.17)$$

emerges immediately, while the magnetoelectric 3×3 DGFs

$$\underline{\underline{G}}^{\mathrm{em}}_{\mathrm{iso}}(\underline{r} - \underline{r}', \omega) = -\underline{\underline{G}}^{\mathrm{me}}_{\mathrm{iso}}(\underline{r} - \underline{r}', \omega) = -\nabla \times g_{\mathrm{iso}}(\underline{r} - \underline{r}', \omega)\,\underline{\underline{I}} \qquad (5.18)$$

follow from Eqs. (5.8) and (5.9). On considering the transposes of the DGFs, the following symmetries are revealed:

$$\left. \begin{aligned} \left[\underline{\underline{G}}^{\mathrm{ee}}_{\mathrm{iso}}(\underline{r} - \underline{r}', \omega)\right]^{\mathrm{T}} &= \underline{\underline{G}}^{\mathrm{ee}}_{\mathrm{iso}}(\underline{r}' - \underline{r}, \omega) \\[4pt] \left[\underline{\underline{G}}^{\mathrm{em}}_{\mathrm{iso}}(\underline{r} - \underline{r}', \omega)\right]^{\mathrm{T}} &= -\underline{\underline{G}}^{\mathrm{me}}_{\mathrm{iso}}(\underline{r}' - \underline{r}, \omega) \\[4pt] \left[\underline{\underline{G}}^{\mathrm{me}}_{\mathrm{iso}}(\underline{r} - \underline{r}', \omega)\right]^{\mathrm{T}} &= -\underline{\underline{G}}^{\mathrm{em}}_{\mathrm{iso}}(\underline{r}' - \underline{r}, \omega) \\[4pt] \left[\underline{\underline{G}}^{\mathrm{mm}}_{\mathrm{iso}}(\underline{r} - \underline{r}', \omega)\right]^{\mathrm{T}} &= \underline{\underline{G}}^{\mathrm{mm}}_{\mathrm{iso}}(\underline{r}' - \underline{r}, \omega) \end{aligned} \right\}, \qquad (5.19)$$

whereas interchanging $\underline{r}$ and $\underline{r}'$ delivers

$$\left. \begin{aligned} \underline{\underline{G}}^{\mathrm{ee}}_{\mathrm{iso}}(\underline{r} - \underline{r}', \omega) &= \underline{\underline{G}}^{\mathrm{ee}}_{\mathrm{iso}}(\underline{r}' - \underline{r}, \omega) \\[4pt] \underline{\underline{G}}^{\mathrm{em}}_{\mathrm{iso}}(\underline{r} - \underline{r}', \omega) &= -\underline{\underline{G}}^{\mathrm{em}}_{\mathrm{iso}}(\underline{r}' - \underline{r}, \omega) \\[4pt] \underline{\underline{G}}^{\mathrm{me}}_{\mathrm{iso}}(\underline{r} - \underline{r}', \omega) &= -\underline{\underline{G}}^{\mathrm{me}}_{\mathrm{iso}}(\underline{r}' - \underline{r}, \omega) \\[4pt] \underline{\underline{G}}^{\mathrm{mm}}_{\mathrm{iso}}(\underline{r} - \underline{r}', \omega) &= \underline{\underline{G}}^{\mathrm{mm}}_{\mathrm{iso}}(\underline{r}' - \underline{r}, \omega) \end{aligned} \right\}. \qquad (5.20)$$

5.2.1.2 *Isotropic chiral mediums*

Fields and sources in an isotropic chiral medium, as characterized by the Tellegen constitutive relations (2.5), are efficiently studied via the introduction of the Beltrami fields $\mathbb{Q}_1(\underline{r}, \omega)$ and $\mathbb{Q}_2(\underline{r}, \omega)$, as defined in Eqs. (4.28), and the corresponding Beltrami source current densities [9]

$$\left. \begin{aligned} \underline{\mathbb{W}}_1(\underline{r}, \omega) &= \frac{k_1}{2\omega\,[\epsilon(\omega)\,\mu(\omega)]^{1/2}}\left\{ i\left[\frac{\mu(\omega)}{\epsilon(\omega)}\right]^{1/2}\underline{J}_{\mathrm{e}}(\underline{r}, \omega) - \underline{J}_{\mathrm{m}}(\underline{r}, \omega)\right\} \\[8pt] \underline{\mathbb{W}}_2(\underline{r}, \omega) &= \frac{k_2}{2\omega\,[\epsilon(\omega)\,\mu(\omega)]^{1/2}}\left\{ \underline{J}_{\mathrm{e}}(\underline{r}, \omega) - i\left[\frac{\epsilon(\omega)}{\mu(\omega)}\right]^{1/2}\underline{J}_{\mathrm{m}}(\underline{r}, \omega)\right\} \end{aligned} \right\}, $$

$$(5.21)$$

where the wavenumbers k_1 and k_2 are defined in Eqs. (4.27). The Maxwell curl postulates may then be set down as

$$\nabla \times \underline{Q}_\ell(\underline{r}, \omega) + (-1)^\ell \, k_\ell \, \underline{Q}_\ell(\underline{r}, \omega) = \underline{W}_\ell(\underline{r}, \omega), \qquad (\ell = 1, 2). \qquad (5.22)$$

The solution of Eq. (5.22) is

$$\underline{Q}_\ell(\underline{r}, \omega) = \underline{Q}_\ell^{\text{cf}}(\underline{r}, \omega) + \int_{V'} \underline{\underline{G}}_\ell(\underline{r} - \underline{r}', \omega) \bullet \underline{W}_\ell(\underline{r}', \omega) \, d^3r', \qquad (\ell = 1, 2), \qquad (5.23)$$

where $\underline{Q}_\ell^{\text{cf}}(\underline{r}, \omega)$ is the complementary function satisfying the equation

$$\left[\nabla \times \underline{\underline{I}} + (-1)^\ell \, k_\ell \underline{\underline{I}} \right] \bullet \underline{Q}_\ell^{\text{cf}}(\underline{r}, \omega) = \underline{0}, \qquad (\ell = 1, 2). \qquad (5.24)$$

The dyadic Beltrami–Green functions

$$\underline{\underline{G}}_\ell(\underline{r} - \underline{r}', \omega) = \left[\nabla \times \underline{\underline{I}} - (-1)^\ell \, k_\ell \underline{\underline{I}} \right] \bullet \left(\underline{\underline{I}} + \frac{1}{k_\ell^2} \nabla \nabla \right) g_\ell(\underline{r} - \underline{r}', \omega),$$
$$(\ell = 1, 2) \qquad (5.25)$$

incorporate the scalar Green functions

$$g_\ell(\underline{r} - \underline{r}', \omega) = \frac{\exp\left(i k_\ell \, |\underline{r} - \underline{r}'| \right)}{4\pi \, |\underline{r} - \underline{r}'|}, \qquad (\ell = 1, 2), \qquad (5.26)$$

which are isomorphic to $g_{\text{iso}}(\underline{R}, \omega)$ defined in Eq. (5.15). The following symmetries may be deduced from Eq. (5.25) [9]:

$$\underline{\underline{G}}_\ell^{\text{T}}(\underline{r} - \underline{r}', \omega) = \underline{\underline{G}}_\ell(\underline{r}' - \underline{r}, \omega), \qquad (\ell = 1, 2). \qquad (5.27)$$

A similar procedure is useful for constructing DGFs for biisotropic mediums characterized by Eqs. (2.7) [10].

5.2.2 *Uniaxial dielectric–magnetic mediums*

We now consider anisotropic dielectric–magnetic mediums characterized by a single distinguished axis, oriented arbitrarily along the direction of the unit vector $\hat{\underline{u}}$. These mediums are characterized by the constitutive relations (2.14) with permittivity and permeability dyadics given by Eqs. (2.9) and (2.12), respectively.

By a process involving the diagonalization of the linear differential operator $\underline{L}_e$ (or, equivalently, $\underline{L}_m$) combined with dyadic factorizations, the 3×3 DGFs for the uniaxial dielectric–magnetic medium are established as [11]

$$
\begin{aligned}
\underline{\underline{G}}_{\text{uni}}^{\text{ee}}(\underline{r} - \underline{r}', \omega) &= i\omega\,\mu(\omega)\bigg\{ -\underline{\underline{B}}(\underline{r} - \underline{r}', \omega) \\
&\quad + \left[\epsilon_u(\omega)\,\underline{\underline{\epsilon}}_{\text{uni}}^{-1}(\omega) + \frac{\nabla\nabla}{\omega^2\epsilon(\omega)\,\mu(\omega)}\right] g_{\text{uni}}^{\epsilon}(\underline{r} - \underline{r}', \omega) \bigg\} \\
\underline{\underline{G}}_{\text{uni}}^{\text{em}}(\underline{r} - \underline{r}', \omega) &= -\epsilon(\omega)\,\underline{\underline{\epsilon}}_{\text{uni}}^{-1}(\omega) \bullet (\nabla \times \underline{\underline{I}}) \\
&\quad \bullet \left[\mu_u(\omega)\,g_{\text{uni}}^{\mu}(\underline{r} - \underline{r}', \omega)\,\underline{\underline{\mu}}_{\text{uni}}^{-1}(\omega) + \underline{\underline{B}}(\underline{r} - \underline{r}', \omega)\right] \\
\underline{\underline{G}}_{\text{uni}}^{\text{me}}(\underline{r} - \underline{r}', \omega) &= -\mu(\omega)\,\underline{\underline{\mu}}_{\text{uni}}^{-1}(\omega) \bullet (\nabla \times \underline{\underline{I}}) \\
&\quad \bullet \left[-\epsilon_u(\omega)\,g_{\text{uni}}^{\epsilon}(\underline{r} - \underline{r}', \omega)\,\underline{\underline{\epsilon}}_{\text{uni}}^{-1}(\omega) + \underline{\underline{B}}(\underline{r} - \underline{r}', \omega)\right] \\
\underline{\underline{G}}_{\text{uni}}^{\text{mm}}(\underline{r} - \underline{r}', \omega) &= i\omega\,\epsilon(\omega)\bigg\{ \underline{\underline{B}}(\underline{r} - \underline{r}', \omega) \\
&\quad + \left[\mu_u(\omega)\,\underline{\underline{\mu}}_{\text{uni}}^{-1}(\omega) + \frac{\nabla\nabla}{\omega^2\epsilon(\omega)\,\mu(\omega)}\right] g_{\text{uni}}^{\mu}(\underline{r} - \underline{r}', \omega) \bigg\}
\end{aligned}
$$
(5.28)

The scalar Green functions $g_{\text{uni}}^{\epsilon,\mu}(\underline{r} - \underline{r}', \omega)$ in Eqs. (5.28) are defined by

$$
g_{\text{uni}}^{\eta}(\underline{R}, \omega) = \frac{\exp\left\{ i\omega\,[\epsilon(\omega)\,\mu(\omega)]^{1/2} \left[\eta_u(\omega)\,\underline{R} \bullet \underline{\underline{\eta}}_{\text{uni}}^{-1}(\omega) \bullet \underline{R}\right]^{1/2} \right\}}{4\pi \left[\eta_u(\omega)\,\underline{R} \bullet \underline{\underline{\eta}}_{\text{uni}}^{-1}(\omega) \bullet \underline{R}\right]^{1/2}},
$$
$$(\eta = \epsilon, \mu), \qquad (5.29)$$

while the 3×3 dyadic $\underline{\underline{B}}(\underline{r} - \underline{r}', \omega)$ is specified as

$$
\begin{aligned}
\underline{\underline{B}}(\underline{R}, \omega) &= \frac{(\underline{R} \times \hat{\underline{u}})\,(\underline{R} \times \hat{\underline{u}})}{(\underline{R} \times \hat{\underline{u}})^2} \left[\frac{\epsilon_u(\omega)}{\epsilon(\omega)}\,g_{\text{uni}}^{\epsilon}(\underline{R}, \omega) - \frac{\mu_u(\omega)}{\mu(\omega)}\,g_{\text{uni}}^{\mu}(\underline{R}, \omega)\right] \\
&\quad + \frac{1}{i\omega\,[\epsilon(\omega)\,\mu(\omega)]^{1/2}\,(\underline{R} \times \hat{\underline{u}})^2} \left[\underline{\underline{I}} - \hat{\underline{u}}\,\hat{\underline{u}} - \frac{2\,(\underline{R} \times \hat{\underline{u}})\,(\underline{R} \times \hat{\underline{u}})}{(\underline{R} \times \hat{\underline{u}})^2}\right] \\
&\quad \times \bigg\{ g_{\text{uni}}^{\epsilon}(\underline{R}, \omega) \left[\epsilon_u(\omega)\,\underline{R} \bullet \underline{\underline{\epsilon}}_{\text{uni}}^{-1}(\omega) \bullet \underline{R}\right]^{1/2} \\
&\quad - g_{\text{uni}}^{\mu}(\underline{R}, \omega) \left[\mu_u(\omega)\,\underline{R} \bullet \underline{\underline{\mu}}_{\text{uni}}^{-1}(\omega) \bullet \underline{R}\right]^{1/2} \bigg\}.
\end{aligned}
$$
(5.30)

The DGFs $\underline{\underline{G}}_{\text{uni}}^{\alpha\beta}(\underline{r} - \underline{r}', \omega)$ ($\alpha\beta = $ ee, em, me, mm) exhibit the same symmetries as those specified in Eqs. (5.19) and (5.20) for isotropic dielectric–magnetic mediums.

The specializations appropriate to uniaxial dielectric mediums and uniaxial magnetic mediums follow straightforwardly from Eqs. (5.28), by implementing the substitutions $\underline{\underline{\mu}}_{\text{uni}}(\omega) = \mu_0 \underline{\underline{I}}$ and $\underline{\underline{\epsilon}}_{\text{uni}}(\omega) = \epsilon_0 \underline{\underline{I}}$, respectively.

5.2.3 *More complex mediums*

For gyrotropic dielectric mediums and gyrotropic magnetic mediums, i.e., those with permittivity and permeability dyadics of the form

$$\underline{\underline{\eta}}_{\text{gyro}}(\omega) = \eta(\omega)\left(\underline{\underline{I}} - \hat{\underline{u}}\,\hat{\underline{u}}\right) + i\,\eta_g(\omega)\,\hat{\underline{u}} \times \underline{\underline{I}} + \eta_u(\omega)\,\hat{\underline{u}}\,\hat{\underline{u}}, \qquad (\eta = \epsilon, \mu),$$
$$(5.31)$$

respectively, DGFs in terms of one–dimensional integrals involving cylindrical functions are available, but closed–form representations are not [12]. Also, some analytical progress towards closed–form DGFs for general uniaxial bianisotropic mediums described by the Tellegen constitutive dyadics

$$\underline{\underline{\eta}}_{\text{uni}}(\omega) = \eta(\omega)\left(\underline{\underline{I}} - \hat{\underline{u}}\,\hat{\underline{u}}\right) + \eta_u(\omega)\,\hat{\underline{u}}\,\hat{\underline{u}}, \qquad (\eta = \epsilon, \xi, \zeta, \mu), \qquad (5.32)$$

has been reported, but a satisfactory degree of conclusion has yet to be achieved [5, 13, 14].

A notable example of a bianisotropic scenario for which a closed–form DGF is available occurs within the context of gravitationally affected vacuum. Let us recall from Sec. 2.4.2 and Sec. 4.7.2 that the electromagnetic properties of vacuum in generally curved spacetime are equivalent to those of a fictitious, instantaneously responding, nonhomogeneous, bianisotropic medium, according to a noncovariant formalism. The nonhomogeneous bianistropic medium may be approximated by the homogeneous bianisotropic medium with time–domain constitutive relations (4.82), within a spacetime neighbourhood which is small compared to the curvature of spacetime.

On using the frequency–domain counterpart of the constitutive relations (4.82), the corresponding DGFs may be established by the following two-step procedure. First, by applying the field–source transformations

$$\left.\begin{array}{l}\underline{F}^{\sharp}(\underline{r}, \omega) = \underline{F}(\underline{r}, \omega)\,\exp\left[-i\omega\,(\epsilon_0\mu_0)^{1/2}\,\bar{\underline{\Gamma}}_{\text{GAV}} \bullet \underline{r}\right] \\[2mm] \underline{Q}^{\sharp}(\underline{r}, \omega) = \underline{Q}(\underline{r}, \omega)\,\exp\left[-i\omega\,(\epsilon_0\mu_0)^{1/2}\,\bar{\underline{\Gamma}}_{\text{GAV}} \bullet \underline{r}\right]\end{array}\right\}, \qquad (5.33)$$

the problem simplifies to that for an orthorhombic biaxial dielectric–magnetic medium. Second, under the affine transformations

$$
\left.
\begin{aligned}
\mathbf{F}^{\sharp\sharp}(\underline{r},\omega) &= \begin{bmatrix} \underline{\underline{\bar\gamma}}^{1/2}_{\text{GAV}} & \underline{0} \\ \underline{0} & \underline{\underline{\bar\gamma}}^{1/2}_{\text{GAV}} \end{bmatrix} \bullet \mathbf{F}^{\sharp}\left(\underline{\underline{\bar\gamma}}^{1/2}_{\text{GAV}} \bullet \underline{r},\omega \right) \\[2mm]
\mathbf{Q}^{\sharp\sharp}(\underline{r},\omega) &= \begin{bmatrix} \text{adj}\,\underline{\underline{\bar\gamma}}^{1/2}_{\text{GAV}} & \underline{0} \\ \underline{0} & \text{adj}\,\underline{\underline{\bar\gamma}}^{1/2}_{\text{GAV}} \end{bmatrix} \bullet \mathbf{Q}^{\sharp}\left(\underline{\underline{\bar\gamma}}^{1/2}_{\text{GAV}} \bullet \underline{r},\omega \right)
\end{aligned}
\right\}, \tag{5.34}
$$

wherein $\underline{\underline{\bar\gamma}}^{1/2}_{\text{GAV}} \bullet \underline{\underline{\bar\gamma}}^{1/2}_{\text{GAV}} = \underline{\underline{\bar\gamma}}_{\text{GAV}}$, the problem further simplifies to that for an isotropic dielectric–magnetic medium. Thus, the 3×3 DGFs emerge as [15]

$$
\left.
\begin{aligned}
\underline{\underline{G}}^{ee}_{\text{GAV}}(\underline{r}-\underline{r}',\omega) &= i\omega\mu_0 \left(\text{adj}\,\underline{\underline{\bar\gamma}}^{1/2}_{\text{GAV}} \right) \\
&\quad \bullet \left(\underline{\underline{I}} + \frac{1}{\omega^2\epsilon_0\mu_0\det\underline{\underline{\bar\gamma}}_{\text{GAV}}}\, \underline{\underline{\bar\gamma}}_{\text{GAV}} \bullet \nabla\nabla \right) g_{\text{GAV}}(\underline{r}-\underline{r}',\omega) \\
\underline{\underline{G}}^{em}_{\text{GAV}}(\underline{r}-\underline{r}',\omega) &= -\underline{\underline{\bar\gamma}}^{-1}_{\text{GAV}} \bullet \left[\nabla\times\underline{\underline{I}} - \omega\,(\epsilon_0\mu_0)^{1/2}\,\underline{\underline{\bar\Gamma}}_{\text{GAV}}\times\underline{\underline{I}} \right] \\
&\quad \bullet \underline{\underline{G}}^{ee}_{\text{GAV}}(\underline{r}-\underline{r}',\omega) \\
\underline{\underline{G}}^{me}_{\text{GAV}}(\underline{r}-\underline{r}',\omega) &= -\underline{\underline{G}}^{em}_{\text{GAV}}(\underline{r}-\underline{r}',\omega) \\
\underline{\underline{G}}^{mm}_{\text{GAV}}(\underline{r}-\underline{r}',\omega) &= \frac{\epsilon_0}{\mu_0}\,\underline{\underline{G}}^{ee}_{\text{GAV}}(\underline{r}-\underline{r}',\omega)
\end{aligned}
\right\},
$$
$$\tag{5.35}$$

with the scalar Green function $g_{\text{GAV}}(\underline{r}-\underline{r}',\omega)$ being given by

$$
g_{\text{GAV}}(\underline{R},\omega) = \frac{1}{4\pi\left| \underline{\underline{\bar\gamma}}^{-1/2}_{\text{GAV}} \bullet \underline{R} \right|} \exp\left\{ i\omega\,(\epsilon_0\mu_0)^{1/2}\left[\underline{\underline{\bar\Gamma}}_{\text{GAV}} \bullet \underline{R} \right.\right.
$$
$$
\left.\left. + \left(\det\underline{\underline{\bar\gamma}}_{\text{GAV}} \right)^{1/2}\left| \underline{\underline{\bar\gamma}}^{-1/2}_{\text{GAV}} \bullet \underline{R} \right| \right] \right\}. \tag{5.36}
$$

The symmetries of the DGFs $\underline{\underline{G}}^{\alpha\beta}_{\text{iso}}(\underline{r}-\underline{r}',\omega)$ ($\alpha\beta$ = ee, em, me, mm) given in Eqs. (5.19) and (5.20) are also exhibited by $\underline{\underline{G}}^{\alpha\beta}_{\text{GAV}}(\underline{r}-\underline{r}',\omega)$.

5.3 Huygens principle

The Huygens principle represents a keystone of electromagnetic scattering theory [16]. By means of this principle, the electromagnetic fields in a source–free region may be related to the DGF and the tangential field components on a closed surface enclosing the source region. The Huygens principle may be applied, for example, to the analysis of diffraction from an aperture, wherein the aperture is formally represented as an equivalent source [7]. It can also be used to formulate the Ewald–Oseen extinction theorem [17] and the T–matrix method for scattering [9].   Due to the general scarcity of DGFs in closed form, exact formulations of the Huygens principle are available only for isotropic mediums [9] and relatively simple anisotropic mediums [1, 18–20]. We note that a formulation of the Huygens principle for a generally anisotropic dielectric medium is available, but it does not yield explicit expressions as it employs a spectral representation of the dyadic Green functions [21] (see Sec. 5.4.1).

5.3.1   *Uniaxial dielectric–magnetic mediums*

The mathematical formulations of the Huygens principle are essentially the same for isotropic dielectric–magnetic, uniaxial dielectric, uniaxial magnetic and uniaxial dielectric–magnetic mediums. The differences between them are restricted to the form of the DGFs that are involved. Therefore, here we focus on the uniaxial dielectric–magnetic scenario.

Suppose that the source–free region V_{SF} is filled with a uniaxial dielectric–magnetic medium, as described by the constitutive relations (2.14) with permittivity and permeability dyadics given by Eqs. (2.9) and (2.12), respectively. The region V_{SF} is enclosed by the finite surfaces S_{in} and S_{out}, with the unit vector $\hat{\underline{n}}(\underline{r})$ on $S_{\mathrm{in}} \cup S_{\mathrm{out}}$ directed into V_{SF}, as illustrated in Fig. 5.1. The source phasor $\mathbf{Q}(\underline{r}, \omega) \neq \underline{\mathbf{0}}$ only for $\underline{r}$ inside S_{in}. Now we introduce the 3×3 dyadic function [20]

$$
\begin{aligned}
\underline{\underline{M}}(\underline{r},\underline{r}',\omega) = &-\underline{E}(\underline{r},\omega) \times \left\{ \underline{\underline{\mu}}_{\mathrm{uni}}^{-1}(\omega) \boldsymbol{\cdot} \left[\nabla \times \underline{\underline{G}}_{\mathrm{uni}}^{\mathrm{ee}}(\underline{r}-\underline{r}',\omega) \right] \right\} \\
&- \left\{ \underline{\underline{\mu}}_{\mathrm{uni}}^{-1}(\omega) \boldsymbol{\cdot} \left[\nabla \times \underline{E}(\underline{r},\omega) \right] \right\} \times \underline{\underline{G}}_{\mathrm{uni}}^{\mathrm{ee}}(\underline{r}-\underline{r}',\omega),
\end{aligned} \tag{5.37}
$$

and observe that

$$
\begin{aligned}
\nabla \boldsymbol{\cdot} \underline{\underline{M}}(\underline{r},\underline{r}',\omega) = &\ \underline{E}(\underline{r},\omega) \boldsymbol{\cdot} \left(\nabla \times \left\{ \underline{\underline{\mu}}_{\mathrm{uni}}^{-1}(\omega) \boldsymbol{\cdot} \left[\nabla \times \underline{\underline{G}}_{\mathrm{uni}}^{\mathrm{ee}}(\underline{r}-\underline{r}',\omega) \right] \right\} \right) \\
&- \left(\nabla \times \left\{ \underline{\underline{\mu}}_{\mathrm{uni}}^{-1}(\omega) \boldsymbol{\cdot} \left[\nabla \times \underline{E}(\underline{r},\omega) \right] \right\} \right) \boldsymbol{\cdot} \underline{\underline{G}}_{\mathrm{uni}}^{\mathrm{ee}}(\underline{r}-\underline{r}',\omega),
\end{aligned} \tag{5.38}
$$

by virtue of the symmetry of $\underline{\underline{\mu}}^{-1}(\omega)$. Application of the dyadic divergence theorem [1, 22] to $\nabla \bullet \underline{\underline{M}}(\underline{r},\underline{r}',\omega)$ yields

$$\int\int\int_{V_{\rm SF}} \left[\underline{E}(\underline{r},\omega) \bullet \left(\nabla \times \left\{\underline{\underline{\mu}}_{\rm uni}^{-1}(\omega) \bullet \left[\nabla \times \underline{\underline{G}}_{\rm uni}^{\rm ee}(\underline{r}-\underline{r}',\omega)\right]\right\}\right)\right.$$
$$\left. - \left(\nabla \times \left\{\underline{\underline{\mu}}_{\rm uni}^{-1}(\omega) \bullet [\nabla \times \underline{E}(\underline{r},\omega)]\right\}\right) \bullet \underline{\underline{G}}_{\rm uni}^{\rm ee}(\underline{r}-\underline{r}',\omega)\right] d^3\underline{r}$$
$$= \int\int_{S_{\rm in}\cup S_{\rm out}} \left[[\hat{n}(\underline{r}) \times \underline{E}(\underline{r},\omega)] \bullet \left\{\underline{\underline{\mu}}_{\rm uni}^{-1}(\omega) \bullet \left[\nabla \times \underline{\underline{G}}_{\rm uni}^{\rm ee}(\underline{r}-\underline{r}',\omega)\right]\right\}\right.$$
$$\left. + \left(\hat{n}(\underline{r}) \times \left\{\underline{\underline{\mu}}_{\rm uni}^{-1}(\omega) \bullet [\nabla \times \underline{E}(\underline{r},\omega)]\right\}\right) \bullet \underline{\underline{G}}_{\rm uni}^{\rm ee}(\underline{r}-\underline{r}',\omega)\right] d^2\underline{r}. \quad (5.39)$$

If $S_{\rm out}$ is taken to be be sufficiently distant from $S_{\rm in}$, then the integral on the surface $S_{\rm out}$ can be neglected owing to the satisfaction of the Sommerfeld radiation condition by the DGFs [3]. Then, by exploiting the Maxwell curl postulates (1.42), along with the relations (5.6) and (5.8) and symmetries (5.19) and (5.20) satisfied by the DGFs, the Huygens principle for the electric field phasor emerges as

$$\underline{E}(\underline{r},\omega) = \int\int_{S_{\rm in}} \left\{\underline{\underline{G}}_{\rm uni}^{\rm ee}(\underline{r}-\underline{r}',\omega) \bullet [\hat{n}(\underline{r}') \times \underline{H}(\underline{r}',\omega)]\right.$$
$$\left. - \underline{\underline{G}}_{\rm uni}^{\rm em}(\underline{r}-\underline{r}',\omega) \bullet [\hat{n}(\underline{r}') \times \underline{E}(\underline{r}',\omega)]\right\} d^2\underline{r}', \qquad \underline{r} \in V_{\rm SF}.$$
$$(5.40)$$

The corresponding expression for the magnetic field phasor follows from taking the curl of both sides of Eq. (5.40), and then exploiting the Maxwell curl postulates (1.42), along with the DGF relations (5.6), (5.8) and (5.9). Thus, we find

$$\underline{H}(\underline{r},\omega) = \int\int_{S_{\rm in}} \left\{\underline{\underline{G}}_{\rm uni}^{\rm me}(\underline{r}-\underline{r}',\omega) \bullet [\hat{n}(\underline{r}') \times \underline{H}(\underline{r}',\omega)]\right.$$
$$\left. - \underline{\underline{G}}_{\rm uni}^{\rm mm}(\underline{r}-\underline{r}',\omega) \bullet [\hat{n}(\underline{r}') \times \underline{E}(\underline{r}',\omega)]\right\} d^2\underline{r}', \qquad \underline{r} \in V_{\rm SF}, \underline{r} \notin S_{\rm in}.$$
$$(5.41)$$

Formulations of the Huygens principle for isotropic dielectric–magnetic mediums are obtained by replacing $\underline{\underline{G}}_{\rm uni}^{\alpha\beta}(\underline{r}-\underline{r}',\omega)$ in Eqs. (5.40) and (5.41) by $\underline{\underline{G}}_{\rm iso}^{\alpha\beta}(\underline{r}-\underline{r}',\omega)$ ($\alpha\beta$ = ee, em, me, mm). Similarly, the corresponding results for uniaxial dielectric and uniaxial magnetic mediums follow from Eqs. (5.40) and (5.41) by substitution of the appropriate DGFs [19]. By first utilizing the field–source transformations (5.34), the derivation presented here can be implemented to develop a formulation of the Huygens principle for a uniaxial dielectric–magnetic medium which exhibits gyrotropic–like

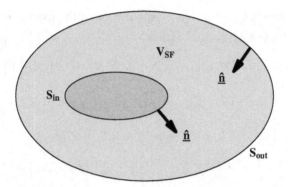

Figure 5.1   The source–free bounded region V_{SF} enclosed by the surfaces S_{in} and S_{out}. The unit vector $\hat{\underline{n}}$ on $S_{in} \cup S_{out}$ is directed into V_{SF}.

magnetoelectric properties [20]. A special case of that is the formulation of the Huygens principle for a simply moving isotropic dielectric medium, based on the Minkowski constitutive relations [1]; however, we note that these constitutive relations describe only nondispersive mediums.

5.3.2   *Isotropic chiral mediums*

In a similar vein to that outlined in Sec. 5.3.1, the Huygens principle for isotropic chiral mediums can be formulated. Keeping Fig. 5.1 in view, and using Green's first identity in lieu of the dyadic divergence theorem, we find that the Beltrami fields for $\underline{r} \in V_{SF}$ are related to those on the surface S_{in} as [9]

$$
\underline{Q}_\ell(\underline{r}, \omega) = (-1)^{\ell+1} \int\!\!\int_{S_{in}} \underline{\underline{G}}_\ell(\underline{r} - \underline{r}', \omega)
$$
$$
\bullet \left[\hat{\underline{n}}(\underline{r}') \times \underline{Q}_\ell(\underline{r}', \omega) \right] d^2\underline{r}', \quad \underline{r} \in V_{SF}, \quad (\ell = 1, 2).
$$

$$(5.42)$$

5.4   Eigenfunction representations

5.4.1   *Homogeneous mediums*

DGFs may be expressed in terms of expansions of eigenfunction solutions of Eq. (5.3), as an alternative to the closed–form representations considered in Sec. 5.2. Thereby, the *spectral representation* [24]

$$\underline{\underline{\mathbf{G}}}(\underline{R},\omega) = \left(\frac{1}{2\pi}\right)^3 \int_{\underline{q}} \underline{\underline{\check{\mathbf{G}}}}(\underline{q},\omega)\,\exp(i\,\underline{q}\bullet\underline{R})\,d^3\underline{q} \qquad (5.43)$$

arises. While closed–form representations of DGFs for anisotropic and bian-isotropic mediums are relatively scarce, the spatial Fourier transform[1] [24]

$$\underline{\underline{\check{\mathbf{G}}}}(\underline{q},\omega) = \int_{\underline{R}} \underline{\underline{\mathbf{G}}}(\underline{R},\omega)\,\exp(-i\,\underline{q}\bullet\underline{R})\,d^3\underline{R} \qquad (5.44)$$

can always be found. Thus, for the homogeneous bianisotropic medium characterized by the Tellegen 6×6 constitutive dyadic $\underline{\underline{\mathbf{K}}}_{\mathrm{EH}}(\omega)$, the spatial Fourier transform of Eq. (5.3) delivers

$$\underline{\underline{\check{\mathbf{G}}}}(\underline{q},\omega) = \frac{1}{i\omega}\frac{\mathrm{adj}\,\underline{\underline{\check{\mathbf{A}}}}(\underline{q},\omega)}{\det\,\underline{\underline{\check{\mathbf{A}}}}(\underline{q},\omega)}, \qquad (5.45)$$

where

$$\underline{\underline{\check{\mathbf{A}}}}(\underline{q},\omega) = \begin{bmatrix} \underline{\underline{0}} & (\underline{q}/\omega)\times\underline{\underline{I}} \\ -(\underline{q}/\omega)\times\underline{\underline{I}} & \underline{\underline{0}} \end{bmatrix} + \underline{\underline{\mathbf{K}}}_{\mathrm{EH}}(\omega). \qquad (5.46)$$

The analytic properties of the inverse Fourier transform

$$\underline{\underline{\mathbf{G}}}(\underline{R},\omega) = \left(\frac{1}{2\pi}\right)^3 \frac{1}{i\omega} \int_{\underline{q}} \frac{\mathrm{adj}\left[\underline{\underline{\tilde{\mathbf{A}}}}(\underline{q},\omega)\right]}{\det\left[\underline{\underline{\tilde{\mathbf{A}}}}(\underline{q},\omega)\right]} \exp(i\,\underline{q}\bullet\underline{R})\,d^3\underline{q} \qquad (5.47)$$

are established [25], but numerical techniques are generally needed to ex-plicitly evaluate the right side of Eq. (5.47).

Equation (5.43) may be regarded as a spectral representation in terms of plane waves. In a similar fashion, other sets of eigenfunctions can be used to establish representations of DGFs. Most notably, expansions in terms of cylindrical vector wavefunctions [26, 27] have been developed for certain uniaxial bianisotropic mediums [28, 29, 23, 30–32]. Also, formu-lations are available in terms of spherical harmonics [10]. However, these DGF representations are exceedingly cumbersome to handle and restricted in scope, and generally require numerical implementation.

---

[1]We follow the convention that the spatial Fourier transform (5.44) incorporates a $-i$ factor in the exponential term whereas the temporal Fourier transform (1.25) incorpo-rates a $+i$ factor in the exponential term.

5.4.2   *Nonhomogeneous mediums*

Nonhomogeneity increases the mathematical complexity, so much so that
DGFs for nonhomogeneous mediums are quite painful to even contemplate.
A noteworthy exception is the matrix Green function of a HBM, which may
be established as follows. We recall from Sec. 4.7.1 that for a HBM with
its helicoidal axis aligned with the Cartesian z axis, as characterized by
the Tellegen constitutive relations (2.64), with constitutive dyadics given
by Eqs. (2.65) and (2.66), planewave propagation in source–free regions is
governed by the 4×4 matrix differential Eq. (4.77). A modification of this
equation may be used to derive the matrix Green function for a HBM.

We consider the Oseen–transformed fields (cf. Eq. (4.74))

$$\left.\begin{array}{l} \underline{E}'(\underline{r},\omega) = \underline{\underline{S}}_z^{-1}(z) \bullet \underline{E}(\underline{r},\omega) \\ \underline{H}'(\underline{r},\omega) = \underline{\underline{S}}_z^{-1}(z) \bullet \underline{H}(\underline{r},\omega) \end{array}\right\} \tag{5.48}$$

and Oseen–transformed source terms

$$\underline{J}'_{\,\mathrm{e,m}}(\underline{r},\omega) = \underline{\underline{S}}_z^{-1}(z) \bullet \underline{J}_{\,\mathrm{e,m}}(\underline{r},\omega). \tag{5.49}$$

In a manner similar to Eq. (4.76), the Fourier representations

$$\underline{A}'(\underline{r},\omega) = \underline{A}'_0(z,\omega)\,\exp\left(i\kappa_{\mathrm{HBM}}x\right), \qquad (A=E,H,J_{\mathrm{e,m}}) \tag{5.50}$$

are introduced, where κ_{HBM} may be interpreted as a wavenumber in the xy
plane. In view of Eq. (4.77), the governing equation here may be expressed
as [33]

$$\frac{\partial}{\partial z}\underline{F}'_0(z,\omega) = \underline{\underline{M}}'(z,\omega) \bullet \underline{F}'_0(z,\omega) + \underline{Q}'_0(z,\omega), \tag{5.51}$$

wherein the 4–vector field term

$$\underline{F}'_0(z,\omega) = \begin{bmatrix} \hat{\underline{x}} \bullet \underline{E}'_0(z,\omega) \\ \hat{\underline{y}} \bullet \underline{E}'_0(z,\omega) \\ \hat{\underline{x}} \bullet \underline{H}'_0(z,\omega) \\ \hat{\underline{y}} \bullet \underline{H}'_0(z,\omega) \end{bmatrix}, \tag{5.52}$$

the 4–vector source term

$$\underline{Q}'_0(z,\omega) = \begin{bmatrix} \hat{\underline{x}} \bullet \underline{J}'_{\mathrm{e}0}(z,\omega) \\ \hat{\underline{y}} \bullet \underline{J}'_{\mathrm{e}0}(z,\omega) \\ \hat{\underline{x}} \bullet \underline{J}'_{\mathrm{m}0}(z,\omega) \\ \hat{\underline{y}} \bullet \underline{J}'_{\mathrm{m}0}(z,\omega) \end{bmatrix} \tag{5.53}$$

and the 4×4 matrix function $\underline{\underline{M}}'(z,\omega)$ has the form given in Eq. (4.79).

The solution of Eq. (5.51) is compactly stated as

$$\underline{F}'_0(z,\omega) = \underline{\underline{\breve{G}}}'_{HBM}(z,\omega) \bullet \underline{F}'_0(0,\omega) + \int_0^z \underline{\underline{\breve{G}}}'_{HBM}(z - z_s,\omega) \bullet \underline{Q}'_0(z_s,\omega)\,dz_s,$$
(5.54)

in terms of the spectral matrix Green function

$$\underline{\underline{\breve{G}}}'_{HBM}(z - z_s,\omega) = \underline{\underline{Z}}'(z,\omega) \bullet \left[\underline{\underline{Z}}'(z_s,\omega)\right]^{-1}.$$
(5.55)

The *matrizant* $\underline{\underline{Z}}'(z,\omega)$ in Eq. (5.55) is the 4×4 matrix which satisfies the differential equation [34]

$$\frac{\partial}{\partial z}\underline{\underline{Z}}'(z,\omega) = \underline{\underline{M}}'(z,\omega) \bullet \underline{\underline{Z}}'(z,\omega),$$
(5.56)

with boundary condition

$$\underline{\underline{Z}}'(0,\omega) = \underline{\underline{I}},$$
(5.57)

where $\underline{\underline{I}}$ is the 4×4 matrix identity.

5.5 Depolarization dyadics

In practical applications, it is often sufficient to construct approximate solutions of restricted validity. An important example — which is presented in detail in Chap. 6 — arises in the homogenization of particulate composite materials [35, 36], wherein the scattering response of an electrically small, homogeneous particle embedded in a homogeneous ambient medium is sought.

Let the 6×6 Tellegen constitutive dyadics for the particle and the ambient mediums be denoted by $\underline{\underline{K}}_p(\omega)$ and $\underline{\underline{K}}_{amb}(\omega)$, respectively. The corresponding Maxwell curl postulates (1.42) may then be expressed for all $\underline{r}$ as

$$\left[\underline{\underline{L}}(\nabla) + i\omega\underline{\underline{K}}_{amb}(\omega)\right] \bullet \underline{F}(\underline{r},\omega) = \underline{Q}_{equiv}(\underline{r},\omega).$$
(5.58)

Herein, the particle is represented by an equivalent source current density

$$\underline{Q}_{equiv}(\underline{r},\omega) = \begin{cases} i\omega\left[\underline{\underline{K}}_{amb}(\omega) - \underline{\underline{K}}_p(\omega)\right] \bullet \underline{F}(\underline{r},\omega), & \underline{r} \in V' \\ \underline{0}, & \underline{r} \notin V' \end{cases},$$
(5.59)

which is assumed to be uniform within the region V' occupied by the particle. As the particle is sufficiently small relative to all relevant electromagnetic wavelengths, the Rayleigh approximation may be implemented to estimate the uniform field $\underline{F}(\underline{r}, \omega)$ in V' as (cf. Eq. (5.1))

$$\underline{F}(\underline{r}, \omega) \approx \underline{F}^{\text{cf}}(\underline{r}, \omega) + \left[\int_{V'} \underline{\underline{G}}_{\text{amb}}(\underline{r} - \underline{r}', \omega) \, d^3\underline{r}' \right] \bullet \underline{Q}_{\text{equiv}}(\underline{r}, \omega),$$

$$(\underline{r} \in V'). \qquad (5.60)$$

For convenience, the particle is taken to be centred at the origin of the coordinate system. At the origin, the equivalent source is linearly mapped onto the field by the 6×6 *depolarization dyadic* $\underline{\underline{D}}_{\text{amb}}(\omega)$ of a region of the same shape, orientation and size as the particle [36]. That is,

$$\underline{F}(\underline{0}, \omega) \approx \underline{F}^{\text{cf}}(\underline{0}, \omega) + \underline{\underline{D}}_{\text{amb}}(\omega) \bullet \underline{Q}_{\text{equiv}}(\underline{0}, \omega), \qquad (5.61)$$

wherein

$$\underline{\underline{D}}_{\text{amb}}(\omega) = \int_{V'} \underline{\underline{G}}_{\text{amb}}(-\underline{r}', \omega) \, d^3\underline{r}'. \qquad (5.62)$$

Care is required in evaluating the depolarization dyadic due to the singularity of the ambient medium DGF $\underline{\underline{G}}_{\text{amb}}(\underline{r} - \underline{r}', \omega)$ within the source region V' [8].

5.5.1   *Ellipsoidal shape*

Let us consider an ellipsoidal particle, centred at the origin of the coordinate system, with surface parameterized as

$$\underline{r}_e(\theta, \phi) = \rho \, \underline{\underline{U}} \bullet \hat{\underline{r}}(\theta, \phi), \qquad (5.63)$$

where $\hat{\underline{r}}(\theta, \phi)$ is the radial unit vector specified by the spherical polar coordinates θ and ϕ, and $\rho > 0$ is a measure of the particle's linear dimensions. The shape dyadic $\underline{\underline{U}}$ is a real–valued 3×3 dyadic with positive eigenvalues and unit determinant [37], the normalized lengths of the ellipsoid semi–axes being specified by the eigenvalues of $\underline{\underline{U}}$. The particle is embedded within a homogeneous ambient medium described by the 3×3 Tellegen constitutive dyadics $\underline{\underline{\epsilon}}_{\text{amb}}(\omega)$, $\underline{\underline{\xi}}_{\text{amb}}(\omega)$, $\underline{\underline{\zeta}}_{\text{amb}}(\omega)$ and $\underline{\underline{\mu}}_{\text{amb}}(\omega)$. After using the spectral representation (5.43), the depolarization dyadic for the ellipsoidal particle given in Eq. (5.62) may be expressed as [38, 39]

$$\underline{\underline{D}}_{U/\text{amb}}(\rho, \omega) = \frac{\rho}{2\pi^2} \int_q \frac{1}{q^2} \left[\frac{\sin(q\rho)}{q\rho} - \cos(q\rho) \right] \underline{\underline{\check{G}}}_{\text{amb}}(\underline{\underline{U}}^{-1} \bullet \underline{q}, \omega) \, d^3\underline{q},$$

$$(5.64)$$

where $q^2 = \underline{q} \bullet \underline{q}$. Our notation highlights the depolarization dyadic's dependency on the shape and size of the ellipsoidal particle. Notice that while the depolarization dyadic depends upon the constitutive parameters of the ambient medium, it is independent of the constitutive parameters of the material from which the ellipsoidal particle itself is constituted.

Let us consider the depolarization dyadic as the sum [40, 41]

$$\underline{\underline{\mathbf{D}}}_{U/\mathrm{amb}}(\rho,\omega) = \underline{\underline{\mathbf{D}}}^0_{U/\mathrm{amb}}(\omega) + \underline{\underline{\mathbf{D}}}^+_{U/\mathrm{amb}}(\rho,\omega), \qquad (5.65)$$

wherein

$$\left.\begin{array}{l}\underline{\underline{\mathbf{D}}}^0_{U/\mathrm{amb}}(\omega) = \dfrac{\rho}{2\pi^2} \displaystyle\int_q \dfrac{1}{q^2}\left[\dfrac{\sin(q\rho)}{q\rho} - \cos(q\rho)\right] \underline{\underline{\tilde{\mathbf{G}}}}^{\infty}_{\mathrm{amb}}(\underline{\underline{U}}^{-1}\bullet\hat{\underline{q}},\omega)\,d^3\underline{q} \\[4mm] \underline{\underline{\mathbf{D}}}^+_{U/\mathrm{amb}}(\rho,\omega) = \dfrac{\rho}{2\pi^2} \displaystyle\int_q \dfrac{1}{q^2}\left[\dfrac{\sin(q\rho)}{q\rho} - \cos(q\rho)\right] \underline{\underline{\tilde{\mathbf{G}}}}^{+}_{\mathrm{amb}}(\underline{\underline{U}}^{-1}\bullet\underline{q},\omega)\,d^3\underline{q}\end{array}\right\},$$
$$(5.66)$$

with

$$\left.\begin{array}{l}\underline{\underline{\check{\mathbf{G}}}}^{\infty}_{\mathrm{amb}}(\hat{\underline{q}},\omega) = \lim_{q\to\infty} \underline{\underline{\check{\mathbf{G}}}}_{\mathrm{amb}}(\underline{q},\omega) \\[3mm] \underline{\underline{\check{\mathbf{G}}}}^{+}_{\mathrm{amb}}(\underline{q},\omega) = \underline{\underline{\check{\mathbf{G}}}}_{\mathrm{amb}}(\underline{q},\omega) - \underline{\underline{\check{\mathbf{G}}}}^{\infty}_{\mathrm{amb}}(\hat{\underline{q}},\omega)\end{array}\right\}, \qquad (5.67)$$

and the unit vector

$$\hat{\underline{q}} = \frac{1}{q}\underline{q} = \hat{\underline{x}}\sin\theta\cos\phi + \hat{\underline{y}}\sin\theta\sin\phi + \hat{\underline{z}}\cos\theta. \qquad (5.68)$$

For later convenience, $\underline{\underline{\mathbf{D}}}^0_{U/\mathrm{amb}}(\omega)$ and $\underline{\underline{\mathbf{D}}}^+_{U/\mathrm{amb}}(\rho,\omega)$ are written in terms of their 3×3 dyadic components as

$$\left.\begin{array}{l}\underline{\underline{\mathbf{D}}}^0_{U/\mathrm{amb}}(\omega) = \left[\begin{array}{cc} {}^{ee}\underline{\underline{D}}^0_{U/\mathrm{amb}}(\omega) & {}^{em}\underline{\underline{D}}^0_{U/\mathrm{amb}}(\omega) \\[3mm] {}^{me}\underline{\underline{D}}^0_{U/\mathrm{amb}}(\omega) & {}^{mm}\underline{\underline{D}}^0_{U/\mathrm{amb}}(\omega) \end{array}\right] \\[7mm] \underline{\underline{\mathbf{D}}}^+_{U/\mathrm{amb}}(\rho,\omega) = \left[\begin{array}{cc} {}^{ee}\underline{\underline{D}}^+_{U/\mathrm{amb}}(\rho,\omega) & {}^{em}\underline{\underline{D}}^+_{U/\mathrm{amb}}(\rho,\omega) \\[3mm] {}^{me}\underline{\underline{D}}^+_{U/\mathrm{amb}}(\rho,\omega) & {}^{mm}\underline{\underline{D}}^+_{U/\mathrm{amb}}(\rho,\omega) \end{array}\right]\end{array}\right\}. \qquad (5.69)$$

The dyadic $\underline{\underline{\mathbf{D}}}^0_{U/\mathrm{amb}}(\omega)$ represents the depolarization contribution that derives from the vanishingly small ellipsoidal region in the limit $\rho \to 0$. Its owes its existence to the singularity of the DGF of the ambient medium. In contrast, the dyadic $\underline{\underline{\mathbf{D}}}^+_{U/\mathrm{amb}}(\rho,\omega)$ provides the depolarization contribution arising from the ellipsoidal region of nonzero volume. Often in homogenization studies the contribution of $\underline{\underline{\mathbf{D}}}^+_{U/\mathrm{amb}}(\rho,\omega)$ is neglected and $\underline{\underline{\mathbf{D}}}^0_{U/\mathrm{amb}}(\omega)$ alone is taken as the depolarization dyadic [42]. However, the significance of the spatial extent of depolarization regions has been emphasized in studies of isotropic [43–47], anisotropic [40] and bianisotropic [41] HCMs.

The integral representation of $\underline{\underline{\mathbf{D}}}^0_{U/\mathrm{amb}}(\omega)$ in Eq. (5.66)$_1$ has been extensively reported upon [42]. The volume integral therein reduces to the ρ–independent surface integral [38, 39]

$$\underline{\underline{\mathbf{D}}}^0_{U/\mathrm{amb}}(\omega) = \int_{\phi=0}^{2\pi}\int_{\theta=0}^{\pi} \underline{\underline{\check{\mathbf{G}}}}^{\infty}(\underline{\underline{U}}^{-1}\bullet\hat{q},\omega)\,\sin\theta\,d\theta\,d\phi. \qquad (5.70)$$

Using Eq. (5.45) to expand the integrands in Eq. (5.70), we have

$$^{\alpha\beta}\underline{\underline{D}}^0_{U/\mathrm{amb}}(\omega) = \underline{\underline{U}}^{-1}\bullet\,^{\alpha\beta}\underline{\underline{d}}^0_{U/\mathrm{amb}}(\omega)\bullet\underline{\underline{U}}^{-1}, \qquad (\alpha\beta = \mathrm{ee},\,\mathrm{em},\,\mathrm{me},\,\mathrm{mm}), \qquad (5.71)$$

where the 3×3 dyadics

$$\left.\begin{aligned}
^{\mathrm{ee}}\underline{\underline{d}}^0_{U/\mathrm{amb}}(\omega) &= \int_{\phi=0}^{2\pi}\int_{\theta=0}^{\pi} \frac{\hat{q}\bullet\underline{\underline{U}}^{-1}\bullet\underline{\underline{\mu}}_{\mathrm{amb}}(\omega)\bullet\underline{\underline{U}}^{-1}\bullet\hat{q}}{4\pi i\omega\,b_{U/\mathrm{amb}}(\theta,\phi,\omega)}\,\hat{q}\,\hat{q}\,\sin\theta\,d\theta\,d\phi \\[6pt]
^{\mathrm{em}}\underline{\underline{d}}^0_{U/\mathrm{amb}}(\omega) &= \int_{\phi=0}^{2\pi}\int_{\theta=0}^{\pi} -\frac{\hat{q}\bullet\underline{\underline{U}}^{-1}\bullet\underline{\underline{\xi}}_{\mathrm{amb}}(\omega)\bullet\underline{\underline{U}}^{-1}\bullet\hat{q}}{4\pi i\omega\,b_{U/\mathrm{amb}}(\theta,\phi,\omega)}\,\hat{q}\,\hat{q}\,\sin\theta\,d\theta\,d\phi \\[6pt]
^{\mathrm{me}}\underline{\underline{d}}^0_{U/\mathrm{amb}}(\omega) &= \int_{\phi=0}^{2\pi}\int_{\theta=0}^{\pi} -\frac{\hat{q}\bullet\underline{\underline{U}}^{-1}\bullet\underline{\underline{\zeta}}_{\mathrm{amb}}(\omega)\bullet\underline{\underline{U}}^{-1}\bullet\hat{q}}{4\pi i\omega\,b_{U/\mathrm{amb}}(\theta,\phi,\omega)}\,\hat{q}\,\hat{q}\,\sin\theta\,d\theta\,d\phi \\[6pt]
^{\mathrm{mm}}\underline{\underline{d}}^0_{U/\mathrm{amb}}(\omega) &= \int_{\phi=0}^{2\pi}\int_{\theta=0}^{\pi} \frac{\hat{q}\bullet\underline{\underline{U}}^{-1}\bullet\underline{\underline{\epsilon}}_{\mathrm{amb}}(\omega)\bullet\underline{\underline{U}}^{-1}\bullet\hat{q}}{4\pi i\omega\,b_{U/\mathrm{amb}}(\theta,\phi,\omega)}\,\hat{q}\,\hat{q}\,\sin\theta\,d\theta\,d\phi
\end{aligned}\right\} \qquad (5.72)$$

and the scalar function

$$\begin{aligned}
b_{U/\mathrm{amb}}(\theta,\phi,\omega) = &\left\{\left[\hat{q}\bullet\underline{\underline{U}}^{-1}\bullet\underline{\underline{\epsilon}}_{\mathrm{amb}}(\omega)\bullet\underline{\underline{U}}^{-1}\bullet\hat{q}\right]\right.\\[4pt]
&\times\left.\left[\hat{q}\bullet\underline{\underline{U}}^{-1}\bullet\underline{\underline{\mu}}_{\mathrm{amb}}(\omega)\bullet\underline{\underline{U}}^{-1}\bullet\hat{q}\right]\right\}\\[4pt]
&-\left\{\left[\hat{q}\bullet\underline{\underline{U}}^{-1}\bullet\underline{\underline{\xi}}_{\mathrm{amb}}(\omega)\bullet\underline{\underline{U}}^{-1}\bullet\hat{q}\right]\right.\\[4pt]
&\times\left.\left[\hat{q}\bullet\underline{\underline{U}}^{-1}\bullet\underline{\underline{\zeta}}_{\mathrm{amb}}(\omega)\bullet\underline{\underline{U}}^{-1}\bullet\hat{q}\right]\right\}. \qquad (5.73)
\end{aligned}$$

Thus, Eqs. (5.72) provide the components of the integrated singularity of the DGF. Since each 3×3 constitutive dyadic term in Eqs. (5.72) is pre– and post–multiplied by $\underline{\underline{U}}^{-1}$, the evaluations of $\underline{\underline{\mathbf{D}}}^0_{U/\mathrm{amb}}(\omega)$ for an ellipsoidal particle embedded in an isotropic ambient medium are isomorphic to the evaluations for certain anisotropic and bianisotropic ambient mediums when $\underline{\underline{U}} = \underline{\underline{I}}$. Similarly, the evaluations of $\underline{\underline{\mathbf{D}}}^0_{U/\mathrm{amb}}(\omega)$ for a spheroidal particle embedded in a uniaxial ambient medium are isomorphic to the evaluations for certain anisotropic and bianisotropic ambient mediums when $\underline{\underline{U}} = \underline{\underline{I}}$. Therefore, we postpone our discussion of explicit representations of $\underline{\underline{D}}^0_{U/\mathrm{amb}}(\omega)$ to Sec. 5.5.2 wherein the case of spherical depolarization regions is considered. Parenthetically, expressions for $\underline{\underline{D}}^0_{U/\mathrm{amb}}(\omega)$ associated with certain cylindrical shapes — which may be viewed as limiting cases of the ellipsoidal depolarization region characterized by $\underline{\underline{U}}$ — have also been derived [48, 49].

Now we turn to the depolarization contribution given by $\underline{\underline{D}}^+_{U/\mathrm{amb}}(\rho, \omega)$, via which the spatial extent of the $\underline{\underline{U}}$-shaped particle is taken into account. A helpful simplification occurs in the case of Lorentz–reciprocal ambient mediums (for which $\underline{\underline{\epsilon}}_{\mathrm{amb}}(\omega) = \underline{\underline{\epsilon}}^{\mathrm{T}}_{\mathrm{amb}}(\omega)$, $\underline{\underline{\xi}}_{\mathrm{amb}}(\omega) = -\underline{\underline{\zeta}}^{\mathrm{T}}_{\mathrm{amb}}(\omega)$, and $\underline{\underline{\mu}}_{\mathrm{amb}}(\omega) = \underline{\underline{\mu}}^{\mathrm{T}}_{\mathrm{amb}}(\omega)$). Then $\underline{\underline{D}}^+_{U/\mathrm{amb}}(\rho, \omega)$ has the surface integral representation [50]

$$
\underline{\underline{D}}^+_{U/\mathrm{amb}}(\rho, \omega) = \frac{\omega^4}{4\pi} \int_{\phi=0}^{2\pi} \int_{\theta=0}^{\pi} \frac{1}{b_{U/\mathrm{amb}}(\theta, \phi, \omega)} \left[\frac{1}{\kappa_+(\omega) - \kappa_-(\omega)} \right.
$$

$$
\times \left(\frac{\exp(i\rho q)}{2q^2} (1 - i\rho q) \left\{ \det\left[\underline{\underline{\check{A}}}_{\mathrm{amb}}(\underline{\underline{U}}^{-1} \cdot \underline{q}, \omega) \right] \right. \right.
$$

$$
\times \underline{\underline{\check{G}}}^+_{\mathrm{amb}}(\underline{\underline{U}}^{-1} \cdot \underline{q}, \omega) + \det\left[\underline{\underline{\check{A}}}_{\mathrm{amb}}(-\underline{\underline{U}}^{-1} \cdot \underline{q}, \omega) \right]
$$

$$
\left. \times \underline{\underline{\check{G}}}^+_{\mathrm{amb}}(-\underline{\underline{U}}^{-1} \cdot \underline{q}, \omega) \right\} \Bigg)_{\substack{q=[\kappa_+(\omega)]^{1/2} \\ q=[\kappa_-(\omega)]^{1/2}}}
$$

$$
\left. + \frac{\det\left[\underline{\underline{\check{A}}}_{\mathrm{amb}}(\underline{0}, \omega) \right]}{\kappa_+(\omega)\, \kappa_-(\omega)} \underline{\underline{\check{G}}}^+_{\mathrm{amb}}(\underline{0}, \omega) \right] \sin\theta \, d\theta \, d\phi, \qquad (5.74)
$$

wherein

$$
\underline{\underline{\check{A}}}_{\mathrm{amb}}(\underline{q}, \omega) = \frac{1}{i\omega} \left[\underline{\underline{\check{G}}}_{\mathrm{amb}}(\underline{q}, \omega) \right]^{-1} \qquad (5.75)
$$

and $\kappa_\pm(\omega)$ are the q^2 roots of $\det\left[\underline{\underline{\check{A}}}_{\mathrm{amb}}(\underline{\underline{U}}^{-1} \cdot \underline{q}, \omega) \right] = 0$. As outlined in Sec. 5.5.2, explicit representations of $\underline{\underline{D}}^+_{U/\mathrm{amb}}(\rho, \omega)$ are available for certain isotropic and uniaxial ambient mediums when $\underline{\underline{U}} = \underline{\underline{I}}$, but numerical methods are generally required to evaluate the integrals on the right side of Eq. (5.74). In those cases where explicit representations of $\underline{\underline{D}}^+_{U/\mathrm{amb}}(\rho, \omega)$ are available, their derivation exploits the fact that the $\exp(i\rho q)(1 - i\rho q)$ term in the integrand on the right side of Eq. (5.74) can be approximated by its asymptotic expansion $1+(\rho q)^2/2+i(\rho q)^3/3$ in the long wavelength regime, since therein $\rho\,[\kappa_\pm(\omega)]^{1/2}\ll 1$ [50].

5.5.2   Spherical shape

In view of their extensive use in the homogenization formalisms discussed in Chap. 6, let us move on to depolarization dyadics for the spherical shape (i.e., $\underline{\underline{U}}=\underline{\underline{I}}$).

5.5.2.1   Isotropic ambient mediums

When the ambient medium is isotropic dielectric–magnetic, i.e., described by the constitutive relations (2.3), the corresponding ρ–independent contributions to the depolarization dyadic reduce to the well–known form [8]

$$\left.\begin{aligned}
{}^{ee}\underline{\underline{D}}^{\,0}_{I/\mathrm{amb}}(\omega) &= \frac{1}{3i\omega\epsilon(\omega)}\underline{\underline{I}}, & {}^{em}\underline{\underline{D}}^{\,0}_{I/\mathrm{amb}}(\omega) &= \underline{\underline{0}}\\[6pt]
{}^{me}\underline{\underline{D}}^{\,0}_{I/\mathrm{amb}}(\omega) &= \underline{\underline{0}}, & {}^{mm}\underline{\underline{D}}^{\,0}_{I/\mathrm{amb}}(\omega) &= \frac{1}{3i\omega\mu(\omega)}\underline{\underline{I}}
\end{aligned}\right\}, \quad (5.76)$$

while ρ–dependent contributions are given as [50]

$$\left.\begin{aligned}
{}^{ee}\underline{\underline{D}}^{+}_{I/\mathrm{amb}}(\rho,\omega) &= \frac{i\omega\rho^2\mu(\omega)}{9}\left\{3+i2\rho\omega\,[\epsilon(\omega)\mu(\omega)]^{1/2}\right\}\underline{\underline{I}}\\[4pt]
{}^{em}\underline{\underline{D}}^{+}_{I/\mathrm{amb}}(\rho,\omega) &= {}^{me}\underline{\underline{D}}^{+}_{I/\mathrm{amb}}(\rho,\omega) = \underline{\underline{0}}\\[4pt]
{}^{mm}\underline{\underline{D}}^{+}_{I/\mathrm{amb}}(\rho,\omega) &= \frac{i\omega\rho^2\epsilon(\omega)}{9}\left\{3+i2\rho\omega\,[\epsilon(\omega)\mu(\omega)]^{1/2}\right\}\underline{\underline{I}}
\end{aligned}\right\}. \quad (5.77)$$

When the ambient medium is isotropic chiral, i.e., described by the constitutive relations (2.5), the 3×3 dyadic contributions to the depolarization dyadic are provided by the following scalar multiples of the 3×3 identity matrix [50]:

$$\left.\begin{aligned}
{}^{ee}\underline{\underline{D}}^{\,0}_{I/\mathrm{amb}}(\omega) &= \frac{\mu(\omega)}{3i\omega\,\Upsilon_+(\omega)}\underline{\underline{I}}, & {}^{em}\underline{\underline{D}}^{\,0}_{I/\mathrm{amb}}(\omega) &= -\frac{\chi(\omega)}{3i\omega\,\Upsilon_+(\omega)}\underline{\underline{I}}\\[6pt]
{}^{me}\underline{\underline{D}}^{\,0}_{I/\mathrm{amb}}(\omega) &= \frac{\chi(\omega)}{3i\omega\,\Upsilon_+(\omega)}\underline{\underline{I}}, & {}^{mm}\underline{\underline{D}}^{\,0}_{I/\mathrm{amb}}(\omega) &= \frac{\epsilon(\omega)}{3i\omega\,\Upsilon_+(\omega)}\underline{\underline{I}}
\end{aligned}\right\}$$

$$(5.78)$$

and

$$
\left.
\begin{aligned}
{}^{\text{ee}}\underline{\underline{D}}^+_{I/\text{amb}}(\rho,\omega) &= \frac{i\omega\rho^2}{3}\left\{\mu(\omega) + \frac{2i\rho\,\omega\,\Upsilon_-(\omega)}{3}\left[\frac{\mu(\omega)}{\epsilon(\omega)}\right]^{1/2}\right\}\underline{\underline{I}} \\
{}^{\text{em}}\underline{\underline{D}}^+_{I/\text{amb}}(\rho,\omega) &= \frac{i\omega\rho^2}{3}\left\{\chi(\omega) + \frac{4i\rho\,\omega\,\chi(\omega)}{3}\left[\epsilon(\omega)\,\mu(\omega)\right]^{1/2}\right\}\underline{\underline{I}} \\
{}^{\text{me}}\underline{\underline{D}}^+_{I/\text{amb}}(\rho,\omega) &= -\frac{i\omega\rho^2}{3}\left\{\chi(\omega) + \frac{4i\rho\,\omega\,\chi(\omega)}{3}\left[\epsilon(\omega)\,\mu(\omega)\right]^{1/2}\right\}\underline{\underline{I}} \\
{}^{\text{mm}}\underline{\underline{D}}^+_{I/\text{amb}}(\rho,\omega) &= \frac{i\omega\rho^2}{3}\left\{\epsilon(\omega) + \frac{2i\rho\,\omega\,\Upsilon_-(\omega)}{3}\left[\frac{\epsilon(\omega)}{\mu(\omega)}\right]^{1/2}\right\}\underline{\underline{I}}
\end{aligned}
\right\}, \quad (5.79)
$$

where

$$
\Upsilon_{\pm}(\omega) = \epsilon(\omega)\,\mu(\omega) \pm \chi^2(\omega). \tag{5.80}
$$

5.5.2.2 *Anisotropic mediums*

We now turn to the uniaxial dielectric–magnetic scenario wherein the constitutive dyadics of the ambient medium are

$$
\left.
\begin{aligned}
\underline{\underline{\eta}}_{\text{amb}}(\omega) &= \underline{\underline{\eta}}_{\text{uni}}(\omega), \qquad (\eta = \epsilon, \mu) \\
\underline{\underline{\xi}}_{\text{amb}}(\omega) &= \underline{\underline{\zeta}}_{\text{amb}}(\omega) = \underline{\underline{0}}
\end{aligned}
\right\}, \tag{5.81}
$$

with $\underline{\underline{\epsilon}}_{\text{uni}}(\omega)$ and $\underline{\underline{\mu}}_{\text{uni}}(\omega)$ as specified in Eqs. (2.9) and (2.12), respectively. The ρ–independent contributions to the depolarization dyadic are provided by [38]

$$
\left.
\begin{aligned}
{}^{\text{ee}}\underline{\underline{D}}^0_{I/\text{uni}}(\omega) &= D^0_\epsilon(\omega)\left(\underline{\underline{I}} - \hat{u}\,\hat{u}\right) + D^{0,u}_\epsilon(\omega)\,\hat{u}\,\hat{u} \\
{}^{\text{em}}\underline{\underline{D}}^0_{I/\text{uni}}(\omega) &= {}^{\text{me}}\underline{\underline{D}}^0_{I/\text{uni}}(\omega) = \underline{\underline{0}} \\
{}^{\text{mm}}\underline{\underline{D}}^0_{I/\text{uni}}(\omega) &= D^0_\mu(\omega)\left(\underline{\underline{I}} - \hat{u}\,\hat{u}\right) + D^{0,u}_\mu(\omega)\,\hat{u}\,\hat{u}
\end{aligned}
\right\}, \tag{5.82}
$$

where the form that explicit representations of $D^0_\eta(\omega)$ and $D^{0,u}_\eta(\omega)$ $(\eta = \epsilon, \mu)$ take depends upon the nature of the the dimensionless scalar

$$
\gamma_\eta(\omega) = \frac{\eta_u(\omega)}{\eta(\omega)}, \qquad (\eta = \epsilon, \mu). \tag{5.83}
$$

If $0 < \gamma_\eta(\omega) < 1$ then we have [38, 51]

$$
\left.
\begin{aligned}
D_\eta^0(\omega) &= \frac{1}{2i\omega\,[\,\eta(\omega) - \eta_u(\omega)\,]} \\
&\quad \times \left\{ 1 - \frac{\gamma_\eta(\omega)\sinh^{-1}\left[\dfrac{1 - \gamma_\eta(\omega)}{\gamma_\eta(\omega)}\right]^{1/2}}{[\,1 - \gamma_\eta(\omega)\,]^{1/2}} \right\} \\[2em]
D_\eta^{0,u}(\omega) &= \frac{1}{i\omega\,[\,\eta(\omega) - \eta_u(\omega)\,]} \left\{ \frac{\sinh^{-1}\left[\dfrac{1 - \gamma_\eta(\omega)}{\gamma_\eta(\omega)}\right]^{1/2}}{[\,1 - \gamma_\eta(\omega)\,]^{1/2}} - 1 \right\}
\end{aligned}
\right\}, \quad (\eta = \epsilon, \mu); \quad (5.84)
$$

whereas

$$
\left.
\begin{aligned}
D_\eta^0(\omega) &= \frac{1}{2i\omega\,[\,\eta(\omega) - \eta_u(\omega)\,]} \\
&\quad \times \left\{ 1 - \frac{\gamma_\eta(\omega)\sec^{-1}[\,\gamma_\eta(\omega)\,]^{1/2}}{[\,\gamma_\eta(\omega) - 1\,]^{1/2}} \right\} \\[2em]
D_\eta^{0,u}(\omega) &= \frac{1}{i\omega\,[\,\eta(\omega) - \eta_u(\omega)\,]} \left\{ \frac{\sec^{-1}[\,\gamma_\eta(\omega)\,]^{1/2}}{[\,\gamma_\eta(\omega) - 1\,]^{1/2}} - 1 \right\}
\end{aligned}
\right\}, \quad (\eta = \epsilon, \mu)
$$

$$(5.85)$$

are appropriate if $\gamma_\eta(\omega) > 1$. In the general case wherein $\gamma_\eta(\omega)$ are complex–valued (with nonzero imaginary part) then either of Eqs. (5.84) or (5.85) can be used. Notice that $\gamma_\eta(\omega) < 0$ — which conditions hold true for nondissipative uniaxial ambient mediums with indefinite constitutive dyadics [52–54] — the components of $\underline{\underline{D}}^0(\omega)$ are undefined.

Dyadics of the form (5.82) also arise for a spheroidal depolarization region, with its rotational axis aligned with $\hat{u}$, immersed in an isotropic dielectric–magnetic ambient medium [51]. The contributions to the depolarization dyadic appropriate to uniaxial dielectric mediums and uniaxial magnetic ambient mediums follow immediately from Eqs. (5.82), upon substituting $\underline{\underline{\mu}}_{\text{uni}}(\omega) = \mu_0\,\underline{\underline{I}}$ and $\underline{\underline{\epsilon}}_{\text{uni}}(\omega) = \epsilon_0\,\underline{\underline{I}}$, respectively.

In the case of uniaxial dielectric ambient mediums, we have the corresponding ρ–dependent contributions [50]

$$
{}^{ee}\underline{\underline{D}}^+_{I/\text{uni}}(\rho, \omega) = D_\epsilon^+(\rho, \omega)\left(\underline{\underline{I}} - \hat{u}\,\hat{u}\right) + D_\epsilon^{+,u}(\rho, \omega)\,\hat{u}\,\hat{u}, \quad (5.86)
$$

where the forms that explicit representations of $D_\epsilon^+(\rho,\omega)$ and $D_\epsilon^{+,u}(\rho,\omega)$ take depend upon the nature of the the dimensionless scalar

$$\tau_\epsilon(\omega) = \frac{\epsilon(\omega) - \epsilon_u(\omega)}{\epsilon(\omega)}. \tag{5.87}$$

If $0 < \tau_\epsilon(\omega) < 1$, we have

$$
\left.
\begin{aligned}
D_\epsilon^+(\rho,\omega) &= \frac{i\omega\mu_0\rho^2}{8}\left\{ \frac{1-\tau_\epsilon(\omega)}{\tau_\epsilon(\omega)} + i\rho\frac{4\omega\left[3\epsilon(\omega)+\epsilon_u(\omega)\right]}{9}\left[\frac{\mu_0}{\epsilon(\omega)}\right]^{1/2} \right. \\
&\qquad \left. -\left[\frac{\epsilon_u(\omega)}{\tau_\epsilon(\omega)\,\epsilon(\omega)}\right]^2 \left[\tau_\epsilon(\omega)\right]^{1/2} \tanh^{-1}\left[\tau_\epsilon(\omega)\right]^{1/2} \right\} \\[6pt]
D_\epsilon^{+,u}(\rho,\omega) &= \frac{i\omega\mu_0\rho^2}{4}\left(i\rho\frac{4\omega\left[\epsilon(\omega)\,\mu_0\right]^{1/2}}{9} \right. \\
&\qquad \left. +\frac{1}{\tau_\epsilon(\omega)}\left\{ \frac{1+\tau_\epsilon(\omega)}{\tau_\epsilon(\omega)}\left[\tau_\epsilon(\omega)\right]^{1/2}\tanh^{-1}\left[\tau_\epsilon(\omega)\right]^{1/2} - 1\right\} \right)
\end{aligned}
\right\};
\tag{5.88}
$$

whereas the appropriate expressions for $\tau_\epsilon(\omega) < 0$ are as follows:

$$
\left.
\begin{aligned}
D_\epsilon^+(\rho,\omega) &= \frac{i\omega\mu_0\rho^2}{8}\left\{ \frac{1-\tau_\epsilon(\omega)}{\tau_\epsilon(\omega)} + i\rho\frac{4\omega\left[3\epsilon(\omega)+\epsilon_u(\omega)\right]}{9}\left[\frac{\mu_0}{\epsilon(\omega)}\right]^{1/2} \right. \\
&\qquad \left. +\left[\frac{\epsilon_u(\omega)}{\tau_\epsilon(\omega)\,\epsilon(\omega)}\right]^2 \left[-\tau_\epsilon(\omega)\right]^{1/2} \tan^{-1}\left[-\tau_\epsilon(\omega)\right]^{1/2} \right\} \\[6pt]
D_\epsilon^{+,u}(\rho,\omega) &= \frac{i\omega\mu_0\rho^2}{4}\left(i\rho\frac{4\omega\left[\epsilon(\omega)\,\mu_0\right]^{1/2}}{9} \right. \\
&\qquad \left. +\frac{1}{\tau_\epsilon(\omega)}\left\{ 1-\frac{1+\tau_\epsilon(\omega)}{\tau_\epsilon(\omega)}\left[-\tau_\epsilon(\omega)\right]^{1/2}\tan^{-1}\left[-\tau_\epsilon(\omega)\right]^{1/2} \right\} \right)
\end{aligned}
\right\}.
\tag{5.89}
$$

As is the case for the components of $\underline{\underline{D}}^0(\omega)$, the components of $\underline{\underline{D}}^+(\rho,\omega)$ are not defined for the $\tau_\epsilon(\omega) > 1$ regime which corresponds to nondissipative uniaxial ambient mediums with indefinite constitutive dyadics [52–54]. For the general case in which $\tau_\epsilon(\omega)$ is complex–valued (with nonzero imaginary part), either representation (5.88) or (5.89) is appropriate. We note that the corresponding expressions for a uniaxial magnetic medium are precisely analogous to those given in Eqs. (5.86)–(5.89).

If the ambient medium is either gyrotropic dielectric, i.e.,

$$\left.\begin{array}{l} \underline{\underline{\epsilon}}_{amb}(\omega) = \underline{\underline{\epsilon}}_{gyro}(\omega) \\ \underline{\underline{\xi}}_{amb}(\omega) = \underline{\underline{\zeta}}_{amb}(\omega) = \underline{\underline{0}} \\ \underline{\underline{\mu}}_{amb}(\omega) = \mu_0 \underline{\underline{I}} \end{array}\right\}, \tag{5.90}$$

or gyrotropic magnetic, i.e.,

$$\left.\begin{array}{l} \underline{\underline{\epsilon}}_{amb}(\omega) = \epsilon_0 \underline{\underline{I}} \\ \underline{\underline{\xi}}_{amb}(\omega) = \underline{\underline{\zeta}}_{amb}(\omega) = \underline{\underline{0}} \\ \underline{\underline{\mu}}_{amb}(\omega) = \underline{\underline{\mu}}_{gyro}(\omega) \end{array}\right\}, \tag{5.91}$$

with $\underline{\underline{\epsilon}}_{gyro}(\omega)$ and $\underline{\underline{\mu}}_{gyro}(\omega)$ specified in Eqs. (5.31), the skew–symmetric dyadic components make no contribution to $\underline{\underline{D}}^0(\omega)$.[2] Hence, the ρ–independent contribution to the depolarization dyadics for gyrotropic dielectric mediums and gyrotropic magnetic mediums are the same as those for the corresponding uniaxial dielectric mediums and uniaxial magnetic mediums, respectively.

Finally, suppose that the ambient medium is orthorhombic dielectric–magnetic; i.e.,

$$\left.\begin{array}{l} \underline{\underline{\eta}}_{amb}(\omega) = \underline{\underline{\eta}}^{ortho}_{bi}(\omega), \quad (\eta = \epsilon, \mu) \\ \underline{\underline{\xi}}_{amb}(\omega) = \underline{\underline{\zeta}}_{amb}(\omega) = \underline{\underline{0}} \end{array}\right\}, \tag{5.92}$$

with

$$\underline{\underline{\eta}}^{ortho}_{bi}(\omega) = \eta_x(\omega)\,\hat{\underline{x}}\,\hat{\underline{x}} + \eta_y(\omega)\,\hat{\underline{y}}\,\hat{\underline{y}} + \eta_z(\omega)\,\hat{\underline{z}}\,\hat{\underline{z}}, \quad (\eta = \epsilon, \mu). \tag{5.93}$$

The corresponding ρ–independent contributions to the depolarization dyadics for the spherical shape are given as [55]

$$\left.\begin{array}{l} ^{ee}\underline{\underline{D}}^0_{I/ortho}(\omega) = \dfrac{1}{i\omega\epsilon_0}\left[D^\epsilon_x(\omega)\,\hat{\underline{x}}\,\hat{\underline{x}} + D^\epsilon_y(\omega)\,\hat{\underline{y}}\,\hat{\underline{y}} + D^\epsilon_z(\omega)\,\hat{\underline{z}}\,\hat{\underline{z}}\right] \\[2mm] ^{em}\underline{\underline{D}}^0_{I/ortho} = {}^{me}\underline{\underline{D}}^0_{I/ortho} = \underline{\underline{0}} \\[2mm] ^{mm}\underline{\underline{D}}^0_{I/ortho} = \dfrac{1}{i\omega\mu_0}\left[D^\mu_x(\omega)\,\hat{\underline{x}}\,\hat{\underline{x}} + D^\mu_y(\omega)\,\hat{\underline{y}}\,\hat{\underline{y}} + D^\mu_z(\omega)\,\hat{\underline{z}}\,\hat{\underline{z}}\right] \end{array}\right\}. \tag{5.94}$$

---

[2] In fact, it follows from the expressions (5.72) that the skew–symmetric components of any constitutive dyadic vanish from $\underline{\underline{D}}^0(\omega)$.

Here,

$$
\left.
\begin{aligned}
D_x^\eta &= \frac{\eta_y^{1/2}(\omega)\,[F(\lambda_1,\lambda_2) - E(\lambda_1,\lambda_2)]}{[\eta_y(\omega) - \eta_x(\omega)]\,[\eta_z(\omega) - \eta_x(\omega)]^{1/2}} \\[2ex]
D_y^\eta &= \frac{1}{\eta_y(\omega) - \eta_x(\omega)} \left\{ \frac{\eta_x(\omega) - \eta_y(\omega)}{\eta_z(\omega) - \eta_y(\omega)} - \left[\frac{\eta_z(\omega) - \eta_x(\omega)}{\eta_y(\omega)} \right]^{1/2} \right. \\[2ex]
&\quad \times \left[\frac{\eta_x(\omega)}{\eta_z(\omega) - \eta_x(\omega)} F(\lambda_1,\lambda_2) \right. \\[2ex]
&\quad \left. \left. - \frac{\eta_y(\omega)}{\eta_z(\omega) - \eta_y(\omega)} E(\lambda_1,\lambda_2) \right] \right\} \\[2ex]
D_z^\eta &= \frac{1}{\eta_z(\omega) - \eta_y(\omega)} \left\{ 1 - \left[\frac{\eta_y(\omega)}{\eta_z(\omega) - \eta_x(\omega)} \right]^{1/2} E(\lambda_1,\lambda_2) \right\}
\end{aligned}
\right\},
$$
$$(\eta = \epsilon, \mu), \qquad (5.95)$$

involve $F(\lambda_1,\lambda_2)$ and $E(\lambda_1,\lambda_2)$ as elliptic integrals of the first and second kinds [56], respectively, with arguments

$$
\left.
\begin{aligned}
\lambda_1 &= \tan^{-1}\left[\frac{\eta_z(\omega) - \eta_x(\omega)}{\eta_x(\omega)} \right]^{1/2} \\[2ex]
\lambda_2 &= \left\{ \frac{\eta_z(\omega)\,[\eta_y(\omega) - \eta_x(\omega)]}{\eta_y(\omega)\,[\eta_z(\omega) - \eta_x(\omega)]} \right\}^{1/2}
\end{aligned}
\right\},
\qquad (\eta = \epsilon, \mu). \qquad (5.96)
$$

The corresponding contributions to the depolarization dyadic for orthorhombic dielectric mediums and orthorhombic magnetic mediums follow immediately from Eqs. (5.94), upon substituting $\underline{\underline{\mu}}_{\mathrm{bi}}^{\,\mathrm{ortho}}(\omega) = \mu_0\,\underline{\underline{I}}$ and $\underline{\underline{\epsilon}}^{\,\mathrm{ortho}}(\omega) = \epsilon_0\,\underline{\underline{I}}$, respectively. The form of $\underline{\underline{D}}^0(\omega)$ represented in Eqs. (5.94) also arises for an ellipsoidal shape, with rotational axes aligned with $\hat{x}$, $\hat{y}$ and $\hat{z}$, in an isotropic dielectric–magnetic ambient medium [55, 57].

5.5.3 *Bianisotropic mediums*

Closed–form representations of depolarization dyadics are not available for a general linear, homogeneous bianisotropic medium, but analytical progress has been reported for certain uniaxial bianisotropic mediums [39]. When the ambient medium is characterized by the constitutive relations (4.82), the skew–symmetric components of the magnetoelectric constitutive dyadics do not contribute to the associated ρ–independent depolarization terms; therefore, the dyadic $\underline{\underline{D}}^0(\omega)$ is simply the same as that for the corresponding anisotropic dielectric–magnetic mediums.

5.6   Connection to plasmonics

Physical insight into the role of the depolarization dyadic may be gained by considering the related quantity

$$
\underline{\underline{\alpha}}_{\ell/\text{amb}}(\omega) = \left[\underline{\underline{K}}_\ell(\omega) - \underline{\underline{K}}_{\text{amb}}(\omega) \right] \cdot \left\{ \underline{\underline{I}} + i\omega \underline{\underline{D}}^0_{U^\ell/\text{amb}}(\omega) \right.
$$
$$
\left. \cdot \left[\underline{\underline{K}}_\ell(\omega) - \underline{\underline{K}}_{\text{amb}}(\omega) \right] \right\}^{-1}, \tag{5.97}
$$

which is the *polarizability density* dyadic of a $\underline{\underline{U}}^\ell$-shaped ellipsoid of medium ℓ embedded in an ambient medium [42]. The medium ℓ and the ambient medium are described by the Tellegen constitutive dyadics $\underline{\underline{K}}_\ell(\omega)$ and $\underline{\underline{K}}_{\text{amb}}(\omega)$, respectively. Both mediums are homogeneous.

The polarizability density dyadic provides a measure of the electromagnetic response of the ellipsoidal particle to an applied field. As the determinant of the dyadic term enclosed in curly brackets on the right side of Eq. (5.97) becomes vanishingly small, the prospect looms of the response represented by $\underline{\underline{\alpha}}_{\ell/\text{amb}}(\omega)$ becoming exceeding large. To see this more clearly, let us consider the isotropic dielectric–magnetic specialization of the polarizability density dyadic with $\underline{\underline{U}} = \underline{\underline{I}}$, by exploiting the depolarization dyadics (5.76): If the medium ℓ and the ambient medium are both isotropic dielectric–magnetic mediums, characterized by the permittivities $\epsilon_{\ell,\text{amb}}(\omega)$ and permeabilities $\mu_{\ell,\text{amb}}(\omega)$, respectively, in the Tellegen constitutive relations (2.3), then the corresponding polarizability density dyadic for a spherical particle takes the form

$$
\underline{\underline{\alpha}}_{\ell/\text{amb}}(\omega) = \begin{bmatrix} \alpha^\epsilon_{\ell/\text{amb}}(\omega)\,\underline{\underline{I}} & \underline{\underline{0}} \\ \underline{\underline{0}} & \alpha^\mu_{\ell/\text{amb}}(\omega)\,\underline{\underline{I}} \end{bmatrix}, \tag{5.98}
$$

with

$$
\alpha^\eta_{\ell/\text{amb}}(\omega) = \frac{3\,\eta_{\text{amb}}(\omega)\,[\,\eta_\ell(\omega) - \eta_{\text{amb}}(\omega)\,]}{2\eta_{\text{amb}}(\omega) + \eta_\ell(\omega)}, \qquad (\eta = \epsilon, \mu). \tag{5.99}
$$

Clearly, the scalar polarizability densities $\alpha^\eta_{\ell/\text{amb}}(\omega)$ become unbounded in the limit $\eta_\ell(\omega) \to -2\eta_{\text{amb}}(\omega)$. While this limit cannot be realized for realistic (dissipative) materials, $\alpha^\eta_{\ell/\text{amb}}(\omega)$ can become exceedingly large for the case where medium ℓ and the ambient medium are weakly nondissipative and $\text{Re}\{\eta_\ell(\omega)\} \approx -2\,\text{Re}\{\eta_{\text{amb}}(\omega)\}$. This phenomenon underpins the emergence of localized surface plasmon resonance (LSPR) shown by electrically small particles [58]. The excitation of LSPR is acutely sensitive to the constitutive properties of the ambient medium — a property which may be usefully exploited in various technological devices, particularly at the nanoscale [59, 60].

References

[1] C.T. Tai, *Dyadic Green functions in electromagnetic theory, 2nd ed*, IEEE Press, Piscataway, NJ, USA, 1994.

[2] W.S. Weiglhofer, Frequency–dependent dyadic Green functions for bianisotropic media, *Advanced electromagnetism: Foundations, theory and applications*, T.W. Barrett and D.M. Grimes (eds), World Scientific, Singapore, 1995, 376–389.

[3] L.B. Felsen and N. Marcuvitz, *Radiation and scattering of waves*, IEEE Press, Piscataway, NJ, USA, 1994.

[4] W.S. Weiglhofer, Analytic methods and free–space dyadic Green's functions, *Radio Sci* **28** (1993), 847–857.

[5] F. Olyslager and I.V. Lindell, Electromagnetics and exotic media: a quest for the holy grail, *IEEE Antennas Propagat Mag* **44** (2002), 48–58.

[6] W.S. Weiglhofer, The connection between factorization properties and closed–form solutions of certain linear dyadic differential operators, *J Phys A: Math Gen* **33** (2000), 6253–6261.

[7] H.C. Chen, *Theory of electromagnetic waves*, McGraw–Hill, New York, NY, USA, 1983.

[8] J. Van Bladel, *Singular electromagnetic fields and sources*, Oxford University Press, Oxford, UK, 1991 (reissued in association with IEEE Press, New York, NY, USA, 1995).

[9] A. Lakhtakia, *Beltrami fields in chiral media*, World Scientific, Singapore, 1994.

[10] J.C. Monzon, Three–dimensional field expansion in the most general rotationally symmetric anisotropic material: application to scattering by a sphere, *IEEE Trans Antennas Propagat* **37** (1989), 728–735.

[11] W.S. Weiglhofer, Dyadic Green's functions for general uniaxial media, *IEE Proc, Part H* **137** (1990), 5–10.

[12] W.S. Weiglhofer, A dyadic Green's functions representation in electrically gyrotropic media, *Arch Elektron Übertrag* **48** (1994), 125–130.

[13] W.S. Weiglhofer and I.V. Lindell, Analytic solution for the dyadic Green function of a nonreciprocal uniaxial bianisotropic medium, *Arch Elektron Übertrag* **48** (1994), 116–119.

[14] A. Lakhtakia and W.S. Weiglhofer, On electromagnetic fields in a linear medium with gyrotropic–like magnetoelectric properties, *Microw Opt Technol Lett* **15** (1997), 168–170.

[15] A. Lakhtakia and T.G. Mackay, Dyadic Green function for an elecromagnetic medium inspired by general relativity, *Chinese Phys Lett* **23** (2006), 832–833.

[16] C.A. Brau, *Modern problems in classical electrodynamics*, Oxford University Press, New York, NY, USA, 2004.

[17] A. Lakhtakia, On the Huygens's principles and the Ewald–Oseen extinction theorems for, and the scattering of, Beltrami fields, *Optik* **91** (1992), 35–40.

[18] L. Bergstein and T. Zachos, A Huygens' principle for uniaxially anisotropic media, *J Opt Soc Am* **56** (1966), 931–937.

[19] A. Lakhtakia, V.K. Varadan and V.V. Varadan, A note on Huygens's principle for uniaxial dielectric media, *J Wave–Mater Interact* **4** (1989), 339–343.

[20] T.G. Mackay and A. Lakhtakia, The Huygens principle for a uniaxial dielectric–magnetic medium with gyrotropic–like magnetoelectric properties, *Electromagnetics* **29** (2009), 143–150.

[21] N.R. Ogg, A Huygen's principle for anisotropic media, *J Phys A: Math Gen* **4** (1971), 382–388.

[22] J. Van Bladel, *Electromagnetic fields*, Hemisphere, Washington, DC, USA, 1985.

[23] X.B. Wu and K. Yasumoto, Cylindrical vector–wave–function representations of fields in a biaxial Ω–medium, *J Electromagn Waves Applics* **11** (1997), 1407-1423.

[24] J.A. Kong, Theorems of bianisotropic media, *Proc IEEE* **60** (1972), 1036–1046.

[25] P.G. Cottis and G.D. Kondylis, Properties of the dyadic Green's function for an unbounded anisotropic medium, *IEEE Trans Antennas Propagat* **43** (1995), 154–161.

[26] W.C. Chew, *Waves and fields in inhomogenous media*, IEEE Press, Piscataway, NJ, USA, 1999.

[27] L.-W. Li, X.-K. Kang and M.-S. Leong, *Spheroidal wave functions in electromagnetic theory*, Wiley, Hoboken, NJ, USA, 2001.

[28] D. Cheng, Eigenfunction expansion of the dyadic Green's function in a gyroelectric chiral medium by cylindrical vector wave functions, *Phys Rev E* **55** (1997), 1950–1958.

[29] D. Cheng, W. Ren and Y.-Q. Jin, Green dyadics in uniaxial bianisotropic–ferrite medium by cylindrical vector wavefunctions, *J Phys A: Math Gen* **30** (1997), 573–585.

[30] E.L. Tan and S.Y. Tan, On the eigenfunction expansions of dyadic Green's functions for bianisotropic media, *Prog Electromagn Res* **20** (1998), 227–247.

[31] L.-W. Li, M.-S. Leong, P.-S. Kooi and T.-S. Yeo, Comment on "Eigenfunction expansion of the dyadic Green's function in a gyroelectric chiral medium by cylindrical vector wave functions", *Phys Rev E* **59** (1999), 3767–3771.

[32] W. Ren, Comment on "Eigenfunction expansion of the dyadic Green's function in a gyroelectric chiral medium by cylindrical vector wave functions", *Phys Rev E* **59** (1999), 3772–3773.

[33] A. Lakhtakia and W.S. Weiglhofer, Green function for radiation and propagation in helicoidal bianisotropic mediums, *IEE Proc–Microw Antennas Propagate* **144** (1997), 57–59.

[34] V.A. Yakubovich and V.M. Starzhinskii, *Linear differential equations with periodic coefficients*, Wiley, New York, NY, USA, 1975.

[35] A. Lakhtakia (ed), *Selected papers on linear optical composite materials*, SPIE Optical Engineering Press, Bellingham, WA, USA, 1996.

[36] A. Lakhtakia, On direct and indirect scattering approaches for homogenization of particulate composites, *Microw Opt Technol Lett* **25** (2000), 53–56.

[37] A. Lakhtakia, Orthogonal symmetries of polarizability dyadics of bianisotropic ellipsoids, *Microw Opt Technol Lett*, **27** (2000), 175–177.

[38] B. Michel, A Fourier space approach to the pointwise singularity of an anisotropic dielectric medium, *Int J Appl Electromagn Mech* **8** (1997), 219–227.

[39] B. Michel and W.S. Weiglhofer, Pointwise singularity of dyadic Green function in a general bianisotropic medium, *Arch Elektron Übertrag* **51** (1997), 219–223. Corrections: **52** (1998), 310.

[40] T.G. Mackay, Depolarization volume and correlation length in the homogenization of anisotropic dielectric composites, *Waves Random Media* **14** (2004), 485–498. Corrections: **16** (2006), 85.

[41] J. Cui and T.G. Mackay, Depolarization regions of nonzero volume in bianisotropic homogenized composites, *Waves Random Complex Media* **17** (2007), 269–281.

[42] B. Michel, Recent developments in the homogenization of linear bianisotropic composite materials, in *Electromagnetic fields in unconventional materials and structures* (O.N. Singh and A. Lakhtakia, eds), Wiley, New York, NY, USA, 2000, 39–82.

[43] W.T. Doyle, Optical properties of a suspension of metal spheres, *Phys Rev B* **39** (1989), 9852–9858.

[44] C.E. Dungey and C.F. Bohren, Light scattering by nonspherical particles: a refinement to the coupled–dipole method, *J Opt Soc Am A* **8** (1991), 81–87.

[45] B. Shanker and A. Lakhtakia, Extended Maxwell Garnett model for chiral-in–chiral composites, *J Phys D: Appl Phys* **26** (1993), 1746–1758.

[46] M.T. Prinkey, A. Lakhtakia and B. Shanker, On the extended Maxwell–Garnett and the extended Bruggeman approaches for dielectric-in-dielectric composites, *Optik* **96** (1994), 25–30.

[47] B. Shanker, The extended Bruggeman approach for chiral–in–chiral mixtures, *J Phys D: Appl Phys* **29** (1996), 281–288.

[48] P.G. Cottis, C.N. Vazouras and C. Spyrou, Green's function for an unbounded biaxial medium in cylindrical coordinates, *IEEE Trans Antennas Propagat* **47** (1999), 195–199.

[49] W.S. Weiglhofer and T.G. Mackay, Needles and pillboxes in anisotropic mediums, *IEEE Trans Antennas Propagat* **50** (2002), 85–86.

[50] T.G. Mackay, On extended homogenization formalisms for nanocomposites, *J Nanophoton* **2** 2008, 021850.

[51] T.G. Mackay and A. Lakhtakia, Anisotropic enhancement of group velocity in a homogenized dielectric composite medium, *J Opt A: Pure Appl Opt* **7** (2005), 669–674.

[52] T.G. Mackay, A. Lakhtakia and R.A. Depine, Uniaxial dielectric media with hyperbolic dispersion relations, *Microw Opt Technol Lett* **48** (2006), 363–367.

[53] G.X. Li, H.L. Tam, F.Y. Wang and K.W. Cheah, Superlens from complementary anisotropic metamaterials, *J Appl Phys* **102** (2007), 116101.

[54] D.R. Smith and D. Schurig, Electromagnetic wave propagation in media with indefinite permittivity and permeability tensors, *Phys Rev Lett* **90** (2003), 077405.

[55] W.S. Weiglhofer, Electromagnetic depolarization dyadics and elliptic integrals, *J Phys A: Math Gen* **31** (1998), 7191–7196.

[56] I.S. Gradshteyn and I.M. Ryzhik, *Table of integrals, series, and products,* *7th ed*, Academic Press, London, UK, 2007.

[57] H. Fricke, The Maxwell–Wagner dispersion in a suspension of ellipsoids, *J Phys Chem* **57** (1953), 934–937.

[58] S.A. Kalele, N.R. Tiwari, S.W. Gosavi and S.K. Kulkarni, Plasmon–assisted photonics at the nanoscale, *J Nanophoton* **1** (2007), 012501.

[59] E. Hutter and J.H. Fendler, Exploitation of localized surface plasmon resonance, *Adv Mater* **16** (2004), 1685–1706.

[60] S.A. Maier, *Plasmonics: Fundamentals and applications*, Springer, New York, NY, USA, 2007.

Chapter 6

Homogenization

A mixture of two (or more) different mediums may be viewed as being effectively homogeneous, provided that wavelengths are much longer than the length–scales of the mixture's nonhomogeneities [1–3]. In fact, the process of homogenization implicitly underpins our descriptions of anisotropy and bianisotropy, since the constitutive relations relate macroscopic electromagnetic fields which represent the volume–averages of their microscopic counterparts [4, 5]. On a more practical level, homogenization is important in the interpretation of experimental measurements and in material design. This latter topic, in particular, continues to motivate much research. For example, bianisotropic mediums may be readily conceptualized as homogenized composite mediums (HCMs), arising from constituent mediums which are themselves not bianisotropic (or even anisotropic in certain instances).

HCMs, and anisotropic and bianisotropic HCMs in particular, furnish prime examples of *metamaterials* [6]. These may be regarded as artificial composite materials which exhibit properties that are either not exhibited at all by their constituent materials or not exhibited to the same extent by their constituent materials. The notion that complex mediums may be realized through the homogenization of relatively simple constituent mediums, was promoted initially for isotropic chiral HCMs [7, 8], and subsequently extended to anisotropic and bianisotropic HCMs [9].

The constitutive parameters of HCMs are estimated using homogenization formalisms. Numerous formalisms have been proposed, some dating back to the earliest years of electromagnetic theory, but only a few of these have been rigorously established for bianisotropic HCMs. For an historical perspective, see the introduction to an anthology of milestone papers on HCMs [10]. Two of the most widely used formalisms today — the Maxwell Garnett formalism [11] and the Bruggeman formalism [12] — were first de-

veloped for isotropic dielectric–magnetic HCMs. Generalizations of these
formalisms appropriate to anisotropic and bianisotropic HCMs emerged
much more recently [13], following the development of expressions for the
corresponding depolarization dyadics [14].

In the conventional approaches to homogenization, as exemplified by the
Maxwell Garnett and Bruggeman formalisms, the distributional statistics
of the HCM constituents are described solely in terms of their respective
volume fractions. A more sophisticated approach is taken in the strong–
property–fluctuation theory (SPFT), wherein spatial correlation functions
of arbitrarily high order may be accommodated [15]. The SPFT for bianisotropic HCMs was established, at a practical level, at the beginning of
the twenty–first century [16], building upon its development for isotropic
[15, 17] and anisotropic [18, 19] HCMs. Furthermore, the range of the
SPFT was recently extended to orthorhombic $mm2$ piezoelectric HCMs
[20].

The estimation of constitutive dyadics of linear bianisotropic HCMs is
described in the following subsections, using the Maxwell Garnett, Bruggeman and SPFT formalisms. These approaches are rigorously established,
in contrast to extrapolations of isotropic formalisms [21–23].

6.1   Constituent mediums

Let us consider an HCM arising from a random mixture of two particulate
constituent mediums, labelled 'a' and 'b', with each constituent medium
itself being homogeneous. The Tellegen constitutive dyadics of the con-
stituent mediums are denoted by $\underline{\underline{\mathbf{K}}}_a(\omega)$ and $\underline{\underline{\mathbf{K}}}_b(\omega)$; i.e., in general, the
constituent mediums are bianisotropic.

The constituent particles are taken to be generally ellipsoidal in shape.
The ellipsoids of each constituent medium are conformal and have the same
orientation, but they are randomly distributed. The shapes of the ellipsoids
are characterized by the real–symmetric 3×3 dyadics $\underline{\underline{U}}^a$ and $\underline{\underline{U}}^b$, as introduced in Eq. (5.63); and their linear dimensions must be much smaller than
the electromagnetic wavelength(s) in both constituent mediums [24, 25].

Let V denote the unbounded space occupied by the composite medium.
This space is partitioned into the disjoint regions V_a and V_b which contain
the constituent mediums 'a' and 'b', respectively. In order to completely
fill V with ellipsoidal particles, a fractal–like distribution of the constituent
particles may be envisaged, as schematically illustrated in Fig. 6.1. The

distributions of the two constituent mediums throughout V are specified in terms of the characteristic functions

$$\Phi_\ell(\underline{r}) = \begin{cases} 1, & \underline{r} \in V_\ell, \\ 0, & \underline{r} \notin V_\ell, \end{cases} \qquad (\ell = \text{a,b}). \qquad (6.1)$$

The nth moment of $\Phi_\ell(\underline{r})$ is the ensemble–average $\langle \Phi_\ell(\underline{r}_1) \cdots \Phi_\ell(\underline{r}_n) \rangle_e$, which represents the probability of $\underline{r}_1, \dots, \underline{r}_n \in V_\ell$ ($\ell = $ a,b). The volume fraction of constituent medium ℓ is given by $f_\ell = \langle \Phi_\ell(\underline{r}) \rangle_e$; clearly, $f_a + f_b = 1$. Only the first moments of $\Phi_{a,b}(\underline{r})$ are used in the Maxwell Garnett and Bruggeman formalisms, whereas arbitrarily high–order moments can be accommodated in the SPFT [15, 16].

Figure 6.1 Schematic representation of the randomly mixed constituent mediums 'a' and 'b'. The shapes and orientations of the ellipsoidal particles comprising constituent medium 'a' are specified by the shape dyadic $\underline{\underline{U}}^a$, while those of constituent medium 'b' are specified by $\underline{\underline{U}}^b$, where $\underline{\underline{U}}^a \neq \underline{\underline{U}}^b$ in general.

The electromagnetic response of an electrically small ellipsoid of medium ℓ, ($\ell = $ a,b), embedded in a homogeneous ambient medium described by the Tellegen constitutive dyadic $\underline{\underline{K}}_{\text{amb}}(\omega)$, is encapsulated by the polarizability density dyadic, as defined in Eq. (5.97) [9]. This dyadic $\underline{\underline{\alpha}}_{\ell/\text{amb}}(\omega)$ is central to the Maxwell Garnett and Bruggeman formalisms. In Eq. (5.97), the conventional practice of adopting $\underline{\underline{\mathbf{D}}}^0_{U^\ell/\text{amb}}(\omega)$ — which is specified by Eq. (5.70) — as the depolarization dyadic is adhered to. As outlined in Sec. 5.5, this is an approximation which involves neglecting the spatial extent of the constituent particles. *Extended* homogenization formalisms may be developed which do take the particle size into account, by adopting $\underline{\underline{\mathbf{D}}}_{U^\ell/\text{amb}}(\rho,\omega)$ as the depolarization dyadic (or by other means). We postpone a discussion of such extended formalisms until Sec. 6.5, and for now concentrate on the more widely encountered ρ–independent formulations.

6.2 Maxwell Garnett formalism

The Maxwell Garnett homogenization formalism[1] has been much used, despite its applicability being limited to dilute composites [10]. Faxén established a rigorous basis for this formalism for isotropic dielectric–in–dielectric composite mediums [26], and its standing was further bolstered by the development of the influential Hashin–Shtrikman bounds, which are closely related to the Maxwell Garnett estimate of the HCM constitutive parameters [27].

In the Maxwell Garnett formalism, the mixture of the two constituent mediums may be envisaged as a collection of well–separated particles (of medium 'a', say) randomly dispersed in a simply connected host medium (medium 'b', say). The Maxwell Garnett estimate of the Tellegen constitutive dyadic of the HCM is [13]

$$\underline{\underline{\mathbf{K}}}_{\text{MG}}(\omega) = \underline{\underline{\mathbf{K}}}_{\text{b}}(\omega) + f_\text{a}\,\underline{\underline{\boldsymbol{\alpha}}}_{\text{a/b}}(\omega) \bullet \left(\underline{\underline{\mathbf{I}}} - i\omega f_\text{a}\,\underline{\underline{\mathbf{D}}}^0_{I/\text{b}}(\omega) \bullet \underline{\underline{\boldsymbol{\alpha}}}_{\text{a/b}}(\omega) \right)^{-1}. \quad (6.2)$$

Herein, the depolarization dyadic $\underline{\underline{\mathbf{D}}}^0_{I/\text{b}}(\omega)$ for a spherical region in constituent medium 'b' indicates the incorporation of a spherical Lorentzian cavity in the Maxwell Garnett formalism. Notice that if $\underline{\underline{\mathbf{D}}}^0_{I/\text{b}}(\omega)$ in Eq. (6.2) were to be replaced by $\underline{\underline{\mathbf{D}}}^0_{U^\text{a}/\text{b}}(\omega)$, then the Bragg–Pippard formalism would arise [28].

In Eq. (6.2), the constitutive parameters of the HCM are estimated in terms of the constitutive parameters of constituent mediums 'a' and 'b', and the shape of the particles comprising constituent medium 'a'. For certain applications it may be desirable to invert the homogenization procedure so

---

[1]It should have been called the Garnett formalism, after James Clerk Maxwell Garnett who propounded it [11]. The Maxwell Garnett formula for isotropic dielectric composite materials — which had been discovered by several other researchers before Garnett — is also called the Lorenz-Lorentz formula and the Mossotti-Clausius formula.

that the constituent parameters of constituent medium 'a' are estimated in terms of the constitutive parameters of constituent medium 'b' and the HCM, and the shape of the particles comprising constituent medium 'a'. Thus, we find [29]

$$\underline{\underline{K}}_a(\omega) = \underline{\underline{K}}_b(\omega) + \frac{1}{f_a}\left[\underline{\underline{I}} - i\omega\frac{1}{f_a}\,\underline{\underline{\alpha}}^{\dagger}_{MG/b}(\omega)\cdot\underline{\underline{D}}^0_{U^a/b}(\omega)\right]^{-1}\cdot\underline{\underline{\alpha}}^{\dagger}_{MG/b}(\omega),$$
(6.3)

where

$$\underline{\underline{\alpha}}^{\dagger}_{MG/b}(\omega) = \left\{\underline{\underline{I}} + i\omega\left[\underline{\underline{K}}_{MG}(\omega) - \underline{\underline{K}}_b(\omega)\right]\cdot\underline{\underline{D}}^0_{I/b}(\omega)\right\}^{-1}$$
$$\cdot\left[\underline{\underline{K}}_{MG}(\omega) - \underline{\underline{K}}_b(\omega)\right].$$
(6.4)

The Maxwell Garnett estimate represented by Eq. (6.2) is valid for $f_a \lesssim 0.3$ only. In order to overcome this restriction, the following two refinements of the Maxwell Garnett formalism have been established.

In the *incremental* Maxwell Garnett homogenization formalism, the estimate of the HCM constitutive dyadic is constructed incrementally, by adding the constituent medium 'a' to the constituent medium 'b' not all at once but in a fixed number N of stages [30]. After each increment, the composite is homogenized using the Maxwell Garnett formalism. Thereby, the iteration scheme

$$\underline{\underline{K}}(\omega, n+1) = \underline{\underline{K}}(\omega, n) + \left(1 - f_b^{1/N}\right)\underline{\underline{\alpha}}_{a/n}(\omega)$$
$$\cdot\left[\underline{\underline{I}} - i\omega\left(1 - f_b^{1/N}\right)\underline{\underline{D}}^0_{I/n}(\omega)\cdot\underline{\underline{\alpha}}_{a/n}(\omega)\right]^{-1}$$
(6.5)

emerges for $n = 0, 1, \ldots, N-1$, where $\underline{\underline{K}}(\omega, 0) = \underline{\underline{K}}_b(\omega)$. In Eq. (6.5), $\underline{\underline{\alpha}}_{a/n}(\omega)$ is the polarizability density dyadic of a particle of constituent material 'a', of shape specified by $\underline{U}^a$, relative to the homogeneous medium characterized by the Tellegen constitutive dyadic $\underline{\underline{K}}(\omega, n)$. The incremental Maxwell Garnett estimate of the Tellegen constitutive dyadic of the HCM can be stated as

$$\underline{\underline{K}}_{IMG}(\omega) = \underline{\underline{K}}(\omega, N).$$
(6.6)

In the limit $N \to \infty$, the incremental Maxwell Garnett formalism gives rise to the *differential* Maxwell Garnett homogenization formalism [31]. In other words, the differential Maxwell Garnett estimate $\underline{\underline{K}}_{DMG}(\omega)$ of the Tellegen constitutive dyadic of the HCM is the solution of the ordinary differential equation

$$\frac{\partial}{\partial s}\underline{\underline{K}}(\omega, s) = \frac{1}{1-v}\,\underline{\underline{\alpha}}_{a/s}(\omega),$$
(6.7)

with initial value $\underline{\underline{K}}(\omega, 0) = \underline{\underline{K}}_b(\omega)$. The continuous variable represents the volume fraction of constituent medium 'a' added to medium 'b', and $\underline{\underline{\alpha}}_{a/s}(\omega)$ is the polarizability density dyadic of a particle made of constituent medium 'a', and of shape given by $\underline{U}^a$, immersed in a medium with constitutive dyadic $\underline{\underline{K}}(\omega, s)$. The particular solution of Eq. (6.7) satisfying $s = f_a$ is the differential Maxwell Garnett estimate; i.e.,

$$\underline{\underline{K}}_{DMG}(\omega) = \underline{\underline{K}}(\omega, f_a). \tag{6.8}$$

We note that the recursive approach that underpins the incremental and differential Maxwell Garnett formalisms has been applied quite widely within the realm of isotropic dielectric HCMs [32–35].

6.3   Bruggeman formalism

A key characteristic of the Bruggeman homogenization formalism is that the constituent mediums 'a' and 'b' are treated symmetrically [36, 37]. Consequently, the Bruggeman formalism is applicable for all volume fractions $f_a \in (0, 1)$, unlike the conventional Maxwell Garnett formalism. The symmetrical treatment of the constituent mediums enables the phenomenon of percolation threshold to be predicted by the Bruggeman formalism, albeit not always correctly [38–41][2]. Several variations on the Bruggeman formalism have been developed for isotropic dielectric–magnetic mediums [43–45]. Furthermore, by a microscopic consideration of the local field effects, the Bruggeman formalism is found to arise naturally from the Maxwell Garnett formalism [46]. The rigorous establishment of the Bruggeman formalism for the most general linear scenarios follows from the SPFT [16].

Within the Bruggeman formalism, the two populations of ellipsoidal particles which make up the two constituent mediums need not have the same shape; i.e., $\underline{U}^a \neq \underline{U}^b$, in general. The polarizability density dyadics implemented in the Bruggeman formalism, for each constituent medium, are calculated relative to the HCM itself; i.e., the polarizability density dyadic $\underline{\underline{\alpha}}_{\ell/Br}(\omega)$, $(\ell = a,b)$, is defined in terms of the Bruggeman estimate $\underline{\underline{K}}_{Br}(\omega)$ of the Tellegen constitutive dyadic of the HCM.

The assertion that the net polarizability density is zero throughout the HCM underlies the Bruggeman homogenization formalism. Thus, the Bruggeman estimate of the HCM constitutive dyadic is provided implicitly

---

[2]Percolation threshold may be more accurately treated using more numerically intensive techniques, based on Monte Carlo simulations for example [42].

by the nonlinear equation[3] [13]

$$f_a \underline{\underline{\alpha}}_{a/Br}(\omega) + f_b \underline{\underline{\alpha}}_{b/Br}(\omega) = \underline{\underline{0}}, \qquad (6.9)$$

from which $\underline{\underline{K}}_{Br}(\omega)$ can be extracted by applying the simple Jacobi technique [50]. Hence, the recursive solution

$$\underline{\underline{K}}_{Br}(\omega, n) = \mathcal{S}\left\{ \underline{\underline{K}}_{Br}(\omega, n-1) \right\}, \qquad (n = 1, 2, \ldots) \qquad (6.10)$$

is developed, with the initial value $\underline{\underline{K}}_{Br}(\omega, 0) = \underline{\underline{K}}_{MG}(\omega)$. The action of the operator $\mathcal{S}$ is defined by

$$\begin{aligned}
\mathcal{S}\left\{ \underline{\underline{K}}_{Br}(\omega) \right\} = &\left(f_a \underline{\underline{K}}_a(\omega) \cdot \left\{ \underline{\underline{I}} + i\omega \underline{\underline{D}}^0_{U^a/Br}(\omega) \right.\right. \\
&\left. \cdot \left[\underline{\underline{K}}_a(\omega) - \underline{\underline{K}}_{Br}(\omega) \right] \right\}^{-1} + f_b \underline{\underline{K}}_b(\omega) \\
&\left. \cdot \left\{ \underline{\underline{I}} + i\omega \underline{\underline{D}}^0_{U^b/Br}(\omega) \cdot \left[\underline{\underline{K}}_b(\omega) - \underline{\underline{K}}_{Br}(\omega) \right] \right\}^{-1} \right) \\
&\cdot \left(f_a \left\{ \underline{\underline{I}} + i\omega \underline{\underline{D}}^0_{U^a/Br}(\omega) \cdot \left[\underline{\underline{K}}_a(\omega) - \underline{\underline{K}}_{Br}(\omega) \right] \right\}^{-1} \right. \\
&\left. + f_b \left\{ \underline{\underline{I}} + i\omega \underline{\underline{D}}^0_{U^b/Br}(\omega) \cdot \left[\underline{\underline{K}}_b(\omega) - \underline{\underline{K}}_{Br}(\omega) \right] \right\}^{-1} \right)^{-1}.
\end{aligned}$$
$$(6.11)$$

In a similar manner to the Maxwell Garnett equation presented in Sec. 6.2, the Bruggeman equation (6.9) may be inverted in order to deliver an estimate of the constituent parameters of constituent medium 'a' in terms of the constitutive parameters of constituent medium 'b' and the HCM, and the shapes of the particles of the constituent mediums 'a' and 'b'; we obtain [29]

$$\underline{\underline{K}}_a(\omega) = \underline{\underline{K}}_{Br}(\omega) - \left[\frac{f_a}{1-f_a} \underline{\underline{I}} + i\omega \underline{\underline{\alpha}}_{a/Br}(\omega) \cdot \underline{\underline{D}}^0_{I/b}(\omega) \right]^{-1} \cdot \underline{\underline{\alpha}}_{a/Br}(\omega).$$
$$(6.12)$$

Notice that the inverted formalism yields an explicit estimate of $\underline{\underline{K}}_a(\omega)$, in contrast to the implicit estimate of $\underline{\underline{K}}_{Br}(\omega)$ which may be extracted from the nonlinear Bruggeman equation (6.9).

---

[3]The Bruggeman equation for isotropic dielectric HCMs precedes and is equivalent to an equation derived by Polder & van Santen [47], as shown later by Bohren and Battan [48], as well as to Böttcher's equation [49].

6.4   Strong–property–fluctuation theory

6.4.1   *Background*

The provenance of the strong–property–fluctuation theory[4] (SPFT) lies in wave–propagation studies for continuous random mediums [51, 52]. It was subsequently adapted to estimate the constitutive parameters of HCMs [15]. In contrast to the conventional Maxwell Garnett and Bruggeman formalisms, the SPFT accommodates a comprehensive description of the distributional statistics of the constituent mediums. Thereby, coherent scattering losses may be accounted for. Within the SPFT regime, the electrically small ellipsoidal particles comprising the two constituent mediums are assumed to be specified by the same shape dyadic; i.e., $\underline{\underline{U}}^{\mathrm{a}} = \underline{\underline{U}}^{\mathrm{b}} = \underline{\underline{U}}$, as illustrated in Fig. 6.2.

The SPFT approach to homogenization is based upon the recursive refinement of a comparison medium; i.e., the homogeneous medium specified by the Tellegen constitutive dyadic $\underline{\underline{K}}_{\mathrm{comp}}(\omega)$. The refinement process involves applying a Feynman–diagrammatic technique to ensemble average a Born–series representation of the electromagnetic fields. The straightforward approach is limited to weak spatial fluctuations in the infinity norm of

$$\underline{\underline{K}}_{\mathrm{a}}(\omega)\,\Phi_{\mathrm{a}}(\underline{r}) + \underline{\underline{K}}_{\mathrm{b}}(\omega)\,\Phi_{\mathrm{b}}(\underline{r}) - \underline{\underline{K}}_{\mathrm{comp}}(\omega), \tag{6.13}$$

but a careful consideration of the singularity of the DGF $\underline{\underline{G}}_{\mathrm{comp}}(\underline{r}-\underline{r}',\omega)$ of the comparison medium allows a reformulation that is applicable for strong fluctuations.

When electromagnetic wavelengths are much larger than the length–scales of nonhomogeneities, the mixture of constituent mediums may viewed as being effectively homogeneous. Within this long–wavelength regime, the nth–order estimate $\underline{\underline{K}}_{\mathrm{SPFT}}^{[n]}(\omega)$ of the Tellegen constitutive dyadic of the HCM is provided by the SPFT as [16, 53]

$$\underline{\underline{K}}_{\mathrm{SPFT}}^{[n]}(\omega) = \underline{\underline{K}}_{\mathrm{comp}}(\omega) - \frac{1}{i\omega}\left[\underline{\underline{I}} + \underline{\underline{\Sigma}}^{[n]}(\omega) \bullet \underline{\underline{D}}_{U/\mathrm{comp}}^{0}(\omega)\right]^{-1} \bullet \underline{\underline{\Sigma}}^{[n]}(\omega), \tag{6.14}$$

wherein the *mass operator* term $\underline{\underline{\Sigma}}^{[n]}(\omega)$ — which has an infinite series representation — is defined in terms of the DGF for the comparison medium, together with the corresponding polarizability density dyadics. The condition

$$\left\langle \Phi_{\mathrm{a}}(\underline{r})\,\underline{\underline{\alpha}}_{\mathrm{a/comp}}(\omega) + \Phi_{\mathrm{b}}(\underline{r})\,\underline{\underline{\alpha}}_{\mathrm{b/comp}}(\omega) \right\rangle_{\mathrm{e}} = \underline{\underline{0}} \tag{6.15}$$

---

[4]It is otherwise known as the strong–*permittivity*–fluctuation theory in the context of dielectric mediums.

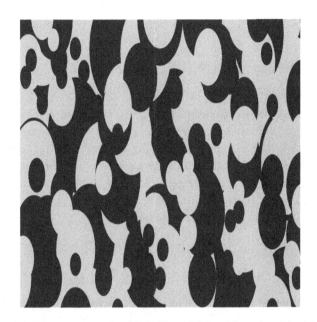

Figure 6.2   As Fig. 6.1 except that the ellipsoidal particles comprising the two constituent mediums are specified by the same shape dyadic; i.e., $\underline{\underline{U}}^a = \underline{\underline{U}}^b$.

is imposed in order to eliminate secular terms from the Born–series representation [54].

6.4.2   *Estimates of constitutive parameters*

6.4.2.1   *Zeroth and first order*

Both the zeroth–order and first–order mass operator are null–valued; i.e.,

$$\underline{\underline{\Sigma}}^{[0]}(\omega) = \underline{\underline{\Sigma}}^{[1]}(\omega) = \underline{\underline{0}}. \qquad (6.16)$$

The estimates $\underline{\underline{K}}^{[0]}_{\mathrm{SPFT}}(\omega) = \underline{\underline{K}}^{[1]}_{\mathrm{SPFT}}(\omega) = \underline{\underline{K}}_{\mathrm{comp}}(\omega)$ may be extracted from the condition (6.15). In fact, as the condition (6.15) is equivalent the Bruggeman equation (6.9), the lowest–order SPFT estimates of the Tellegen constitutive dyadic of the HCM are identical to its Bruggeman estimate.

6.4.2.2   *Second order*

Most commonly, the SPFT is implemented at the second–order level of approximation — otherwise known as the bilocal approximation [15, 19, 17, 16, 55, 18]. Therein, the distributional statistics of the constituent mediums 'a' and 'b' are characterized by the two–point covariance function

$$\Lambda(\underline{r} - \underline{r}') = \langle \Phi_a(\underline{r})\Phi_a(\underline{r}')\rangle_e - \langle \Phi_a(\underline{r})\rangle_e \langle \Phi_a(\underline{r}')\rangle_e$$
$$= \langle \Phi_b(\underline{r})\Phi_b(\underline{r}')\rangle_e - \langle \Phi_b(\underline{r})\rangle_e \langle \Phi_b(\underline{r}')\rangle_e, \qquad (6.17)$$

along with its associated correlation length L_{cor}. Within a region of linear cross–sectional dimensions given by L_{cor}, and of shape dictated by the covariance function, the correlated scattering responses of the constituent particles give rise to an attenuation of the macroscopic coherent field. In contrast, the scattering responses of particles separated by distances much greater than L_{cor} are statistically independent.

The second–order mass operator is given by [16]

$$\underline{\underline{\Sigma}}^{[2]}(L_{cor}, \omega) = -\omega^2 \left[\underline{\underline{\alpha}}_{a/comp}(\omega) - \underline{\underline{\alpha}}_{b/comp}(\omega) \right]$$
$$\bullet \underline{\underline{P}}_\Lambda(L_{cor}, \omega) \bullet \left[\underline{\underline{\alpha}}_{a/comp}(\omega) - \underline{\underline{\alpha}}_{b/comp}(\omega) \right], \quad (6.18)$$

where the dependency on the correlation length has been made explicit through

$$\underline{\underline{P}}_\Lambda(L_{cor}, \omega) = \mathsf{P} \int_{\underline{R}} \Lambda(\underline{R}) \underline{\underline{G}}_{comp}(\underline{R}, \omega) \, d^3\underline{R}, \qquad (6.19)$$

with $\mathsf{P} \int \ldots d^3\underline{R}$ denoting principal–value integration.

The principal–value integral specified in Eq. (6.19) has been investigated theoretically and numerically for various physically motivated choices of covariance function [56, 57], and most conspicuously for the step function $\Lambda(\underline{r} - \underline{r}') = \Lambda_{step}(\underline{r} - \underline{r}')$ for which

$$\langle \Phi_\ell(\underline{r})\Phi_\ell(\underline{r}')\rangle_e = \begin{cases} f_\ell, & |\underline{\underline{U}}^{-1} \bullet (\underline{r} - \underline{r}')| \leq L_{cor}, \\ f_\ell^2, & |\underline{\underline{U}}^{-1} \bullet (\underline{r} - \underline{r}')| > L_{cor}, \end{cases} \quad (\ell = \text{a,b}).$$
$$(6.20)$$

These studies have revealed that the form of the covariance function has only a weak influence upon the SPFT estimate $\underline{\underline{K}}^{[2]}_{SPFT}(\omega)$ [57].

Upon adopting the covariance function $\Lambda_{step}(\underline{r} - \underline{r}')$, the principal–value integral in Eq. (6.19) may be expressed in the more tractable form

$$\underline{\underline{P}}_{\Lambda_{step}}(L_{cor}, \omega) = f_a f_b \underline{\underline{D}}^+_{I/comp}(L_{cor}, \omega), \qquad (6.21)$$

with the dyadic $\underline{\underline{D}}^+_{U/comp}(L_{cor}, \omega)$ being defined in Eq. (5.66)$_2$. As described in Sec. 5.5, some analytical progress has been reported towards the simplification of the integral on the right side of Eq. (6.21) on an unbounded domain: For certain anisotropic [18, 19] and bianisotropic [16] comparison mediums, a two–dimensional integral emerges from Eq. (6.21) (which is equivalent to that given in Eq. (5.74)), but numerical methods are generally required to evaluate that integral. Explicit expressions are available for isotropic chiral and uniaxial dielectric (or magnetic) comparison mediums, per Eqs. (5.79) and (5.86).

6.4.2.3 *Third order*

The calculation of the third–order (i.e., trilocally approximated) mass operator term $\underline{\underline{\Sigma}}^{[3]}$ necessitates a three–point covariance function [58–60]. The two–point covariance function represented in Eq. (6.20), when incorporated into $\Lambda_{\text{step}}(\underline{r} - \underline{r}')$, generalizes to give

$$\langle\, \Phi_{\text{a}}(\underline{r})\, \Phi_{\text{a}}(\underline{r}')\, \Phi_{\text{a}}(\underline{r}'')\,\rangle = \begin{cases} f_{\text{a}}^3, & \min\{L_{12}, L_{13}, L_{23}\} > L_{\text{cor}} \\[2mm] f_{\text{a}}, & \max\{L_{12}, L_{13}, L_{23}\} \le L_{\text{cor}} \\[2mm] \frac{1}{3}\,(f_{\text{a}} + 2f_{\text{a}}^3), & \text{one of } L_{12}, L_{13}, L_{23} \le L_{\text{cor}} \\[2mm] \frac{1}{3}\,(2f_{\text{a}} + f_{\text{a}}^3), & \text{two of } L_{12}, L_{13}, L_{23} \le L_{\text{cor}} \end{cases} \tag{6.22}$$

where the lengths

$$\left. \begin{aligned} L_{12} &= |\,\underline{\underline{U}}^{-1} \boldsymbol{\cdot} (\underline{r} - \underline{r}')\,| \\ L_{13} &= |\,\underline{\underline{U}}^{-1} \boldsymbol{\cdot} (\underline{r} - \underline{r}'')\,| \\ L_{23} &= |\,\underline{\underline{U}}^{-1} \boldsymbol{\cdot} (\underline{r}' - \underline{r}'')\,| \end{aligned} \right\}. \tag{6.23}$$

After implementing Eq. (6.22), the third–order mass operator term emerges as [60]

$$\underline{\underline{\Sigma}}^{[3]}(L_{\text{cor}}, \omega) = \underline{\underline{\Sigma}}^{[2]}(L_{\text{cor}}, \omega) + \frac{i\omega^3 f_{\text{a}}(1 - 2f_{\text{a}})}{3(1 - f_{\text{a}})^2}\, \underline{\underline{\alpha}}_{\text{a/comp}}(\omega)$$

$$\boldsymbol{\cdot} \Big[\underline{\underline{V}}(\omega) \boldsymbol{\cdot} \underline{\underline{\alpha}}_{\text{a/comp}}(\omega) \boldsymbol{\cdot} \underline{\underline{P}}_{\Lambda_{\text{step}}}(L_{\text{cor}}, \omega)$$

$$+ \underline{\underline{P}}_{\Lambda_{\text{step}}}(L_{\text{cor}}, \omega) \boldsymbol{\cdot} \underline{\underline{\alpha}}_{\text{a/comp}}(\omega) \boldsymbol{\cdot} \underline{\underline{V}}(\omega)$$

$$+ \underline{\underline{P}}_{\Lambda_{\text{step}}}(L_{\text{cor}}, \omega) \boldsymbol{\cdot} \underline{\underline{\alpha}}_{\text{a/comp}}(\omega) \boldsymbol{\cdot} \underline{\underline{P}}_{\Lambda_{\text{step}}}(L_{\text{cor}}, \omega) \Big]$$

$$\boldsymbol{\cdot} \underline{\underline{\alpha}}_{\text{a/comp}}(\omega), \tag{6.24}$$

wherein the 6×6 dyadic

$$\underline{\underline{V}}(\omega) = \frac{1}{i\omega}\, \underline{\underline{K}}_{\text{comp}}^{-1}(\omega) - \underline{\underline{D}}_{U/\text{comp}}^0(\omega). \tag{6.25}$$

Strictly, Eq. (6.24) holds for isotropic chiral comparison mediums and spherical constituent particles, but it is probable that it also holds for weakly bianisotropic mediums with diagonally–dominant 3×3 constitutive dyadics. Upon adopting Eq. (6.24), SPFT convergence at the level of the bilocal approximation has been demonstrated for certain isotropic chiral HCMs, as well as bianisotropic HCMs of the Faraday chiral type which are both weakly uniaxial and weakly gyrotropic [60].

6.5 Extended homogenization formalisms

The Maxwell Garnett and Bruggeman formalisms, as well as the SPFT approach, rely on depolarization dyadics to represent the scattering responses of the constituent particles that make up the composites. The usual convention was adopted in Secs. 6.2 to 6.4 of taking $\underline{\underline{\mathbf{D}}}^0_{U/\text{comp}}(\omega)$ as the depolarization dyadic, thereby neglecting the spatial extent of the constituent particles. Extended versions of the Maxwell Garnett (including its incremental and differential variants) and Bruggeman formalisms, and the SPFT, can be realized by instead taking $\underline{\underline{\mathbf{D}}}_{U/\text{comp}}(\rho,\omega)$ as the depolarization dyadic, per Eq. (5.65) [61]. In the most general setting, these extended versions are appropriate to bianisotropic composite mediums comprising ellipsoidal constituent particles [62]. Furthermore, convergence at the second–order level of approximation has been established for the extended SPFT applied to certain bianisotropic HCMs [63]. Similar extensions — not deriving from the spectral representation of the depolarization dyadic — have also been developed for the Maxwell Garnett and Bruggeman formalisms for isotropic HCMs [64–69]

It should be emphasized that, while such extended formalisms take into account the spatial extent of the constituent particles, the size parameter ρ must be much smaller than electromagnetic wavelengths in order to be consistent with the notion of homogenization. The constituent particles are, in effect, modelled as electric and magnetic dipoles, via their depolarization dyadic representation. Higher–order multipoles are not relevant to this formulation. For a discussion of some problems connected with higher–order multipole approaches to homogenization, the reader is referred elsewhere [70].

Extended formalisms can be expected to produce more accurate estimates of the HCM constitutive parameters than their unextended counterparts. Whether that is true or not for a specific composite material requires comparison with actual experimental data. The implementation of an extended formalism is usually a more involved process than that of its unextended version, as the former is likely to require the use of numerical methods. As described in Sec. 5.5, explicit representations of $\underline{\underline{\mathbf{D}}}_{U/\text{comp}}(\rho, \omega)$ are available only for isotropic and certain uniaxial scenarios. It is illuminating to present the explicit form of the extended second–order SPFT for the simplest possible homogenization scenario: that where the constituent mediums are isotropic dielectric (or, analogously, isotropic magnetic) mediums with permittivities $\epsilon_\ell(\omega)$ ($\ell = $ a,b), distributed as spherical particles. In consonance, the comparison medium is also an isotropic dielectric medium with permittivity $\epsilon_{\text{comp}}(\omega)$. The corresponding extended second–order SPFT estimate of the HCM permittivity is [71]

$$\epsilon_{\text{SPFT}}^{[2]}(\omega) = \epsilon_{\text{comp}}(\omega) - \frac{1}{i\omega}\left[\frac{\Sigma^{[2]}(\rho, L_{\text{cor}}, \omega)}{1 + \Sigma^{[2]}(\rho, L_{\text{cor}}, \omega)\, d(\rho, \omega)} \right], \qquad (6.26)$$

wherein the depolarization scalar may be expressed as the sum

$$d(\rho, \omega) = d^0(\omega) + d^+(\rho, \omega). \qquad (6.27)$$

According to Eqs. (5.76) and (5.77)$_1$, the ρ–independent contribution to $d(\rho, \omega)$ is given by

$$d^0(\omega) = \frac{1}{i3\omega\epsilon_{\text{comp}}(\omega)}, \qquad (6.28)$$

whereas the ρ–dependent contribution is

$$d^+(\rho, \omega) = \frac{i\omega\rho^2\mu_0}{9}\left\{ 3 + i2\rho\omega\left[\epsilon_{\text{comp}}(\omega)\mu_0\right]^{1/2} \right\}. \qquad (6.29)$$

The scalar mass operator term in Eq. (6.26) follows from the isotropic dielectric specialization of Eq. (6.18) as

$$\Sigma^{[2]}(\rho, L_{\text{cor}}, \omega) = -f_{\text{a}}f_{\text{b}}\omega^2\left[\alpha_{\text{a/comp}}(\rho, \omega) - \alpha_{\text{b/comp}}(\rho, \omega)\right]^2 d^+(L_{\text{cor}}, \omega), \qquad (6.30)$$

with the polarizability density scalars[5]

$$\alpha_{\ell/\text{comp}}(\rho, \omega) = \frac{\epsilon_\ell(\omega) - \epsilon_{\text{comp}}(\omega)}{1 + i\omega\left[\epsilon_\ell(\omega) - \epsilon_{\text{comp}}(\omega)\right] d(\rho, \omega)}, \qquad (\ell = \text{a,b}). \quad (6.31)$$

A noteworthy consequence of the incorporation of the size parameter ρ is that the extended Maxwell Garnett and Bruggeman formalisms predict that dissipative HCMs can arise even though the constituent mediums are

---

[5]The polarizability density scalars defined in Eq. (6.31) are equivalent to the scalar equivalents of those defined in Eq. (5.99) when the depolarization scalar $d(\rho, \omega)$ is replaced by $d^0(\omega)$.

nondissipative. This is immediately apparent from Eqs. (6.28) and (6.29): for a nondissipative comparison medium, $i\,d^0(\omega)$ is real–valued whereas $i\,d^+(\rho, \omega)$ has a nonzero imaginary part. The dissipation may be attributed to scattering losses from the incoherent field. For the second–order (and third–order) SPFT, this phenomenon also occurs for the *unextended* formulation due to the incorporation of the correlation length — again, this is immediately apparent from the complex–valued nature of $d^+(L_{cor}, \omega)$ for a nondissipative comparison medium.

Let us close this section by reflecting upon the following question: For what range of sizes of the constituent particles are the extended homogenization formalisms discussed herein appropriate? The requirement that the particles must be small relative to the electromagnetic wavelength(s), in order to be consistent with the notion of homogenization, immediately establishes an upper bound [24, 25]. At optical wavelengths, the linear dimensions of the constituent particles must therefore be at most 38–78 nm. A lower bound exists because, at sufficiently small length–scales, quantum processes cannot be neglected in the description of the constituent particles and their interactions. At such small length–scales, the concept of macroscopic electromagnetics is no longer appropriate. The lower bound is material–dependent. For example, in a recent study using spectroscopic ellipsometry, it was found that an extended Maxwell Garnett homogenization formalism adequately characterizes a HCM rather well for silver constituent particles as small as 2.3 nm [72]. Parenthetically, this study also highlighted the prospect of utilizing extended homogenization formalisms in particle–sizing applications for nanocomposites. Notice that the constitutive parameters of very small constituent particles may differ significantly from those of the corresponding bulk materials and depend sensitively upon the shape and size of the particles [73]. This is especially true for metallic particles smaller than the mean free path of conduction electrons in the bulk metal, wherein the mean free path may be dominated by collisions at the particle's boundary [74].

6.6   Anisotropy and bianisotropy via homogenization

Manifestations of anisotropy and bianisotropy may be readily conceptualized through the process of homogenization. Since HCMs 'inherit' the symmetries of their constituent mediums, complex HCMs can arise from relatively simple constituent mediums.

In this regard, biaxial HCMs have been comprehensively studied, both in anisotropic [75, 76] and bianisotropic [77] settings. A biaxial HCM generally arises when the constituent mediums present two distinguished axes of symmetry — crucially, the two distinguished axes must not be collinear. The distinguished axes may be either of electromagnetic origin, as in the case of uniaxial constituent mediums for example, or they may originate from the shapes of the constituent particles, as in the case of spheroidal particles for example. Hence, a biaxial HCM — which is anisotropic or bianisotropic — may develop from constituent mediums 'a' and 'b' wherein:

- both constituent mediums are uniaxial and distributed as spherical particles;
- both constituent mediums are isotropic and distributed as spheroidal particles;
- both constituent mediums are isotropic, with one medium distributed as spherical particles and the other as ellipsoidal particles; or
- one of the constituent mediums is uniaxial while the other is isotropic, and one of the constituent mediums is distributed as spheroidal particles while the other as spherical particles.

If the distinguished axes of the constituent mediums are orthogonal, then orthorhombic biaxial HCMs generally develop; otherwise, monoclinic or triclinic biaxial HCMs arise.

Another example of a bianisotropic HCM arising from relatively simple constituents is provided by Faraday chiral mediums (FCMs), as described in Sec. 2.3.4. FCMs have been studies extensively using the Maxwell Garnett and the Bruggeman formalisms [78, 79], as well as the SPFT [16].

6.7 Homogenized composite mediums as metamaterials

Through the process of homogenization, constitutive–parameter regimes may be accessed which would otherwise be inaccessible and material properties may be extended beyond those of the constituent mediums. That is, metamaterials may be realized as HCMs. Anisotropy and bianisotropy generally present greater opportunities for such metamaterials. Some examples of scenarios in which homogenization may give rise to metamaterials are as follows:

(i) HCMs generally exhibit fewer symmetries than do their constituent mediums. Therefore, as outlined in Sec. 6.6, anisotropic HCMs can arise from isotropic constituent mediums, and bianisotropic HCMs can arise from isotropic and/or anisotropic constituent mediums [9].

(ii) HCMs supporting a host of exotic planewave phenomenons that are not supported by their constituent mediums may be conceptualized. With reference to Sec. 4.8, possible scenarios include HCMs which support:

   - NPV propagation [80, 81];
   - hyperbolic dispersion relations [82];
   - Voigt wave propagation [83]; and
   - negative reflection [84].

(iii) The group velocity in a HCM can be higher in magnitude than those in its constituent mediums [85, 86][6]. The scope for group–velocity enhancement is greater in anisotropic HCMs as compared to isotropic HCMs [89]. Furthermore, group–velocity diminishment can also be achieved via homogenization [90].

(iv) The degree of nonlinearity exhibited by a HCM can exceed that exhibited by its constituent materials [91]. We elaborate on this matter in Sec. 7.3.

Caution is required in applying homogenization formalisms in unconventional constitutive–parameter regimes such as those associated with metamaterials. A simple illustration of the problems that can arise is provided by the Bruggeman estimate of the permittivity of an isotropic dielectric HCM, namely $\epsilon_{Br}(\omega)$, arising from two isotropic dielectric constituent mediums with permittivities $\epsilon_{a}(\omega)$ and $\epsilon_{b}(\omega)$, randomly distributed as electrically small spherical particles. The isotropic dielectric specialization of Eq. (6.9)

---

[6]The group velocity $\underline{v}_{gr}(\omega)$ of a wavepacket is conventionally defined in terms of the gradient of the angular frequency ω with respect to the wavevector $\underline{k}$ [87]; i.e.,

$$\underline{v}_{gr}(\omega) = \nabla_{\underline{k}} \, \omega \big|_{\omega=\omega(k_{avg})} \, , \tag{6.32}$$

where k_{avg} denotes the average wavenumber of the wavepacket. Herein the compact notation

$$\nabla_{\underline{k}} \equiv \left(\frac{\partial}{\partial k_x}, \frac{\partial}{\partial k_y}, \frac{\partial}{\partial k_z} \right) \tag{6.33}$$

for the gradient operator with respect to $\underline{k}$ is adopted, where (k_x, k_y, k_z) is the representation of $\underline{k}$ in terms of its Cartesian components, and it is assumed that $\underline{k} \in \mathbb{R}^3$. When $\underline{k} \in \mathbb{C}^3$ (with Im $\{\underline{k}\} \neq \underline{0}$), the wavevector $\underline{k}$ in Eq. (6.32) should be replaced by Re $\{\underline{k}\}$ [88]. For a nondissipative anisotropic dielectric–magnetic medium with $\underline{k} \in \mathbb{R}^3$, the group velocity coincides with the velocity of energy transport [87].

yields the quadratic equation

$$2\left[\epsilon_{Br}(\omega)\right]^2 + \epsilon_{Br}(\omega)\left[\epsilon_a(\omega)\left(f_b - 2f_a\right) + \epsilon_b(\omega)\left(f_a - 2f_b\right)\right] - \epsilon_a(\omega)\epsilon_b(\omega) = 0, \tag{6.34}$$

which has the discriminant

$$\Delta_{Br} = \left[\epsilon_a(\omega)\left(f_b - 2f_a\right) + \epsilon_b(\omega)\left(f_a - 2f_b\right)\right]^2 + 8\epsilon_a(\omega)\epsilon_b(\omega). \tag{6.35}$$

An insight into the applicability of the Bruggeman formalism may be gained by restricting attention to nondissipative constituent mediums (i.e., $\epsilon_{a,b} \in \mathbb{R}$). It is then clear that Δ_{Br} can be negative–valued provided that $\epsilon_a \epsilon_b < 0$. For example, if $f_a = f_b$ then $\Delta_{Br} < 0$ provided that $\left[\epsilon_a(\omega) + \epsilon_b(\omega)\right]^2 < -32\epsilon_a(\omega)\epsilon_b(\omega)$. Thus, the Bruggeman estimate ϵ_{Br} for $\epsilon_a \epsilon_b < 0$ may be complex–valued with nonzero imaginary part, even though neither component medium is dissipative. This is physically suspect, since the Bruggeman formalism — which is equivalent to the lowest–order SPFT — does not explicitly take into account scattering losses nor does it take into account nonzero dimensions of the constituent particles. This problem extends to the weakly dissipative scenario wherein $\epsilon_{a,b} \in \mathbb{C}$ and $\mathrm{Re}\left\{\epsilon_a\right\}\,\mathrm{Re}\left\{\epsilon_b\right\} < 0$ [92]. This situation corresponds, for example, to certain metal–in–insulator dielectric composites which are of interest as potential metamaterials. In particular, silver is the metal of choice for many meta-material applications since it exhibits relatively low dissipation at optical and near–infrared frequencies. This property suggests that the Bruggeman formalism is not suitable for estimating the effective constitutive properties of silver–in–insulator composites at these frequencies [93], which restriction has not always been appreciated in the recent past [94–97].

The limitation of the Bruggeman homogenization formalism for certain metamaterials described here is quite general: the Maxwell Garnett formalism is likewise restricted, as are higher–order SPFT approaches, for both passive and active HCMs [98]. In a similar vein, the Wiener [99], Hashin–Shtrikman [27, 92] and Bergman–Milton [100–103] bounds on HCM constitutive parameters become exceedingly large and thereby lose their meaning for constitutive parameter regimes associated with certain metamaterials.

Not only should the constituent particles of a composite medium be electrically small for the application of homogenization formalisms [24, 25], but the particles must be numerous and distributed statistically uniformly throughout the composite medium. As a singular case, the homogenization of a composite medium containing just one inclusion, howsoever small, is an ill–conceived exercise [104].

References

[1] J.C. Garland and D.B. Tanner, *Electrical transport and optical properties of inhomogeneous media*, American Institute of Physics, New York, NY, USA, 1978.

[2] A. Priou (ed), *Dielectric properties of heterogeneous materials*, Elsevier, New York, NY, USA, 1992.

[3] A. Lakhtakia, On direct and indirect scattering approaches for homogenization of particulate composites, *Microw Opt Technol Lett* **25** (2000), 53–56.

[4] J. Van Kranendonk and J.E. Sipe, Foundations of the macroscopic electromagnetic theory of dielectric media, *Prog Optics* **15** (1977), 245–350.

[5] J.Z. Buchwald, *From Maxwell to microphysics: Aspects of electromagnetic theory in the last quarter of the nineteenth century*, University of Chicago Press, Chicago, IL, USA, 1985.

[6] R.M. Walser, Metamaterials: an introduction, in *Introduction to complex mediums for optics and electromagnetics* (W.S. Weiglhofer and A. Lakhtakia, eds), SPIE Press, Bellingham, WA, USA, 2003, 295–316.

[7] A.H. Sihvola and I.V. Lindell, Chiral Maxwell–Garnett mixing formula, *Electron Lett* **26** (1990), 118–119.

[8] A. Lakhtakia, Frequency–dependent continuum electromagnetic properties of a gas of scattering centers, *Adv Chem Phys* **85**(2) (1993), 312–359.

[9] B. Michel, Recent developments in the homogenization of linear bianisotropic composite materials, in *Electromagnetic fields in unconventional materials and structures* (O.N. Singh and A. Lakhtakia, eds), Wiley, New York, NY, USA, 2000, 39–82.

[10] A. Lakhtakia (ed), *Selected papers on linear optical composite materials*, SPIE Optical Engineering Press, Bellingham, WA, USA, 1996.

[11] J.C. Maxwell Garnett, Colours in metal glasses and in metallic films, *Phil Trans R Soc Lond A* **203** (1904), 385–420.

[12] D.A.G. Bruggeman, Berechnung verschiedener physikalischer Konstanten von heterogenen Substanzen, I. Dielektrizitätskonstanten und Leitfähigkeiten der Mischkörper aus isotropen Substanzen, *Ann Phys Lpz* **24** (1935), 636–679.

[13] W.S. Weiglhofer, A. Lakhtakia and B. Michel, Maxwell Garnett and Bruggeman formalisms for a particulate composite with bianisotropic host medium, *Microw Opt Technol Lett* **15** (1997), 263–266. Corrections: **22** (1999), 221.

[14] B. Michel and W.S. Weiglhofer, Pointwise singularity of dyadic Green function in a general bianisotropic medium, *Arch Elektron Übertrag* **51** (1997), 219–223. Corrections: **52** (1998), 310.

[15] L. Tsang and J.A. Kong, Scattering of electromagnetic waves from random media with strong permittivity fluctuations, *Radio Sci* **16** (1981), 303–320.

[16] T.G. Mackay, A. Lakhtakia and W.S. Weiglhofer, Strong–property–fluctuation theory for homogenization of bianisotropic composites: formulation, *Phys Rev E* **62** (2000), 6052–6064. Corrections: **63** (2001), 049901.

[17] B. Michel and A. Lakhtakia, Strong–property–fluctuation theory for homogenizing chiral particulate composites, *Phys Rev E* **51** (1995), 5701–5707.

[18] Z.D. Genchev, Anisotropic and gyrotropic version of Polder and van Santen's mixing formula, *Waves Random Media* **2** (1992), 99–110.

[19] N.P. Zhuck, Strong–fluctuation theory for a mean electromagnetic field in a statistically homogeneous random medium with arbitrary anisotropy of electrical and statistical properties, *Phys Rev B* **50** (1994), 15636–15645.

[20] A.J. Duncan, T.G. Mackay and A. Lakhtakia, The homogenization of orthorhombic piezoelectric composites by the strong–property–fluctuation theory, *J Phys A: Math Theor* **42** (2009), 165402.

[21] A.H. Sihvola and O.P.M. Pekonen, Effective medium formulae for bi-isotropic mixtures, *J Phys D: Appl Phys* **29** (1996), 514–521.

[22] B. Shanker, A comment on 'Effective medium formulae for bi–anisotropic mixures', *J Phys D: Appl Phys* **30** (1997), 289–290.

[23] A.H. Sihvola and O.P.M. Pekonen, Six–vector mixing formulae defended. Reply to comment on 'Effective medium formulae for bi–isotropic mixtures', *J Phys D: Appl Phys* **30** (1997), 291–292.

[24] H.C. van de Hulst, *Light scattering by small particles*, Dover, New York, NY, USA, 1981.

[25] T.G. Mackay, Lewin's homogenization formula revisited for nanocomposite materials, *J Nanophoton* **2** (2008), 029503.

[26] H. Faxén, Der Zusammenhang zwischen den Maxwellschen Gleichungen für Dielektrika und den atomistischen Ansätzen von H.A. Lorentz u.a., *Z Phys* **2** (1920), 218–229.

[27] Z. Hashin and S. Shtrikman, A variational approach to the theory of the effective magnetic permeability of multiphase materials, *J Appl Phys* **33** (1962), 3125–3131.

[28] J.A. Sherwin and A. Lakhtakia, Bragg–Pippard formalism for bianisotropic particulate composites, *Microw Opt Technol Lett* **33** (2002), 40–44.

[29] W.S. Weiglhofer, On the inverse homogenization problem of linear composite materials, *Microw Opt Technol Lett* **28** (2001), 421–423.

[30] A. Lakhtakia, Incremental Maxwell Garnett formalism for homogenizing particulate composite media, *Microw Opt Technol Lett* **17** (1998), 276–279.

[31] B. Michel, A. Lakhtakia, W.S. Weiglhofer and T.G. Mackay, Incremental and differential Maxwell Garnett formalisms for bi–anisotropic composites, *Compos Sci Technol* **61** (2001), 13–18.

[32] K. Ghosh and R. Fuchs, Critical behaviour in the dielectric properties of random self–similar composites, *Phys Rev B* **44** (1991), 7330–7343.

[33] R.G. Barrera and R. Fuchs, Theory of electron energy loss in a random system of spheres, *Phys Rev B* **52** (1995), 3256–3273.

[34] R. Fuchs, R.G. Barrera and J.L. Carrillo, Spectral representations of the electron energy loss in composite media, *Phys Rev B* **54** (1996), 12824–12834.

[35] I.O. Sosa, C.I. Mendoza and R.G. Barrera, Calculation of electron–energy-loss spectra of composites and self–similar structures, *Phys Rev E* **63** (2001), 144201.

[36] G.W. Milton, The coherent potential approximation is a realizable effective medium scheme, *Commun Math Phys* **99** (1985), 463–500.

[37] L. Ward, *The optical constants of bulk materials and films, 2nd ed*, Institute of Physics, Bristol, UK, 2000.

[38] F. Brouers, Percolation threshold and conductivity in metal–insulator composite mean–field theories, *J Phys C: Solid State Phys* **19** (1986), 7183–7193.

[39] S. Berthier and J. Peiro, Anomalous infra-red absorption of nanocermets in the percolation range, *J Phys: Condens Matter* **10** (1998), 3679–3694.

[40] A.V. Goncharenko and E.F. Venger, Percolation threshold for Bruggeman composites, *Phys Rev E* **70** (2004), 057102.

[41] T.G. Mackay and A. Lakhtakia, Percolation thresholds in the homogenization of spheroidal particles oriented in two direction, *Opt Commun* **259** (2006), 727–737.

[42] X. Zheng, M.G. Forest, R. Vaia, M. Arlen and R. Zhou, A strategy for dimensional percolation in sheared nanorod dispersions, *Adv Mater* **19** (2007), 4038–4043.

[43] T. Hanai, Theory of the dielectric dispersion due to the interfacial polarization and its applications to emulsions, *Kolloid Z* **171** (1960), 23–31.

[44] G.A. Niklasson and C.G. Granqvist, Optical properties and solar selectivity of coevaporated $Co–Al_2O_3$ composite films, *J Appl Phys* **55** (1984), 3382–3410.

[45] A.V. Goncharenko, Generalizations of the Bruggeman equation and a concept of shape–distributed particle composites, *Phys Rev E* **68** (2003), 041108.

[46] D.E. Aspnes, Local–field effects and effective–medium theory: a microscopic perspective, *Am J Phys* **50** (1982), 704–709.

[47] D. Polder and J.H. van Santen, The effective permeability of mixtures of solids, *Physica* **12** (1946), 257–271.

[48] C.F. Bohren and L.J. Battan, Radar backscattering by inhomogeneous precipitation particles, *J Atmos Sci* **37** (1980), 1821–1827.

[49] C.J.F. Böttcher, The dielectric constant of crystalline powders, *Rec Trav Chim Pays–Bas* **64** (1945), 47–51.

[50] B. Michel, A. Lakhtakia and W.S. Weiglhofer, Homogenization of linear bianisotropic particulate composite media — Numerical studies, *Int J Appl Electromag Mech* **9** (1998), 167–178. Corrections: **10** (1999), 537–538.

[51] Yu.A. Ryzhov and V.V. Tamoikin, Radiation and propagation of electromagnetic waves in randomly inhomogeneous media, *Radiophys Quantum Electron* **14** (1970), 228–233.

[52] V.I. Tatarskii and V.U. Zavorotnyi, Strong fluctuations in light propagation in a randomly inhomogeneous medium, *Prog Optics* **18** (1980), 204–256.

[53] T.G. Mackay, A. Lakhtakia and W.S. Weiglhofer, Ellipsoidal topology, orientation diversity and correlation length in bianisotropic composite mediums, *Arch Elektron Übertrag* **55** (2001), 243–251.

[54] U. Frisch, Wave propagation in random media, in *Probabilistic methods in applied mathematics, Vol. 1* (A.T. Bharucha–Reid, ed.), Academic Press, London, UK, 1970, 75–198.

[55] A. Stogryn, The bilocal approximation for the electric field in strong fluctuation theory, *IEEE Trans Antennas Propagat* **31** (1983), 985–986.

[56] L. Tsang, J.A. Kong and R.W. Newton, Application of strong fluctuation random medium theory to scattering of electromagnetic waves from a half–space of dielectric mixture, *IEEE Trans Antennas Propagat* **30** (1982), 292–302.

[57] T.G. Mackay, A. Lakhtakia and W.S. Weiglhofer, Homogenisation of similarly oriented, metallic, ellipsoidal inclusions using the bilocal–approximated strong–property–fluctuation theory, *Opt Commun* **197** (2001), 89–95.

[58] M.N. Miller, Bounds for effective electrical, thermal, and magnetic properties of heterogeneous materials, *J Math Phys* **10** (1969), 1988–2004.

[59] G.W. Milton and N. Phan–Thien, New bounds on effective elastic moduli of two–component materials, *Proc R Soc Lond A* **380** (1982), 305–331.

[60] T.G. Mackay, A. Lakhtakia and W.S. Weiglhofer, Third–order implementation and convergence of the strong–property–fluctuation theory in electromagnetic homogenisation, *Phys Rev E* **64** (2001), 066616.

[61] T.G. Mackay, Depolarization volume and correlation length in the homogenization of anisotropic dielectric composites, *Waves Random Media* **14** (2004), 485–498. Corrections: **16** (2006), 85.

[62] J. Cui and T.G. Mackay, Depolarization regions of nonzero volume in bianisotropic homogenized composites, *Waves Random Complex Media* **17** (2007), 269–281.

[63] J. Cui and T.G. Mackay, On convergence of the extended strong–property–fluctuation theory for bianisotropic homogenized composites, *Electromagnetics* **27**, (2007) 495–506.

[64] W.T. Doyle, Optical properties of a suspension of metal spheres, *Phys Rev B* **39** (1989), 9852–9858.

[65] M.T. Prinkey, A. Lakhtakia and B. Shanker, On the extended Maxwell–Garnett and the extended Bruggeman approaches for dielectric-in-dielectric composites, *Optik* **96** (1994), 25–30.

[66] B. Shanker and A. Lakhtakia, Extended Maxwell Garnett model for chiral–in–chiral composites, *J Phys D: Appl Phys* **26** (1993), 1746–1758.

[67] B. Shanker, The extended Bruggeman approach for chiral–in–chiral mixtures, *J Phys D: Appl Phys* **29** (1996), 281–288.

[68] A. Lakhtakia and B. Shanker, Beltrami fields within continuous source regions, volume integral equations, scattering algorithms, and the extended Maxwell–Garnett model, *Int J Appl Electromagn Mater* **4** (1993), 65–82.

176   *Electromagnetic Anisotropy and Bianisotropy: A Field Guide*

[69] P. Mallet, C.A. Guérin and A. Sentenac, Maxwell–Garnett mixing rule in the presence of multiple–scattering: Derivation and accuracy, *Phys Rev B* **72** (2005), 014205.

[70] C. F. Bohren, Applicability of effective–medium theories to problems of scattering and absorption by nonhomogeneous atmospheric particles. *J Atmos Sci* **43** (1986), 468–475.

[71] T.G. Mackay, On extended homogenization formalisms for nanocomposites, *J Nanophoton* **2** (2008), 021850.

[72] T.W.H. Oates, Real time spectroscopic ellipsometry of nanoparticle growth, *Appl Phys Lett* **88** (2006), 213115.

[73] C.F. Bohren and D.R. Huffman, *Absoption and scattering of light by small particles*, Wiley, New York, NY, USA, 1983.

[74] U. Kreibig and C. von Fragstein, The limitation of electron mean free path in small silver particles, *Z Physik* **224** (1969), 307–323.

[75] T.G. Mackay and W.S. Weiglhofer, Homogenization of biaxial composite materials: dissipative anisotropic properties, *J Opt A: Pure Appl Opt* **2** (2000), 426–432.

[76] T.G. Mackay and W.S. Weiglhofer, Homogenization of biaxial composite materials: nondissipative dielectric properties, *Electromagnetics* **21** (2001), 15–26.

[77] T.G. Mackay and W.S. Weiglhofer, Homogenization of biaxial composite materials: bianisotropic properties, *J Opt A: Pure Appl Opt* **3** (2001), 45–52.

[78] W.S. Weiglhofer, A. Lakhtakia and B. Michel, On the constitutive parameters of a chiroferrite composite medium, *Microw Opt Technol Lett* **18** (1998), 342–345.

[79] W.S. Weiglhofer and T.G. Mackay, Numerical studies of the constitutive parameters of a chiroplasma composite medium, *Arch Elektron Übertrag* **54** (2000), 259–265.

[80] T.G. Mackay and A. Lakhtakia, Correlation length and negative phase velocity in isotropic dielectric–magnetic materials, *J Appl Phys* **100** (2006), 063533.

[81] T.G. Mackay and A. Lakhtakia, Plane wave with negative phase velocity in Faraday chiral mediums, *Phys Rev E* **69** (2004), 026602.

[82] T.G. Mackay, A. Lakhtakia and R.A. Depine, Uniaxial dielectric media with hyperbolic dispersion relations, *Microw Opt Technol Lett* **48** (2006), 363–367.

[83] T.G. Mackay and A. Lakhtakia, Correlation length facilitates Voigt wave propagation, *Waves Random Media* **14** (2004), L1–L11.

[84] T.G. Mackay and A. Lakhtakia, Negative reflection in a Faraday chiral medium, *Microw Opt Technol Lett* **50** (2008), 1368–1371.

[85] K. Sølna and G.W. Milton, Can mixing materials make electromagnetic signals travel faster? *SIAM J Appl Math* **62** (2002), 2064–2091.

[86] T.G. Mackay and A. Lakhtakia, Enhanced group velocity in metamaterials, *J Phys A: Math Gen* **37** (2004), L19–L24.

[87] H.C. Chen, *Theory of electromagnetic waves*, McGraw–Hill, New York, NY, USA, 1983.

[88] K.E. Oughstun and G.C. Sherman, *Electromagnetic pulse propagation in causal dielectrics*, Springer, Berlin, Germany, 1997.

[89] T.G. Mackay and A. Lakhtakia, Anisotropic enhancement of group velocity in a homogenized dielectric composite medium, *J Opt A: Pure Appl Opt* **7** (2005), 669–674.

[90] L. Gao, Decreased group velocity in compositionally graded films, *Phys Rev E* **73** (2006), 036602.

[91] T.G. Mackay, A. Lakhtakia and W.S. Weiglhofer, The strong–property–fluctuation theory for cubically nonlinear, isotropic chiral composite mediums, *Electromagnetics* **23** (2003), 455–479.

[92] T.G. Mackay and A. Lakhtakia, A limitation of the Bruggeman formalism for homogenization, *Opt Commun* **234** (2004), 35–42. Corrections: **282** (2009), 4028.

[93] T.G. Mackay, On the effective permittivity of silver–insulator nanocomposites, *J Nanophoton* **1** (2007), 019501.

[94] K.P. Velikov, W.L. Vos, A. Moroz and A. van Blaaderen, Reflectivity of metallodielectric photonic glasses *Phys Rev B* **69** (2004), 075108 (2004).

[95] S.M. Aouadi, M. Debessai and P. Filip, Zirconium nitride/silver nanocomposite structures for biomedical applications, *J Vac Sci Technol B* **22** (2004), 1134–1140.

[96] R.K. Roy, S.K. Mandal, D. Bhattacharyya and A.K. Pal, An ellipsometric investigation of Ag/SiO_2 nanocomposite thin films, *Euro Phys J B* **34** (2003), 25–32.

[97] J.C.G. de Sande, R. Serna, J. Gonzalo, C.N. Afonso and D.E. Hole, Refractive index of Ag nanocrystals composite films in the neighborhood of the surface plasmon resonance, *J Appl Phys* **91** (2002), 1536–1541.

[98] T.G. Mackay and A. Lakhtakia, On the application of homogenization formalisms to active dielectric composite materials, *Opt Commun* **282** (2009), 2470–2475.

[99] O. Wiener, Die Theorie des Mischkörpers für das Feld der Stationären Strömung, *Abh Math–Phy Kl Sächs* **32**, (1912), 507–604.

[100] D.J. Bergman, Bounds for the complex dielectric constant of a two-component material, *Phys Rev B* **23** (1981), 3058–3065.

[101] G.W. Milton, Bounds on the complex permittivity of a two-component composite material, *J Appl Phys* **52** (1981), 5286–5293.

[102] D.J. Bergman, Exactly solvable microscopic geometries and rigorous bounds for the complex dielectric constant of a two–component composite material, *Phys Rev Lett* **44** (1980), 1285–1287. Corrections: **45** (1980), 148.

[103] A.J. Duncan, T.G. Mackay and A. Lakhtakia, On the Bergman–Milton bounds for the homogenization of dielectric composite materials, *Opt Commun* **271** (2007), 470–474.

[104] M.C.K. Wiltshire, J.B. Pendry and J.V. Hajnal, Chiral Swiss rolls show a negative refractive index, *J Phys: Condens Matter* **21** (2009), 292201.

Chapter 7

Nonlinear Mediums

In the previous six chapters, we have largely concentrated on mediums in which the induction fields are linearly proportional to the primitive fields. We close by focussing on the considerably more complex scenario wherein the proportionality is nonlinear. Mediums described by nonlinear constitutive relations been studied and exploited for technological purposes since the earliest days of electromagnetics. For example, the nonlinear permeability of ferromagnetic mediums was widely harnessed in electrical machines of the nineteenth century [1]. The Luxembourg effect — which involves the transfer of amplitude modulation between radio waves of different frequencies in the ionosphere, as observed by Tellegen in 1933 [2] — is due to the nonlinearity of plasma at sufficiently high wave intensities [3]. Nonlinear properties at optical wavelengths are generally exhibited only at high light intensities; thus, the speciality of nonlinear optics grew rapidly in the 1960's following the invention of the laser [1, 4]. The combination of nonlinearity and anisotropy or bianisotropy presents formidable challenges to theorists, but the vast scope for complex behaviour also presents tantalizing opportunities.

7.1 Constitutive relations

A general description of nonlinearity in a homogeneous bianisotropic setting may be established as follows: We begin with the frequency–domain counterparts of the definitions of the induction fields given in Eqs. (1.6), namely

$$\left. \begin{array}{l} \underline{D}(\underline{r},\omega) = \epsilon_0\,\underline{E}(\underline{r},\omega) + \underline{P}(\underline{r},\omega) \\[2mm] \underline{H}(\underline{r},\omega) = \dfrac{1}{\mu_0}\,\underline{B}(\underline{r},\omega) - \underline{M}(\underline{r},\omega) \end{array} \right\} . \qquad (7.1)$$

Herein the frequency–domain polarization $\underline{P}(\underline{r},\omega)$ and magnetization $\underline{M}(\underline{r},\omega)$ are the temporal Fourier transforms of $\underline{\tilde{P}}(\underline{r},t)$ and $\underline{\tilde{M}}(\underline{r},t)$, respectively, in accordance with Eq. (1.25) with $\mathcal{Z} = \underline{P}, \underline{M}$. We combine these to form the 6–vector

$$\underline{\mathbf{N}}(\underline{r},\omega) = \begin{bmatrix} \underline{P}(\underline{r},\omega) \\ \underline{M}(\underline{r},\omega) \end{bmatrix}. \tag{7.2}$$

The nonlinear constitutive relations may be expressed in the 6–vector/dyadic notation of Sec. 1.5 as [5]

$$\underline{\mathbf{C}}(\underline{r},\omega) = \underline{\underline{\mathbf{K}}}_0 \cdot \underline{\mathbf{F}}(\underline{r},\omega) + \underline{\mathbf{N}}(\underline{r},\omega), \tag{7.3}$$

where $\underline{\underline{\mathbf{K}}}_0$ is the 6×6 Tellegen constitutive dyadic of free space; i.e.,

$$\underline{\underline{\mathbf{K}}}_0 = \begin{bmatrix} \epsilon_0 \underline{\underline{I}} & \underline{\underline{0}} \\ \underline{\underline{0}} & \mu_0 \underline{\underline{I}} \end{bmatrix}. \tag{7.4}$$

We consider the 6–vector $\underline{\mathbf{N}}(\underline{r},\omega)$ as the sum of linear and nonlinear parts

$$\underline{\mathbf{N}}(\underline{r},\omega) = \underline{\mathbf{N}}_{\mathrm{L}}(\underline{r},\omega) + \underline{\mathbf{N}}_{\mathrm{NL}}(\underline{r},\omega). \tag{7.5}$$

The linear part

$$\underline{\mathbf{N}}_{\mathrm{L}}(\underline{r},\omega) = \left[\underline{\underline{\mathbf{K}}}^{\mathrm{L}}(\omega) - \underline{\underline{\mathbf{K}}}_0 \right] \cdot \underline{\mathbf{F}}(\underline{r},\omega), \tag{7.6}$$

involves $\underline{\underline{\mathbf{K}}}^{\mathrm{L}}(\omega)$ as the 6×6 constitutive dyadic to characterize the linear response of the nonlinear medium. The exclusively nonlinear response of the medium, under the simultaneous stimulation by an ensemble of $M > 1$ fields $\underline{\mathbf{F}}(\underline{r},\omega_m)$, $(m = 1, 2, \ldots, M)$, is characterized by the 6–vector $\underline{\mathbf{N}}_{\mathrm{NL}}(\underline{r},\omega)$. At the frequency $\omega = \omega_{\mathrm{NL}}$, the jth element of $\underline{\mathbf{N}}_{\mathrm{NL}}(\underline{r},\omega_{\mathrm{NL}})$ is given by

$$[\underline{\mathbf{N}}_{\mathrm{NL}}(\underline{r},\omega_{\mathrm{NL}})]_j = \sum_{j_1=1}^{6} \sum_{j_2=1}^{6} \cdots \sum_{j_m=1}^{6} \cdots \sum_{j_M=1}^{6} \left\{ \chi_{jj_1j_2\cdots j_m\cdots j_M}^{\mathrm{NL}} (\omega_{\mathrm{NL}}; \mathcal{W}) \right.$$
$$\left. \times \prod_{n=1}^{M} [\underline{\mathbf{F}}(\underline{r},\omega_n)]_{j_n} \right\}, \tag{7.7}$$

for $j \in [1, 6]$. Here, the set of $M > 1$ angular frequencies

$$\mathcal{W} = \{\omega_1, \omega_2, \ldots, \omega_M\}, \tag{7.8}$$

with no requirement that all members of $\mathcal{W}$ be distinct. The angular frequency ω_{NL} is related to the members of $\mathcal{W}$ via the sum

$$\omega_{\mathrm{NL}} = \sum_{m=1}^{M} a_m \omega_m, \qquad a_m = \pm 1; \qquad (7.9)$$

if $a_n = -1$ then $[\underline{\mathbf{F}}(\underline{r}, \omega_n)]_{j_n}$ in Eq. (7.7) should be replaced by its complex conjugate. The nonlinear constitutive properties are delineated by the nonlinear susceptibility tensor $\chi^{\mathrm{NL}}_{jj_1 j_2 \cdots j_m \cdots j_M}(\omega_{\mathrm{NL}}; \mathcal{W})$.

According to the constitutive relation (7.7), electromagnetic fields oscillating at angular frequency $\omega_m \in \mathcal{W}$, that are launched into the nonlinear medium, together give rise to fields oscillating at angular frequency ω_{NL}. A vast range of nonlinear dielectric effects are thereby described, including various frequency–mixing processes [4, 6]. Also, numerous nonlinear magnetic and magnetoelectric effects are characterized by Eq. (7.7) [7–9].

7.2 Homogenization

In principle at least, homogenization offers a means of engineering mediums with certain nonlinear properties in certain directions, and enhancing (or diminishing) nonlinear properties in certain directions (see Sec. 7.3). However, nonlinearity generally leads to intractability in homogenization analyses, the more so in the contexts of anisotropy and bianisotropy. Accordingly, studies are often restricted to cases of weak nonlinearity wherein

$$|\underline{\mathbf{N}}_{\mathrm{L}}(\underline{r}, \omega)| \gg |\underline{\mathbf{N}}_{\mathrm{NL}}(\underline{r}, \omega)|. \qquad (7.10)$$

Satisfaction of this inequality essentially permits the linear and nonlinear contributions to the constitutive relations of the HCM to be treated separately [5, 10, 11]. In the following subsections, we present the most general formulations currently available of the Maxwell Garnett and SPFT approaches for weakly nonlinear HCMs. Since the Bruggeman homogenization formalism is equivalent to the lowest–order SPFT approach, the Bruggeman formulation is effectively included in the SPFT formulations.

Notice that in order to be consistent with the notion of homogenization, the particles comprising the constituent mediums should be small relative to electromagnetic wavelengths at ω_{NL} as well as at all angular frequencies $\omega_m \in \mathcal{W}$.

7.2.1   *Maxwell Garnett formalism*

Let us recall the scenario described in Sec. 6.1. That is, we have two particulate constituent mediums; the particles of constituent medium 'a' are ellipsoids, described by the shape dyadic $\underline{\underline{U}}^{\mathrm{a}}$, while the ellipsoidal particles of constituent medium 'b' are described by $\underline{\underline{U}}^{\mathrm{b}}$. In consonance with the Maxwell Garnett approach, the constituent medium 'a' is dilutely dispersed in the constituent medium 'b'.

Now, the constituent medium 'a' is taken to be a weakly nonlinear bianisotropic medium, described by the constitutive relations of the form (7.3), wherein the linear part of the 6–vector $\underline{N}(\underline{r}, \omega)$ is given by the right side of Eq. (7.6) with $\underline{\underline{K}}^{\mathrm{L}}(\omega) = \underline{\underline{K}}^{\mathrm{L}}_{\mathrm{a}}(\omega)$. The components of the weakly nonlinear part of $\underline{N}(\underline{r}, \omega)$ are given by the right side of Eq. (7.7) with $\chi^{\mathrm{NL}}_{jj_1 j_2 \cdots j_m \cdots j_M}(\omega_{\mathrm{NL}}; \mathcal{W}) = {}^{\mathrm{a}}\chi^{\mathrm{NL}}_{jj_1 j_2 \cdots j_m \cdots j_M}(\omega_{\mathrm{NL}}; \mathcal{W})$. In contrast, the constituent medium 'b' is taken to be a linear bianisotropic medium, characterized by the Tellegen constitutive dyadic $\underline{\underline{K}}_{\mathrm{b}}(\omega)$.

By exploiting the weak nonlinearity condition (7.10), the constitutive relations of the HCM, as delivered by the Maxwell Garnett approach, emerge as [5]

$$\underline{C}(\underline{r}, \omega) = \underline{\underline{K}}^{\mathrm{L}}_{\mathrm{MG}}(\omega) \bullet \underline{F}(\underline{r}, \omega) + s(\omega, \omega_{\mathrm{NL}}) \, \underline{N}^{\mathrm{NL}}_{\mathrm{MG}}(\underline{r}, \omega), \qquad \omega \in \mathcal{W} \cup \{\omega_{\mathrm{NL}}\}, \tag{7.11}$$

where

$$\underline{N}^{\mathrm{NL}}_{\mathrm{MG}}(\underline{r}, \omega_{\mathrm{NL}}) = \left\{ \underline{\underline{I}} - i\omega_{\mathrm{NL}} \left[\underline{\underline{I}} - i\omega_{\mathrm{NL}} f_{\mathrm{a}} \, \underline{\underline{D}}^0_{I/\mathrm{b}}(\omega_{\mathrm{NL}}) \bullet \underline{\underline{\alpha}}^{\mathrm{a}/\mathrm{b}}(\omega_{\mathrm{NL}}) \right]^{-1} \right. $$
$$\left. \bullet \underline{\underline{D}}^0_{I/\mathrm{b}}(\omega_{\mathrm{NL}}) \right\} \bullet \underline{\check{N}}^{\mathrm{NL}}_{\mathrm{MG}}(\underline{r}, \omega) \tag{7.12}$$

and the components of the 6–vector $\underline{\check{N}}^{\mathrm{NL}}_{\mathrm{MG}}(\underline{r}, \omega_{\mathrm{NL}})$ are given as

$$\left[\underline{\check{N}}_{\mathrm{NL}}(\underline{r}, \omega_{\mathrm{NL}}) \right]_j = \sum_{j_1=1}^{6} \sum_{j_2=1}^{6} \cdots \sum_{j_m=1}^{6} \cdots \sum_{j_M=1}^{6} \left({}^{\mathrm{a}}\chi^{\mathrm{NL}}_{jj_1 j_2 \cdots j_m \cdots j_M}(\omega_{\mathrm{NL}}; \mathcal{W}) \right.$$
$$\times \prod_{n=1}^{M} \left\{ \left[\underline{\underline{K}}^{\mathrm{L}}_{\mathrm{a}}(\omega_n) - \underline{\underline{K}}^{\mathrm{L}}_{\mathrm{MG}}(\omega_n) \right]^{-1} \right.$$
$$\left. \left. \bullet \underline{\underline{\alpha}}^{\mathrm{a}/\mathrm{MG}}(\omega_n) \bullet \underline{F}(\underline{r}, \omega_n) \right\}_{j_n} \right). \tag{7.13}$$

Here, $\underline{\underline{K}}^{\mathrm{L}}_{\mathrm{MG}}(\omega_n)$ is the Maxwell Garnett estimate of the linear contribution to the HCM constitutive properties, as yielded by Eq. (6.2) with $\underline{\underline{K}}^{\mathrm{L}}_{\mathrm{a}}(\omega_n)$ taking the place of $\underline{\underline{K}}_{\mathrm{a}}(\omega)$ therein. The polarizability density dyadic

$\underline{\underline{\alpha}}_{a/MG}(\omega_n)$ is as given in Eq. (5.97). The switching function $s(\omega, \omega_{NL})$ is defined by

$$s(\omega, \omega_{NL}) = \begin{cases} 1, \omega = \omega_{NL} \\ \\ 0, \omega \neq \omega_{NL} \end{cases} . \qquad (7.14)$$

Thus, linear processes are taken into account at all angular frequencies in $\mathcal{W} \cup \{\omega_{NL}\}$, but nonlinear processes contribute only when $\omega = \omega_{NL}$.

The derivation of the bianisotropic result represented by Eq. (7.11) follows along lines analogous to the corresponding derivation for the nonlinear isotropic dielectric case, based on spherical constituent particles [11], and builds upon the corresponding results for biisotropic [12] and gyrotropic [13] HCMs. The Maxwell Garnett estimate presented in Eq. (7.11) implicitly utilizes the depolarization dyadic $\underline{\underline{D}}^0(\omega_{NL})$ which is independent of the size of the constituent particles. This estimate may be extended, in order to take into account the sizes of the constituent particles, by substituting $\underline{\underline{D}}(\rho, \omega_{NL})$ for $\underline{\underline{D}}^0(\omega_{NL})$, where ρ is a measure of the average linear dimensions of the particles, in a manner similar to that discussed in Sec. 6.5.

7.2.2 *Strong–property–fluctuation theory*

Formulations of the second–order SPFT have been developed for weakly nonlinear, isotropic [14–16] and anisotropic [17–19] HCMs in which at least one of the constituent mediums is an isotropic dielectric medium with an intensity–dependent permittivity dyadic. That is, in the framework of Sec. 6.1, the permittivity of the constituent medium is given by

$$\epsilon_\ell(\omega) = \epsilon_\ell^L(\omega) + \epsilon_\ell^{NL}(\omega) \, |\underline{E}_\ell(\omega)|^2 , \qquad (\ell = a,b), \qquad (7.15)$$

where $\epsilon_\ell^L(\omega)$ and $\epsilon_\ell^{NL}(\omega)$ are the linear and weakly nonlinear contributions to the permittivity, respectively. The electric field phasor $\underline{E}_\ell(\omega)$ inside the constituent particles may be taken to be uniform, provided that the particles are sufficiently small. Mediums with intensity–dependent permittivities exhibit the optical Kerr effect [4], and are of considerable scientific and technological interest on account of self–focusing and self–trapping of light [20, 21]. Although such nonlinear mediums exist in nature, they are of particular interest within the context of homogenization since HCMs may exhibit nonlinear effects which are stronger than those of naturally occurring mediums with intensity–dependent permittivity [22], as discussed in Sec. 7.3. Even though a medium with intensity–dependent permittivity may be weakly nonlinear, as indicated by the inequality (7.10), it is capable of inducing Brillouin scattering via electrostriction, which is often a strong process [4].

7.2.2.1 *Weakly nonlinear, isotropic dielectric HCM*

Let us begin with the simplest weakly nonlinear homogenization scenario, wherein both constituent mediums are isotropic dielectric mediums. Both mediums are distributed randomly as spherical particles; i.e., the corresponding shape dyadics are $\underline{U}^a = \underline{U}^b = \underline{I}$. Suppose that constituent medium 'a' is a weakly nonlinear medium with intensity–dependent permittivity described by Eq. (7.15), and constituent medium 'b' is a linear medium with scalar permittivity ϵ_b.

The corresponding comparison medium is a weakly nonlinear, isotropic dielectric medium with intensity–dependent permittivity

$$\epsilon_{comp}(\omega) = \epsilon^L_{comp}(\omega) + \epsilon^{NL}_{comp}(\omega) \, | \underline{E}_{HCM}(\omega) |^2 , \qquad (7.16)$$

where $\underline{E}_{HCM}(\omega)$ is the spatially averaged electric field in the HCM. The linear part of $\epsilon_{comp}(\omega)$, namely $\epsilon^L_{comp}(\omega)$, is delivered from the isotropic dielectric version of the Bruggeman equation, which is given by the quadratic Eq. (6.34) with $\epsilon_{Br}(\omega)$ and $\epsilon_a(\omega)$ being replaced by $\epsilon^L_{comp}(\omega)$ and $\epsilon^L_a(\omega)$, respectively. The weakly nonlinear part of $\epsilon_{comp}(\omega)$, namely $\epsilon^{NL}_{comp}(\omega)$, is given by [14]

$$\epsilon^{NL}_{comp}(\omega) = g^{FF}_\epsilon(\omega) \, \epsilon^{NL}_a(\omega) \, \frac{\partial \epsilon^L_{comp}(\omega)}{\partial \epsilon^L_a(\omega)}, \qquad (7.17)$$

wherein the partial derivative

$$\frac{\partial \epsilon^L_{comp}(\omega)}{\partial \epsilon^L_a(\omega)} = \frac{f_a \epsilon^L_{comp}(\omega) \left[\epsilon_b + 2\epsilon^L_{comp}(\omega) \right]^2}{f_a \epsilon^L_a(\omega) \left[\epsilon_b(\omega) + 2\epsilon^L_{comp}(\omega) \right]^2 + f_b \epsilon_b(\omega) \left[\epsilon^L_a(\omega) + 2\epsilon^L_{comp}(\omega) \right]^2}. \qquad (7.18)$$

A suitable estimate of the local field factor

$$g^{FF}_\epsilon(\omega) = \frac{d \, | \underline{E}_a(\omega) |^2}{d \, | \underline{E}_{HCM}(\omega) |^2} \qquad (7.19)$$

is provided by [23]

$$g^{FF}_\epsilon(\omega) = \left| \frac{3\epsilon^L_{comp}(\omega)}{\epsilon^L_a(\omega) + 2\epsilon^L_{comp}(\omega)} \right|^2 , \qquad (7.20)$$

but we note that other evaluations may also be considered [14, 24].

The second–order SPFT estimate of the HCM permittivity $\epsilon^{[2]}_{SPFT}(\omega)$ — which has the form of Eq. (7.16) but with $\epsilon^{[2]}_{SPFT}(\omega)$, $\epsilon^{[2],L}_{SPFT}(\omega)$ and $\epsilon^{[2],NL}_{SPFT}(\omega)$

replacing $\epsilon_{\text{comp}}(\omega)$, $\epsilon^{\text{L}}_{\text{comp}}(\omega)$ and $\epsilon^{\text{NL}}_{\text{comp}}(\omega)$, respectively — is established for the step covariance function $\Lambda_{\text{step}}(\underline{r} - \underline{r}')$ prescribed in Eq. (6.20). Under the assumption of weak nonlinearity, the linear part is [15]

$$
\epsilon^{[2],\text{L}}_{\text{SPFT}}(\omega) = \epsilon^{\text{L}}_{\text{comp}}(\omega) + \mu_0 \left[\omega \, L_{\text{cor}} \, \epsilon^{\text{L}}_{\text{comp}}(\omega) \right]^2 \left\{ 3 + i2\omega L_{\text{cor}} \left[\mu_0 \, \epsilon^{\text{L}}_{\text{comp}}(\omega) \right]^{1/2} \right\}
$$
$$
\times \left\{ f_{\text{a}} \left[\frac{\epsilon^{\text{L}}_{\text{a}}(\omega) - \epsilon^{\text{L}}_{\text{comp}}(\omega)}{\epsilon^{\text{L}}_{\text{a}}(\omega) + 2\epsilon^{\text{L}}_{\text{comp}}(\omega)} \right]^2 + f_{\text{b}} \left[\frac{\epsilon_{\text{b}}(\omega) - \epsilon^{\text{L}}_{\text{comp}}(\omega)}{\epsilon_{\text{b}}(\omega) + 2\epsilon^{\text{L}}_{\text{comp}}(\omega)} \right]^2 \right\},
$$

$$(7.21)$$

while the nonlinear part is

$$
\epsilon^{[2],\text{NL}}_{\text{SPFT}}(\omega) = \epsilon^{\text{NL}}_{\text{comp}}(\omega) + \mu_0 \left(\omega L_{\text{cor}} \right)^2 \epsilon^{\text{L}}_{\text{comp}}(\omega) \Bigg(2\epsilon^{\text{L}}_{\text{comp}}(\omega)
$$
$$
\times \left\{ 3 + i2\omega L_{\text{cor}} \left[\mu_0 \, \epsilon^{\text{L}}_{\text{comp}}(\omega) \right]^{1/2} \right\} \Bigg\{ f_{\text{a}} \left[\frac{\epsilon^{\text{L}}_{\text{a}}(\omega) - \epsilon^{\text{L}}_{\text{comp}}(\omega)}{\epsilon^{\text{L}}_{\text{a}}(\omega) + 2\epsilon^{\text{L}}_{\text{comp}}(\omega)} \right]^2
$$
$$
\times \left[\frac{g^{\text{FF}}_{\epsilon}(\omega) \epsilon^{\text{NL}}_{\text{a}}(\omega) - \epsilon^{\text{NL}}_{\text{comp}}(\omega)}{\epsilon^{\text{L}}_{\text{a}}(\omega) - \epsilon^{\text{L}}_{\text{comp}}(\omega)} - \frac{g^{\text{FF}}_{\epsilon}(\omega) \epsilon^{\text{NL}}_{\text{a}}(\omega) + 2\epsilon^{\text{NL}}_{\text{comp}}(\omega)}{\epsilon^{\text{L}}_{\text{a}}(\omega) + 2\epsilon^{\text{NL}}_{\text{comp}}(\omega)} \right]
$$
$$
+ f_{\text{b}} \left[\frac{\epsilon_{\text{b}}(\omega) - \epsilon^{\text{L}}_{\text{comp}}(\omega)}{\epsilon_{\text{b}}(\omega) + 2\epsilon^{\text{L}}_{\text{comp}}(\omega)} \right]^2
$$
$$
\times \left[\frac{-\epsilon^{\text{NL}}_{\text{comp}}(\omega)}{\epsilon_{\text{b}}(\omega) - \epsilon^{\text{L}}_{\text{comp}}(\omega)} - \frac{2\epsilon^{\text{NL}}_{\text{comp}}(\omega)}{\epsilon_{\text{b}}(\omega) + 2\epsilon^{\text{NL}}_{\text{comp}}(\omega)} \right] \Bigg\}
$$
$$
+ \epsilon^{\text{NL}}_{\text{comp}}(\omega) \left\{ 6 + i5\omega L_{\text{cor}} \left[\mu_0 \, \epsilon^{\text{L}}_{\text{comp}}(\omega) \right]^{1/2} \right\}
$$
$$
\times \left\{ f_{\text{a}} \left[\frac{\epsilon^{\text{L}}_{\text{a}}(\omega) - \epsilon^{\text{L}}_{\text{comp}}(\omega)}{\epsilon^{\text{L}}_{\text{a}}(\omega) + 2\epsilon^{\text{L}}_{\text{comp}}(\omega)} \right]^2 + f_{\text{b}} \left[\frac{\epsilon_{\text{b}}(\omega) - \epsilon^{\text{L}}_{\text{comp}}(\omega)}{\epsilon_{\text{b}}(\omega) + 2\epsilon^{\text{L}}_{\text{comp}}(\omega)} \right]^2 \right\} \Bigg).
$$

$$(7.22)$$

The implementation of a Gaussian covariance function, instead of the step covariance function, makes only a small difference to the SPFT estimate of $\epsilon^{[2]}_{\text{SPFT}}(\omega)$ [14]. Explicit expressions for the corresponding third–order SPFT estimate of the HCM permittivity — based on the three–point covariance function (6.22) — are available [15], but we note that convergence at the second–order level has been demonstrated in numerical studies [15].

7.2.2.2 *Weakly nonlinear, isotropic chiral HCM*

Suppose that the constituent medium 'a' is a weakly nonlinear, isotropic dielectric medium with scalar permittivity given by Eq. (7.15) and the constituent medium 'b' is an isotropic chiral medium characterized by the Tellegen constitutive relations (2.5) with permittivity $\epsilon_b(\omega)$, permeability $\mu_b(\omega)$ and chirality parameter $\chi_b(\omega)$. The constituent mediums are randomly dispersed as electrically small spherical particles; i.e., the corresponding shape dyadics are $\underline{U}^a = \underline{U}^b = \underline{I}$. In the following the key results of the SPFT homogenization procedure are presented, full details being available elsewhere [16].

The HCM 'inherits' the weak nonlinearity of constituent medium 'a' and the chirality of constituent medium 'b'. Hence, the 6×6 Tellegen constitutive dyadic of the comparison medium has the form

$$\underline{\underline{K}}_{\text{comp}}(\omega) = \underline{\underline{K}}^{\text{L}}_{\text{comp}}(\omega) + \underline{\underline{K}}^{\text{NL}}_{\text{comp}}(\omega) \,|\, \underline{E}_{\text{HCM}}(\omega)\,|^2 \tag{7.23}$$

$$= \begin{bmatrix} \epsilon^{\text{L}}_{\text{comp}}(\omega)\,\underline{I} & \chi^{\text{L}}_{\text{comp}}(\omega)\,\underline{I} \\ -\chi^{\text{L}}_{\text{comp}}(\omega)\,\underline{I} & \mu^{\text{L}}_{\text{comp}}(\omega)\,\underline{I} \end{bmatrix}$$

$$+ \begin{bmatrix} \epsilon^{\text{NL}}_{\text{comp}}(\omega)\,\underline{I} & \chi^{\text{NL}}_{\text{comp}}(\omega)\,\underline{I} \\ -\chi^{\text{NL}}_{\text{comp}}(\omega)\,\underline{I} & \mu^{\text{NL}}_{\text{comp}}(\omega)\,\underline{I} \end{bmatrix} |\, \underline{E}_{\text{HCM}}(\omega)\,|^2, \tag{7.24}$$

where $\underline{E}_{\text{HCM}}(\omega)$ is the spatially averaged electric field in the HCM, as in Eq. (7.16). Similarly, the depolarization dyadic for a vanishingly small spherical particle immersed in the comparison medium has the form

$$\underline{\underline{D}}^0_{I/\text{comp}}(\omega) = \underline{\underline{D}}^{0,\text{L}}_{I/\text{comp}}(\omega) + \underline{\underline{D}}^{0,\text{NL}}_{I/\text{comp}}(\omega)\,|\, \underline{E}_{\text{HCM}}(\omega)\,|^2. \tag{7.25}$$

Under the assumption of weak nonlinearity, the linear part of the depolarization dyadic is equivalent to the depolarization dyadic that would arise in the corresponding wholly linear scenario (i.e., when $\epsilon^{\text{NL}}_a(\omega) = 0$). Thus, the four 3×3 dyadics which make up $\underline{\underline{D}}^{0,\text{L}}_{I/\text{comp}}(\omega)$ are those specified in Eq. (5.78), with $\epsilon^{\text{L}}_{\text{comp}}(\omega)$, $\chi^{\text{L}}_{\text{comp}}(\omega)$ and $\mu^{\text{L}}_{\text{comp}}(\omega)$ replacing $\epsilon(\omega)$, $\chi(\omega)$ and $\mu(\omega)$, respectively, therein. The weakly nonlinear part of the depolarization dyadic is given by

$$\underline{\underline{D}}^{0,\text{NL}}_{I/\text{comp}}(\omega) = \frac{1}{3i\omega\,\Upsilon_+(\omega)} \begin{bmatrix} {}^{ee}d^{0,\text{NL}}_{I/\text{comp}}(\omega)\,\underline{I} & {}^{em}d^{0,\text{NL}}_{I/\text{comp}}(\omega)\,\underline{I} \\ {}^{me}d^{0,\text{NL}}_{I/\text{comp}}(\omega)\,\underline{I} & {}^{mm}d^{0,\text{NL}}_{I/\text{comp}}(\omega)\,\underline{I} \end{bmatrix}, \tag{7.26}$$

where the scalar functions

$$\left. \begin{array}{l} {}^{ee}d^{0,\text{NL}}_{I/\text{comp}}(\omega) = \mu^{\text{NL}}_{\text{comp}}(\omega) - \mu^{\text{L}}_{\text{comp}}(\omega)\breve{\Upsilon}(\omega) \\[4pt] {}^{em}d^{0,\text{NL}}_{I/\text{comp}}(\omega) = \chi^{\text{NL}}_{\text{comp}}(\omega)\breve{\Upsilon}(\omega) - \chi^{\text{L}}_{\text{comp}}(\omega) \\[4pt] {}^{me}d^{0,\text{NL}}_{I/\text{comp}}(\omega) = \chi^{\text{NL}}_{\text{comp}}(\omega) - \chi^{\text{NL}}_{\text{comp}}(\omega)\breve{\Upsilon}(\omega) \\[4pt] {}^{mm}d^{0,\text{NL}}_{I/\text{comp}}(\omega) = \epsilon^{\text{NL}}_{\text{comp}}(\omega) - \epsilon^{\text{L}}_{\text{comp}}(\omega)\breve{\Upsilon}(\omega) \end{array} \right\}, \tag{7.27}$$

with

$$\check{\Upsilon}(\omega) = \frac{\epsilon^{\rm L}_{\rm comp}(\omega)\,\mu^{\rm NL}_{\rm comp}(\omega) + 2\,\chi^{\rm L}_{\rm comp}(\omega)\,\chi^{\rm NL}_{\rm comp}(\omega)}{\Upsilon_+(\omega)} \qquad (7.28)$$

and $\Upsilon_+(\omega)$ defined as in Eq. (5.80) but with $\epsilon^{\rm L}_{\rm comp}(\omega)$, $\chi^{\rm L}_{\rm comp}(\omega)$ and $\mu^{\rm L}_{\rm comp}(\omega)$ in lieu of $\epsilon(\omega)$, $\chi(\omega)$ and $\mu(\omega)$, respectively, therein.

The corresponding 6×6 polarizability density dyadics are expressed as

$$\underline{\underline{\alpha}}_{\ell/{\rm comp}}(\omega) = \underline{\underline{\alpha}}^{\rm L}_{\ell/{\rm comp}}(\omega) + \underline{\underline{\alpha}}^{\rm NL}_{\ell/{\rm comp}}(\omega)\,|\underline{E}_{\rm HCM}(\omega)|^2, \qquad (\ell = {\rm a,b}). \quad (7.29)$$

The linear parts $\underline{\underline{\alpha}}^{\rm L}_{\ell/{\rm comp}}(\omega)$ are defined as in Eq. (5.97), but with $\underline{\underline{K}}^{\rm L}_{\rm a}(\omega)$ replacing $\underline{\underline{K}}_{\rm a}(\omega)$, $\underline{\underline{K}}^{\rm L}_{\rm comp}(\omega)$ replacing $\underline{\underline{K}}_{\rm amb}(\omega)$, and $\underline{\underline{D}}^{0,{\rm L}}_{{\rm I}/{\rm comp}}(\omega)$ replacing $\underline{\underline{D}}^{0}_{U^{\ell}/{\rm amb}}(\omega)$ therein. The weakly nonlinear parts of $\underline{\underline{\alpha}}_{\ell/{\rm comp}}(\omega)$ are given by

$$\left.\begin{aligned}
\underline{\underline{\alpha}}^{\rm NL}_{{\rm a}/{\rm comp}}(\omega) &= \left[\, g^{\rm FF}_{\rm ICM}(\omega)\,\underline{\underline{K}}^{\rm NL}_{\rm a}(\omega) - \underline{\underline{K}}^{\rm NL}_{\rm comp}(\omega)\right] \boldsymbol{\cdot} \left[\,\underline{\underline{\aleph}}^{\rm L}_{{\rm a}/{\rm comp}}(\omega)\right]^{-1} \\
&\quad + \left[\,\underline{\underline{K}}^{\rm L}_{\rm a}(\omega) - \underline{\underline{K}}^{\rm L}_{\rm comp}(\omega)\right] \boldsymbol{\cdot} \mathcal{N}_{\rm ICM}\left\{\,\underline{\underline{\aleph}}_{{\rm a}/{\rm comp}}(\omega)\right\} \\
\underline{\underline{\alpha}}^{\rm NL}_{{\rm b}/{\rm comp}}(\omega) &= -\underline{\underline{K}}^{\rm NL}_{\rm comp}(\omega) \boldsymbol{\cdot} \left[\,\underline{\underline{\aleph}}^{\rm L}_{{\rm b}/{\rm comp}}(\omega)\right]^{-1} \\
&\quad + \left[\,\underline{\underline{K}}_{\rm b}(\omega) - \underline{\underline{K}}^{\rm L}_{\rm comp}(\omega)\right] \boldsymbol{\cdot} \mathcal{N}_{\rm ICM}\left\{\,\underline{\underline{\aleph}}_{{\rm b}/{\rm comp}}(\omega)\right\}
\end{aligned}\right\},$$

$$(7.30)$$

wherein the 6×6 dyadic

$$\underline{\underline{\aleph}}_{\ell/{\rm comp}}(\omega) = \underline{\underline{\aleph}}^{\rm L}_{\ell/{\rm comp}}(\omega) + \underline{\underline{\aleph}}^{\rm NL}_{\ell/{\rm comp}}(\omega)\,|\underline{E}_{\rm HCM}(\omega)|^2, \qquad (\ell = {\rm a,b}) \quad (7.31)$$

has linear parts

$$\left.\begin{aligned}
\underline{\underline{\aleph}}^{\rm L}_{{\rm a}/{\rm comp}}(\omega) &= \underline{\underline{I}} + i\omega\underline{\underline{D}}^{0,{\rm L}}_{{\rm I}/{\rm comp}}(\omega) \boldsymbol{\cdot} \left[\,\underline{\underline{K}}^{\rm L}_{\rm a}(\omega) - \underline{\underline{K}}^{\rm L}_{\rm comp}(\omega)\right] \\
\underline{\underline{\aleph}}^{\rm L}_{{\rm b}/{\rm comp}}(\omega) &= \underline{\underline{I}} + i\omega\underline{\underline{D}}^{0,{\rm L}}_{{\rm I}/{\rm comp}}(\omega) \boldsymbol{\cdot} \left[\,\underline{\underline{K}}_{\rm b}(\omega) - \underline{\underline{K}}^{\rm L}_{\rm comp}(\omega)\right]
\end{aligned}\right\} \quad (7.32)$$

and weakly nonlinear parts

$$\left.\begin{aligned}
\underline{\underline{\aleph}}^{\rm NL}_{{\rm a}/{\rm comp}}(\omega) &= i\omega\left\{\underline{\underline{D}}^{0,{\rm L}}_{{\rm I}/{\rm comp}}(\omega) \boldsymbol{\cdot} \left[\, g^{\rm FF}_{\rm ICM}(\omega)\,\underline{\underline{K}}^{\rm NL}_{\rm a}(\omega) - \underline{\underline{K}}^{\rm NL}_{\rm comp}(\omega)\right]\right. \\
&\quad \left. +\underline{\underline{D}}^{0,{\rm NL}}_{{\rm I}/{\rm comp}}(\omega) \boldsymbol{\cdot} \left[\,\underline{\underline{K}}^{\rm L}_{\rm a}(\omega) - \underline{\underline{K}}^{\rm L}_{\rm comp}(\omega)\right]\right\} \\
\underline{\underline{\aleph}}^{\rm NL}_{{\rm b}/{\rm comp}}(\omega) &= i\omega\left\{-\underline{\underline{D}}^{0,{\rm L}}_{{\rm I}/{\rm comp}}(\omega) \boldsymbol{\cdot} \underline{\underline{K}}^{\rm NL}_{\rm comp}(\omega)\right. \\
&\quad \left. +\underline{\underline{D}}^{0,{\rm NL}}_{{\rm I}/{\rm comp}}(\omega) \boldsymbol{\cdot} \left[\,\underline{\underline{K}}_{\rm b}(\omega) - \underline{\underline{K}}^{\rm L}_{\rm comp}(\omega)\right]\right\}
\end{aligned}\right\}. \quad (7.33)$$

The 6×6 dyadic operator $\mathcal{N}_{\mathrm{ICM}}\{\,\bullet\,\}$ introduced in Eqs. (7.30) has the following action:

$$\mathcal{N}_{\mathrm{ICM}}\left\{\underline{\underline{\mathbf{N}}}_{\ell/\mathrm{comp}}(\omega)\right\} = \frac{1}{\det \underline{\underline{\mathbf{N}}}^{\mathrm{L}}_{\ell/\mathrm{comp}}(\omega)}\left\{\operatorname{adj} \underline{\underline{\mathbf{N}}}^{\mathrm{NL}}_{\ell/\mathrm{comp}}(\omega)\right.$$

$$\left. -\mathcal{M}_{\mathrm{ICM}}\left\{\underline{\underline{\mathbf{N}}}_{\ell/\mathrm{comp}}(\omega)\right\}\left[\underline{\underline{\mathbf{N}}}^{\mathrm{L}}_{\ell/\mathrm{comp}}(\omega)\right]^{-1}\right\}. \quad (7.34)$$

It uses the scalar operator

$$\mathcal{M}_{\mathrm{ICM}}\left\{\underline{\underline{\mathbf{N}}}_{\ell/\mathrm{comp}}(\omega)\right\} = {}^{ee}\aleph^{\mathrm{L}}_{\ell/\mathrm{comp}}(\omega)\,{}^{mm}\aleph^{\mathrm{NL}}_{\ell/\mathrm{comp}}(\omega)$$

$$-{}^{em}\aleph^{\mathrm{L}}_{\ell/\mathrm{comp}}(\omega)\,{}^{me}\aleph^{\mathrm{NL}}_{\ell/\mathrm{comp}}(\omega)$$

$$+{}^{ee}\aleph^{\mathrm{NL}}_{\ell/\mathrm{comp}}(\omega)\,{}^{mm}\aleph^{\mathrm{L}}_{\ell/\mathrm{comp}}(\omega)$$

$$-{}^{em}\aleph^{\mathrm{NL}}_{\ell/\mathrm{comp}}(\omega)\,{}^{me}\aleph^{\mathrm{L}}_{\ell/\mathrm{comp}}(\omega), \quad (7.35)$$

wherein

$$\underline{\underline{\mathbf{N}}}^{\mathrm{j}}_{\ell/\mathrm{comp}}(\omega) = \begin{bmatrix} {}^{ee}\aleph^{\mathrm{j}}_{\ell/\mathrm{comp}}(\omega)\,\underline{\underline{I}} & {}^{em}\aleph^{\mathrm{j}}_{\ell/\mathrm{comp}}(\omega)\,\underline{\underline{I}} \\ {}^{me}\aleph^{\mathrm{j}}_{\ell/\mathrm{comp}}(\omega)\,\underline{\underline{I}} & {}^{mm}\aleph^{\mathrm{j}}_{\ell/\mathrm{comp}}(\omega)\,\underline{\underline{I}} \end{bmatrix}, \qquad (\mathrm{j=L,\,NL}). \tag{7.36}$$

A suitable estimate of the local field factor $g^{\mathrm{FF}}_{\mathrm{ICM}}(\omega)$ introduced in Eq. (7.30)$_1$ — which represents the derivative on the side of Eq. (7.19) — is provided by [25, 26]

$$g^{\mathrm{FF}}_{\mathrm{ICM}}(\omega) = \left|\left\{\left[\underline{\underline{\mathbf{N}}}^{\mathrm{L}}_{\mathrm{a/comp}}(\omega)\right]^{-1}\right\}_{11}\right|^{2}. \tag{7.37}$$

The constitutive parameters of the comparison medium are determined by solving the nonlinear equations

$$f_{\mathrm{a}}\,\underline{\underline{\alpha}}^{\mathrm{j}}_{\mathrm{a/comp}}(\omega) + f_{\mathrm{b}}\,\underline{\underline{\alpha}}^{\mathrm{j}}_{\mathrm{b/comp}}(\omega) = \underline{\underline{0}}, \qquad (\mathrm{j=L,\,NL}), \tag{7.38}$$

which represent the weakly nonlinear generalization of Eq. (6.15) (or, equivalently, the Bruggeman equation (6.9)). Just as for the corresponding linear scenario described in Sec. 6.3, the nonlinear Eqs. (7.38) can be solved recursively; i.e.,

$$\underline{\underline{K}}^{\mathrm{j}}_{\mathrm{comp}}[n](\omega) = \mathcal{S}^{\mathrm{j}}_{\mathrm{ICM}}\left\{\underline{\underline{K}}^{\mathrm{j}}_{\mathrm{comp}}[n-1](\omega)\right\}, \qquad (\mathrm{j=L,\,NL};\, n=1,2,\ldots). \tag{7.39}$$

The dyadic operator $\mathcal{S}^{\mathrm{L}}_{\mathrm{ICM}}\{\,\bullet\,\}$ is as given in Eq. (6.11), but with $\underline{\underline{K}}^{\mathrm{L}}(\omega)$ replacing $\underline{\underline{K}}_{\mathrm{a}}(\omega)$, $\underline{\underline{K}}^{\mathrm{L}}_{\mathrm{comp}}(\omega)$ replacing $\underline{\underline{K}}_{\mathrm{Br}}(\omega)$, and $\underline{\underline{D}}^{o,\mathrm{L}}_{\ell/\mathrm{comp}}(\omega)$ replacing

$\mathbf{\underline{D}}^0_{U^a/Br}(\omega)$ and $\mathbf{\underline{D}}^0_{U^b/Br}(\omega)$ therein. The operator providing the weakly nonlinear contribution to $\mathbf{\underline{K}}_{\text{comp}}[n](\omega)$ is specified by

$$\mathcal{S}^{\text{NL}}_{\text{ICM}}\left\{\mathbf{\underline{K}}^{\text{NL}}_{\text{comp}}(\omega)\right\} = \left(f_a\left\{g^{\text{FF}}_{\text{ICM}}(\omega)\,\mathbf{\underline{K}}^{\text{NL}}_{a}(\omega)\bullet\left[\mathbf{\underline{\aleph}}^{\text{L}}_{a/\text{comp}}(\omega)\right]^{-1}\right.\right.$$
$$\left.+\left[\mathbf{\underline{K}}^{\text{L}}_{a}(\omega)-\mathbf{\underline{K}}^{\text{L}}_{\text{comp}}(\omega)\right]\bullet\mathcal{N}_{\text{ICM}}\left\{\mathbf{\underline{\aleph}}_{a/\text{comp}}(\omega)\right\}\right\}$$
$$\left.+f_b\left[\mathbf{\underline{K}}_{b}(\omega)-\mathbf{\underline{K}}^{\text{L}}_{\text{comp}}(\omega)\right]\bullet\mathcal{N}_{\text{ICM}}\left\{\mathbf{\underline{\aleph}}_{b/\text{comp}}(\omega)\right\}\right)$$
$$\bullet\left\{f_a\left[\mathbf{\underline{\aleph}}^{\text{L}}_{a/\text{comp}}(\omega)\right]^{-1}+f_b\left[\mathbf{\underline{\aleph}}^{\text{L}}_{b/\text{comp}}(\omega)\right]^{-1}\right\}^{-1}.\tag{7.40}$$

In order to obtain the mass operator, the linear and weakly nonlinear contributions to the principal–value integral represented by $\mathbf{\underline{P}}_\Lambda(L_{\text{cor}},\omega)$ in Eq. (6.19) must first be considered. We have

$$\mathbf{\underline{P}}_\Lambda(L_{\text{cor}},\omega)=\mathbf{\underline{P}}^{\text{L}}_\Lambda(L_{\text{cor}},\omega)+\mathbf{\underline{P}}^{\text{NL}}_\Lambda(L_{\text{cor}},\omega)\,|\underline{E}_{\text{HCM}}(\omega)|^2.\tag{7.41}$$

Upon implementing the step covariance function $\Lambda_{\text{step}}(\underline{r}-\underline{r}')$, as in Eq. (6.20) with the shape dyadic $\underline{U}=\underline{I}$, the form of $\mathbf{\underline{P}}^{\text{L}}_{\Lambda_{\text{step}}}(L_{\text{cor}},\omega)$ is given by Eq. (6.21), wherein the components of $\mathbf{\underline{D}}^+_{I/\text{comp}}(L_{\text{cor}},\omega)$ are provided by Eqs. (5.79) with $\epsilon^{\text{L}}_{\text{comp}}(\omega)$, $\chi^{\text{L}}_{\text{comp}}(\omega)$ and $\mu^{\text{L}}_{\text{comp}}(\omega)$ replacing $\epsilon(\omega)$, $\chi(\omega)$ and $\mu(\omega)$, respectively. The corresponding weakly nonlinear contribution to $\mathbf{\underline{P}}_\Lambda(L_{\text{cor}},\omega)$ may be expressed as

$$\mathbf{\underline{P}}^{\text{NL}}_{\Lambda_{\text{step}}}(L_{\text{cor}},\omega)=\frac{i\omega L^2_{\text{cor}}f_af_b}{3}\begin{bmatrix}{}^{ee}P^{\text{NL}}_{\Lambda_{\text{step}}}(\omega)\,\underline{I}&{}^{em}P^{\text{NL}}_{\Lambda_{\text{step}}}(\omega)\,\underline{I}\\{}^{me}P^{\text{NL}}_{\Lambda_{\text{step}}}(\omega)\,\underline{I}&{}^{mm}P^{\text{NL}}_{\Lambda_{\text{step}}}(\omega)\,\underline{I}\end{bmatrix},\tag{7.42}$$

where the scalar functions

$$\left.\begin{aligned}{}^{ee}P^{\text{NL}}_{\Lambda_{\text{step}}}(\omega)&=\mu^{\text{NL}}_{\text{comp}}(\omega)+i\frac{2}{3}\omega L_{\text{cor}}\left[\mu^{\text{NL}}_{\text{comp}}(\omega)\,p_1(\omega)+\mu^{\text{L}}_{\text{comp}}(\omega)\,p_2(\omega)\right]\\{}^{em}P^{\text{NL}}_{\Lambda_{\text{step}}}(\omega)&=\chi^{\text{NL}}_{\text{comp}}(\omega)+i\frac{4}{3}\omega L_{\text{cor}}\left[\chi^{\text{NL}}_{\text{comp}}(\omega)\,p_3(\omega)+\chi^{\text{L}}_{\text{comp}}(\omega)\,p_4(\omega)\right]\\{}^{me}P^{\text{NL}}_{\Lambda_{\text{step}}}(\omega)&=-\chi^{\text{NL}}_{\text{comp}}(\omega)-i\frac{4}{3}\omega L_{\text{cor}}\left[\chi^{\text{NL}}_{\text{comp}}(\omega)\,p_3(\omega)+\chi^{\text{L}}_{\text{comp}}(\omega)\,p_4(\omega)\right]\\{}^{mm}P^{\text{NL}}_{\Lambda_{\text{step}}}(\omega)&=\epsilon^{\text{NL}}_{\text{comp}}(\omega)+i\frac{2}{3}\omega L_{\text{cor}}\left[\epsilon^{\text{NL}}_{\text{comp}}(\omega)\,p_1(\omega)+\epsilon^{\text{L}}_{\text{comp}}(\omega)\,p_2(\omega)\right]\end{aligned}\right\}\tag{7.43}$$

contain

$$p_1(\omega) = \frac{\epsilon^L_{comp}(\omega)\,\mu^L_{comp}(\omega) - \left[\xi^L_{comp}(\omega)\right]^2}{p_3(\omega)}$$

$$p_2(\omega) = \left\{1 + \left[\frac{\xi^L_{comp}}{p_3(\omega)}\right]^2\right\} p_4(\omega) - \frac{2\,\xi^L_{comp}(\omega)\,\xi^{NL}_{comp}(\omega)}{p_3(\omega)}$$

$$p_3(\omega) = \left[\epsilon^L_{comp}(\omega)\,\mu^L_{comp}(\omega)\right]^{1/2}$$

$$p_4(\omega) = \frac{\epsilon^L_{comp}(\omega)\,\mu^{NL}_{comp}(\omega) + \epsilon^{NL}_{comp}(\omega)\,\mu^L_{comp}(\omega)}{2\,p_3(\omega)}$$

(7.44)

In consonance with the weak nonlinearity condition (7.10), the second–order mass operator is expressed as

$$\underline{\underline{\Sigma}}^{[2]}(L_{cor},\omega) = \underline{\underline{\Sigma}}^{[2],L}(L_{cor},\omega) + \underline{\underline{\Sigma}}^{[2],NL}(L_{cor},\omega)\,|\underline{E}_{HCM}(\omega)|^2. \quad (7.45)$$

The linear part of the mass operator, namely $\underline{\underline{\Sigma}}^{[2],L}(L_{cor},\omega)$, is delivered from Eq. (6.18) with $\underline{\underline{\alpha}}_{a,b/comp}(\omega)$ therein being replaced by $\underline{\underline{\alpha}}^L_{a,b/comp}(\omega)$, and $\underline{\underline{P}}_\Lambda(L_{cor},\omega)$ therein being replaced by $\underline{\underline{P}}^L_{\Lambda_{step}}(L_{cor},\omega)$. The weakly non-linear part of the mass operator, namely $\underline{\underline{\Sigma}}^{[2],NL}(L_{cor},\omega)$, is given by

$$\underline{\underline{\Sigma}}^{[2],NL}(L_{cor},\omega) = -\omega^2\Bigg(\left\{\left[\underline{\underline{\alpha}}^L_{a/comp}(\omega) - \underline{\underline{\alpha}}^L_{b/comp}(\omega)\right]\cdot\underline{\underline{P}}^L_{\Lambda_{step}}(L_{cor},\omega)\right.$$
$$\cdot\left[\underline{\underline{\alpha}}^{NL}_{a/comp}(\omega) - \underline{\underline{\alpha}}^{NL}_{b/comp}(\omega)\right]\Bigg\}$$
$$+\left\{\left[\underline{\underline{\alpha}}^L_{a/comp}(\omega) - \underline{\underline{\alpha}}^L_{b/comp}(\omega)\right]\cdot\underline{\underline{P}}^{NL}_{\Lambda_{step}}(L_{cor},\omega)\right.$$
$$\cdot\left[\underline{\underline{\alpha}}^L_{a/comp}(\omega) - \underline{\underline{\alpha}}^L_{b/comp}(\omega)\right]\Bigg\}$$
$$+\left\{\left[\underline{\underline{\alpha}}^{NL}_{a/comp}(\omega) - \underline{\underline{\alpha}}^{NL}_{b/comp}(\omega)\right]\cdot\underline{\underline{P}}^L_{\Lambda_{step}}(L_{cor},\omega)\right.$$
$$\cdot\left.\left[\underline{\underline{\alpha}}^L_{a/comp}(\omega) - \underline{\underline{\alpha}}^L_{b/comp}(\omega)\right]\right\}\Bigg). \quad (7.46)$$

Finally, the foregoing expressions come together to deliver the second–order SPFT estimate of the HCM constitutive dyadic, namely

$$\underline{\underline{K}}^{[2]}_{SPFT}(\omega) = \underline{\underline{K}}^{[2],L}_{SPFT}(\omega) + \underline{\underline{K}}^{[2],NL}_{SPFT}(\omega)\,|\underline{E}_{HCM}(\omega)|^2, \quad (7.47)$$

wherein the linear contribution is provided by Eq. (6.14) with $n = 2$; and $\underline{\underline{K}}^L_{comp}(\omega)$, $\underline{\underline{D}}^{0,L}_{1/comp}(\omega)$ and $\underline{\underline{\Sigma}}^{[2],L}(L_{cor},\omega)$ replacing $\underline{\underline{K}}_{comp}(\omega)$, $\underline{\underline{D}}^0_{U/comp}(\omega)$

and $\underline{\underline{\Sigma}}^{[2]}(L_{\mathrm{cor}},\omega)$, respectively. The weakly nonlinear contribution is specified as

$$\underline{\underline{K}}^{[2],\mathrm{NL}}_{\mathrm{SPFT}}(\omega) = \underline{\underline{K}}^{\mathrm{NL}}_{\mathrm{comp}}(\omega) - \frac{1}{i\omega}\left\{ \left[\underline{\underline{e}}^{\mathrm{L}}(\omega)\right]^{-1} \cdot \underline{\underline{\Sigma}}^{[2],\mathrm{NL}}(L_{\mathrm{cor}},\omega) \right.$$

$$\left. +\mathcal{N}_{\mathrm{ICM}}\left\{\underline{\underline{e}}(\omega)\right\} \cdot \underline{\underline{\Sigma}}^{[2],\mathrm{L}}(L_{\mathrm{cor}},\omega) \right\}, \qquad (7.48)$$

wherein the 6×6 dyadic

$$\underline{\underline{e}}(\omega) = \underline{\underline{e}}^{\mathrm{L}}(\omega) + \underline{\underline{e}}^{\mathrm{NL}}(\omega) \, | \, \underline{E}_{\mathrm{HCM}}(\omega) \, |^{2}, \qquad (7.49)$$

with components

$$\left. \begin{aligned} \underline{\underline{e}}^{\mathrm{L}}(\omega) &= \underline{\underline{I}} + \underline{\underline{\Sigma}}^{[2],\mathrm{L}}(L_{\mathrm{cor}},\omega) \cdot \underline{\underline{D}}^{0,\mathrm{L}}_{\mathrm{I/comp}}(\omega) \\ \underline{\underline{e}}^{\mathrm{NL}}(\omega) &= \underline{\underline{\Sigma}}^{[2],\mathrm{L}}(L_{\mathrm{cor}},\omega) \cdot \underline{\underline{D}}^{0,\mathrm{NL}}_{\mathrm{I/comp}}(\omega) + \underline{\underline{\Sigma}}^{[2],\mathrm{NL}}(L_{\mathrm{cor}},\omega) \cdot \underline{\underline{D}}^{0,\mathrm{L}}_{\mathrm{I/comp}}(\omega) \end{aligned} \right\}, \qquad (7.50)$$

has been introduced.

The corresponding third–order SPFT estimate of the HCM constitutive dyadic — derived using the three–point covariance function (6.22) — has been been implemented in numerical studies to demonstrate convergence at the second–order level [16].

7.2.2.3 *Weakly nonlinear, anisotropic dielectric HCM*

We now turn to a scenario where weak nonlinearity is combined with anisotropy. The key results of the second–order SPFT homogenization procedure are presented here, with full details being available elsewhere [18, 19]. These results contain the corresponding results for the Bruggeman homogenization formalism — which is effectively the lowest–order SPFT approach [17].

Suppose that constituent mediums 'a' and 'b' are both weakly nonlinear, isotropic dielectric mediums with scalar permittivities of the form of (7.15). Both mediums are randomly distributed as identically oriented, conformal, ellipsoidal particles described by the shape dyadic $\underline{\underline{U}}$, as illustrated in Fig. 6.2. Without loss of generality, let us take the axes of our coordinate system to be aligned with the principal axes of the ellipsoidal particles, so that $\underline{\underline{U}}$ has a diagonal matrix representation. The corresponding HCM is endowed with anisotropy, on account of the ellipsoidal shape of the constituent particles. Thus, the comparison medium is specified by

$$\underline{\underline{\epsilon}}_{\text{comp}}(\omega) = \underline{\underline{\epsilon}}^{\text{L}}_{\text{comp}}(\omega) + \underline{\underline{\epsilon}}^{\text{NL}}_{\text{comp}}(\omega) \, | \, \underline{E}_{\text{HCM}}(\omega) \, |^2 \qquad (7.51)$$

$$= \begin{pmatrix} \epsilon^{x,\text{L}}_{\text{comp}}(\omega) & 0 & 0 \\ 0 & \epsilon^{y,\text{L}}_{\text{comp}}(\omega) & 0 \\ 0 & 0 & \epsilon^{z,\text{L}}_{\text{comp}}(\omega) \end{pmatrix}$$

$$+ \begin{pmatrix} \epsilon^{x,\text{NL}}_{\text{comp}}(\omega) & 0 & 0 \\ 0 & \epsilon^{y,\text{NL}}_{\text{comp}}(\omega) & 0 \\ 0 & 0 & \epsilon^{z,\text{NL}}_{\text{comp}}(\omega) \end{pmatrix} | \, \underline{E}_{\text{HCM}}(\omega) \, |^2,$$

$$(7.52)$$

with $\underline{E}_{\text{HCM}}(\omega)$ being the spatially averaged electric field in the HCM, as in Eqs. (7.16) and (7.23). Similarly, the corresponding 3×3 depolarization dyadic for a vanishingly small ellipsoidal particle embedded in the comparison medium — denoted by $^{ee}\underline{\underline{D}}^0_{U/\text{comp}}(\omega)$ in the notation of Eq. (5.69)$_1$ — is expressed as the sum of linear and weakly nonlinear parts

$$^{ee}\underline{\underline{D}}^0_{U/\text{comp}}(\omega) = {}^{ee}\underline{\underline{D}}^{0,\text{L}}_{U/\text{comp}}(\omega) + {}^{ee}\underline{\underline{D}}^{0,\text{NL}}_{U/\text{comp}}(\omega) \, | \, \underline{E}_{\text{HCM}}(\omega) \, |^2. \qquad (7.53)$$

By introducing the 3×3 diagonal dyadic

$$\underline{\underline{\Omega}}(\theta, \phi) = \underline{\underline{U}}^{-1} \cdot \begin{pmatrix} \sin^2\theta\cos^2\phi & 0 & 0 \\ 0 & \sin^2\theta\sin^2\phi & 0 \\ 0 & 0 & \cos^2\theta \end{pmatrix} \cdot \underline{\underline{U}}^{-1}, \qquad (7.54)$$

the linear and nonlinear contributions to the depolarization dyadic may be conveniently written as

$$^{ee}\underline{\underline{D}}^{0,\text{L}}_{U/\text{comp}}(\omega) = \frac{1}{4\pi i\omega} \int_{\phi=0}^{2\pi} \int_{\theta=0}^{\pi} \frac{\sin\theta}{\text{tr}\left[\underline{\underline{\epsilon}}^{\text{L}}_{\text{comp}}(\omega) \cdot \underline{\underline{\Omega}}(\theta,\phi)\right]} \underline{\underline{\Omega}}(\theta,\phi) \, d\theta \, d\phi \qquad (7.55)$$

and

$$^{ee}\underline{\underline{D}}^{0,\text{NL}}_{U/\text{comp}}(\omega) = -\frac{1}{4\pi i\omega} \int_{\phi=0}^{2\pi} \int_{\theta=0}^{\pi} \frac{\text{tr}\left[\underline{\underline{\epsilon}}^{\text{NL}}_{\text{comp}}(\omega) \cdot \underline{\underline{\Omega}}(\theta,\phi)\right] \sin\theta}{\left\{\text{tr}\left[\underline{\underline{\epsilon}}^{\text{L}}_{\text{comp}}(\omega) \cdot \underline{\underline{\Omega}}(\theta,\phi)\right]\right\}^2}$$

$$\times \underline{\underline{\Omega}}(\theta,\phi) \, d\theta \, d\phi, \qquad (7.56)$$

respectively. Notice that the surface integral on the right side of Eq. (7.55) may be equivalently expressed in terms of elliptic functions, in accord with Eqs. (5.95).

The corresponding 3×3 polarizability density dyadic has the form

$$
{}^{ee}\underline{\underline{\alpha}}_{\ell/\mathrm{comp}}(\omega) = {}^{ee}\underline{\underline{\alpha}}^{\mathrm{L}}_{\ell/\mathrm{comp}}(\omega) + {}^{ee}\underline{\underline{\alpha}}^{\mathrm{NL}}_{\ell/\mathrm{comp}}(\omega)\,|\underline{E}_{\mathrm{HCM}}(\omega)|^{2}, \qquad (\ell = \mathrm{a,b}),
\tag{7.57}
$$

where the linear contribution

$$
{}^{ee}\underline{\underline{\alpha}}^{\mathrm{L}}_{\ell/\mathrm{comp}}(\omega) = \left[\epsilon^{\mathrm{L}}_{\ell}(\omega)\underline{\underline{I}} - \underline{\underline{\epsilon}}^{\mathrm{L}}_{\mathrm{comp}}(\omega)\right] \bullet \left[{}^{ee}\underline{\underline{N}}_{\ell/\mathrm{comp}}(\omega)\right]^{-1}
\tag{7.58}
$$

and the weakly nonlinear contribution

$$
{}^{ee}\underline{\underline{\alpha}}^{\mathrm{NL}}_{\ell/\mathrm{comp}}(\omega) = \left[g^{\mathrm{FF}}_{\ell,\underline{\underline{\epsilon}}}(\omega)\,\epsilon^{\mathrm{NL}}_{\ell}(\omega)\underline{\underline{I}} - \underline{\underline{\epsilon}}^{\mathrm{NL}}_{\mathrm{comp}}(\omega)\right] \bullet \left[{}^{ee}\underline{\underline{N}}_{\ell/\mathrm{comp}}(\omega)\right]^{-1}
$$
$$
+ \left[\epsilon^{\mathrm{L}}_{\ell}(\omega)\underline{\underline{I}} - \underline{\underline{\epsilon}}^{\mathrm{L}}_{\mathrm{comp}}(\omega)\right] \bullet \mathcal{N}_{\underline{\underline{\epsilon}}}\left\{{}^{ee}\underline{\underline{N}}_{\ell/\mathrm{comp}}(\omega)\right\}.
\tag{7.59}
$$

Herein, the 3×3 dyadic

$$
{}^{ee}\underline{\underline{N}}_{\ell/\mathrm{comp}}(\omega) = {}^{ee}\underline{\underline{N}}^{\mathrm{L}}_{\ell/\mathrm{comp}}(\omega) + {}^{ee}\underline{\underline{N}}^{\mathrm{NL}}_{\ell/\mathrm{comp}}(\omega)\,|\underline{E}_{\mathrm{HCM}}(\omega)|^{2}, \qquad (\ell = \mathrm{a,\ b})
\tag{7.60}
$$

has the linear part

$$
{}^{ee}\underline{\underline{N}}^{\mathrm{NL}}_{\ell/\mathrm{comp}}(\omega) = \underline{\underline{I}} + i\omega\,{}^{ee}\underline{\underline{D}}^{0,\mathrm{L}}_{U/\mathrm{comp}}(\omega) \bullet \left[\epsilon^{\mathrm{L}}_{\ell}(\omega)\underline{\underline{I}} - \underline{\underline{\epsilon}}^{\mathrm{L}}_{\mathrm{comp}}(\omega)\right]
\tag{7.61}
$$

and weakly nonlinear part

$$
{}^{ee}\underline{\underline{N}}^{\mathrm{L}}_{\ell/\mathrm{comp}}(\omega) = i\omega\left\{{}^{ee}\underline{\underline{D}}^{0,\mathrm{L}}_{U/\mathrm{comp}}(\omega) \bullet \left[g^{\mathrm{FF}}_{\ell,\underline{\underline{\epsilon}}}(\omega)\,\epsilon^{\mathrm{NL}}_{\ell}(\omega)\underline{\underline{I}} - \underline{\underline{\epsilon}}^{\mathrm{NL}}_{\mathrm{comp}}(\omega)\right]\right.
$$
$$
\left. + {}^{ee}\underline{\underline{D}}^{0,\mathrm{NL}}_{U/\mathrm{comp}}(\omega) \bullet \left[\epsilon^{\mathrm{L}}_{\ell}(\omega)\underline{\underline{I}} - \underline{\underline{\epsilon}}^{\mathrm{L}}_{\mathrm{comp}}(\omega)\right]\right\},
\tag{7.62}
$$

with ${}^{ee}\underline{\underline{N}}^{\mathrm{j}}_{\ell/\mathrm{comp}}(\omega)$ having the diagonal form

$$
{}^{ee}\underline{\underline{N}}^{\mathrm{j}}_{\ell/\mathrm{comp}}(\omega) = \begin{pmatrix} {}^{ee}N^{x,\mathrm{j}}_{\ell/\mathrm{comp}}(\omega) & 0 & 0 \\ 0 & {}^{ee}N^{y,\mathrm{j}}_{\ell/\mathrm{comp}}(\omega) & 0 \\ 0 & 0 & {}^{ee}N^{z,\mathrm{j}}_{\ell/\mathrm{comp}}(\omega) \end{pmatrix}, \qquad (\mathrm{j = L, NL}).
\tag{7.63}
$$

The 3×3 dyadic operator $\mathcal{N}_{\underline{\underline{\epsilon}}}\{\,\bullet\,\}$ yields

$$
\mathcal{N}_{\underline{\underline{\epsilon}}}\left\{{}^{ee}\underline{\underline{\aleph}}_{\ell/\mathrm{comp}}(\omega)\right\} = \frac{1}{\det\,{}^{ee}\underline{\underline{\aleph}}^{\mathrm{L}}_{\ell/\mathrm{comp}}(\omega)}
$$

$$
\times\left\{\begin{pmatrix} \Pi^{x}_{\ell/\mathrm{comp}}(\omega) & 0 & 0 \\ 0 & \Pi^{y}_{\ell/\mathrm{comp}}(\omega) & 0 \\ 0 & 0 & \Pi^{z}_{\ell/\mathrm{comp}}(\omega) \end{pmatrix}\right.
$$

$$
\left. -\mathcal{M}_{\underline{\underline{\epsilon}}}\left\{{}^{ee}\underline{\underline{\aleph}}_{\ell/\mathrm{comp}}(\omega)\right\}\left[{}^{ee}\underline{\underline{\aleph}}^{\mathrm{L}}_{\ell/\mathrm{comp}}(\omega)\right]^{-1}\right\}, \quad (7.64)
$$

wherein the scalar operator $\mathcal{M}_{\underline{\underline{\epsilon}}}\{\,\bullet\,\}$ delivers

$$
\mathcal{M}_{\underline{\underline{\epsilon}}}\left\{{}^{ee}\underline{\underline{\aleph}}_{\ell/\mathrm{comp}}(\omega)\right\} = {}^{ee}\aleph^{x,\mathrm{L}}_{\ell/\mathrm{comp}}(\omega)\,{}^{ee}\aleph^{y,\mathrm{L}}_{\ell/\mathrm{comp}}(\omega)\,{}^{ee}\aleph^{z,\mathrm{NL}}_{\ell/\mathrm{comp}}(\omega)
$$

$$
+\,{}^{ee}\aleph^{x,\mathrm{L}}_{\ell/\mathrm{comp}}(\omega)\,{}^{ee}\aleph^{y,\mathrm{NL}}_{\ell/\mathrm{comp}}(\omega)\,{}^{ee}\aleph^{z,\mathrm{L}}_{\ell/\mathrm{comp}}(\omega)
$$

$$
+\,{}^{ee}\aleph^{x,\mathrm{NL}}_{\ell/\mathrm{comp}}(\omega)\,{}^{ee}\aleph^{y,\mathrm{L}}_{\ell/\mathrm{comp}}(\omega)\,{}^{ee}\aleph^{z,\mathrm{L}}_{\ell/\mathrm{comp}}(\omega)
$$

$$
(7.65)
$$

and

$$
\left.\begin{aligned}
\Pi^{x}_{\ell/\mathrm{comp}}(\omega) &= {}^{ee}\aleph^{y,\mathrm{NL}}_{\ell/\mathrm{comp}}(\omega)\,{}^{ee}\aleph^{z,\mathrm{L}}_{\ell/\mathrm{comp}}(\omega) + {}^{ee}\aleph^{y,\mathrm{L}}_{\ell/\mathrm{comp}}(\omega)\,{}^{ee}\aleph^{z,\mathrm{NL}}_{\ell/\mathrm{comp}}(\omega) \\
\Pi^{y}_{\ell/\mathrm{comp}}(\omega) &= {}^{ee}\aleph^{z,\mathrm{NL}}_{\ell/\mathrm{comp}}(\omega)\,{}^{ee}\aleph^{x,\mathrm{L}}_{\ell/\mathrm{comp}}(\omega) + {}^{ee}\aleph^{z,\mathrm{L}}_{\ell/\mathrm{comp}}(\omega)\,{}^{ee}\aleph^{x,\mathrm{NL}}_{\ell/\mathrm{comp}}(\omega) \\
\Pi^{z}_{\ell/\mathrm{comp}}(\omega) &= {}^{ee}\aleph^{y,\mathrm{NL}}_{\ell/\mathrm{comp}}(\omega)\,{}^{ee}\aleph^{x,\mathrm{L}}_{\ell/\mathrm{comp}}(\omega) + {}^{ee}\aleph^{y,\mathrm{L}}_{\ell/\mathrm{comp}}(\omega)\,{}^{ee}\aleph^{x,\mathrm{NL}}_{\ell/\mathrm{comp}}(\omega)
\end{aligned}\right\}.
$$

$$
(7.66)
$$

Suitable estimates of the local field factors $g^{\mathrm{FF}}_{\ell,\underline{\underline{\epsilon}}}(\omega)$ ($\ell = $ a,b) in Eq. (7.59) — which arise as the derivative on the right side of Eq. (7.19) with the subscript ℓ replacing 'a' therein — are as follows:

$$
g^{\mathrm{FF}}_{\ell,\underline{\underline{\epsilon}}}(\omega) = \frac{1}{3}\left|\,\mathrm{tr}\left[{}^{ee}\underline{\underline{\aleph}}^{\mathrm{L}}_{\ell/\mathrm{comp}}(\omega)\right]^{-1}\right|, \qquad (\ell = \text{a,b}). \qquad (7.67)
$$

The intensity–dependent permittivity dyadic of the comparison medium, $\underline{\underline{\epsilon}}_{\mathrm{comp}}(\omega)$, is found by solving nonlinear equations equivalent to those specified by Eqs. (7.38), but wherein the 3×3 polarizability density dyadics ${}^{ee}\underline{\underline{\alpha}}^{\mathrm{L,NL}}_{\ell/\mathrm{comp}}(\omega)$ replace the 6×6 polarizability density dyadics $\underline{\underline{\alpha}}^{\mathrm{L,NL}}_{\ell/\mathrm{comp}}(\omega)$. The intensity–dependent permittivity dyadics $\underline{\underline{\epsilon}}^{\mathrm{L,NL}}_{\mathrm{comp}}(\omega)$ may be extracted from these nonlinear equations by means of recurrence; i.e.,

$$
\underline{\underline{\epsilon}}^{\,\mathrm{j}}_{\mathrm{comp}}[n](\omega) = \mathcal{S}^{\mathrm{j}}_{\underline{\underline{\epsilon}}}\left\{\underline{\underline{\epsilon}}_{\mathrm{comp}}[n-1](\omega)\right\}, \qquad (\mathrm{j}=\mathrm{L},\,\mathrm{NL};\; n = 1,2,\ldots).
$$

$$
(7.68)
$$

Here the dyadic operator $\mathcal{S}_{\underline{\underline{\epsilon}}}^{\mathrm{L}}\{\,\bullet\,\}$ is given by the 3×3 dielectric special-ization of Eq. (6.11), with $\underline{\underline{\epsilon}}_{\mathrm{a}}^{\mathrm{L}}(\omega)$ replacing $\underline{\underline{\mathbf{K}}}_{\mathrm{a}}(\omega)$, $\underline{\underline{\epsilon}}_{\mathrm{comp}}^{\mathrm{L}}(\omega)$ replacing $\underline{\underline{\mathbf{K}}}_{\mathrm{Br}}(\omega)$, $^{\mathrm{ee}}\underline{\underline{D}}_{U/\mathrm{comp}}^{0,\mathrm{L}}(\omega)$ replacing $\underline{\underline{\mathbf{D}}}_{U^{\mathrm{a}}/\mathrm{Br}}^{0}(\omega)$ and $\underline{\underline{\mathbf{D}}}_{U^{\mathrm{b}}/\mathrm{Br}}^{0}(\omega)$, and $\underline{\underline{I}}$ replac-ing $\underline{\underline{\mathbf{I}}}$ therein. The weakly nonlinear contribution to $\underline{\underline{\epsilon}}_{\mathrm{comp}}[n](\omega)$ is provided by the operator

$$
\begin{aligned}
\mathcal{S}_{\underline{\underline{\epsilon}}}^{\mathrm{NL}}\left\{\underline{\underline{\epsilon}}_{\mathrm{comp}}^{\mathrm{NL}}(\omega)\right\} = &\left(f_{\mathrm{a}}\left\{g_{\mathrm{a},\underline{\underline{\epsilon}}}^{\mathrm{FF}}(\omega)\,\epsilon_{\mathrm{a}}^{\mathrm{NL}}(\omega)\left[{}^{\mathrm{ee}}\underline{\underline{\aleph}}_{\mathrm{a}/\mathrm{comp}}^{\mathrm{L}}(\omega)\right]^{-1}\right.\right. \\
&\left.+\left[\epsilon_{\mathrm{a}}^{\mathrm{L}}(\omega)\,\underline{\underline{I}}-\underline{\underline{\epsilon}}_{\mathrm{comp}}^{\mathrm{L}}(\omega)\right]\bullet\mathcal{N}_{\underline{\underline{\epsilon}}}\left\{{}^{\mathrm{ee}}\underline{\underline{\aleph}}_{\mathrm{a}/\mathrm{comp}}^{\mathrm{L}}(\omega)\right\}\right\} \\
&+\left(f_{\mathrm{b}}\left\{g_{\mathrm{b},\underline{\underline{\epsilon}}}^{\mathrm{FF}}(\omega)\,\epsilon_{\mathrm{b}}^{\mathrm{NL}}(\omega)\left[{}^{\mathrm{ee}}\underline{\underline{\aleph}}_{\mathrm{b}/\mathrm{comp}}^{\mathrm{L}}(\omega)\right]^{-1}\right.\right. \\
&\left.+\left[\epsilon_{\mathrm{b}}^{\mathrm{L}}(\omega)\,\underline{\underline{I}}-\underline{\underline{\epsilon}}_{\mathrm{comp}}^{\mathrm{L}}(\omega)\right]\bullet\mathcal{N}_{\underline{\underline{\epsilon}}}\left\{{}^{\mathrm{ee}}\underline{\underline{\aleph}}_{\mathrm{b}/\mathrm{comp}}^{\mathrm{L}}(\omega)\right\}\right\} \\
&\bullet\left\{f_{\mathrm{a}}\left[{}^{\mathrm{ee}}\underline{\underline{\aleph}}_{\mathrm{a}/\mathrm{comp}}^{\mathrm{L}}(\omega)\right]^{-1}+f_{\mathrm{b}}\left[{}^{\mathrm{ee}}\underline{\underline{\aleph}}_{\mathrm{b}/\mathrm{comp}}^{\mathrm{L}}(\omega)\right]^{-1}\right\}^{-1}.
\end{aligned}
$$

$$(7.69)$$

We now turn to the calculation of the mass operator, which involves the principal value integral represented by $\underline{\underline{\mathbf{P}}}_{\Lambda}(L_{\mathrm{cor}},\omega)$ in Eq. (6.19). Only the components $\left[\underline{\underline{\mathbf{P}}}_{\Lambda}(L_{\mathrm{cor}},\omega)\right]_{mn}$ with $m,n\in\{1,2,3\}$ need be considered here. These form the 3×3 dyadic $^{\mathrm{ee}}\underline{\underline{P}}_{\Lambda}(L_{\mathrm{cor}},\omega)$ such that $\left[\underline{\underline{\mathbf{P}}}_{\Lambda}(L_{\mathrm{cor}},\omega)\right]_{mn}=\left[{}^{\mathrm{ee}}\underline{\underline{P}}_{\Lambda}(L_{\mathrm{cor}},\omega)\right]_{mn}$ for $m,n\in\{1,2,3\}$. Furthermore, the dyadic $^{\mathrm{ee}}\underline{\underline{P}}_{\Lambda}(L_{\mathrm{cor}},\omega)$ may be viewed as the sum of linear and weakly non-linear parts, as in Eq. (7.41) but with $^{\mathrm{ee}}\underline{\underline{P}}_{\Lambda}(L_{\mathrm{cor}},\omega)$ replacing $\underline{\underline{\mathbf{P}}}_{\Lambda}(L_{\mathrm{cor}},\omega)$. With the step covariance function $\Lambda_{\mathrm{step}}(\underline{r}-\underline{r}')$, as specified in Eq. (6.20), the linear part of $^{\mathrm{ee}}\underline{\underline{P}}_{\Lambda_{\mathrm{step}}}(L_{\mathrm{cor}},\omega)$ has the integral representation

$$
\begin{aligned}
{}^{\mathrm{ee}}\underline{\underline{P}}_{\Lambda_{\mathrm{step}}}^{\mathrm{L}}(L_{\mathrm{cor}},\omega) = &\frac{f_{\mathrm{a}}f_{\mathrm{b}}L_{\mathrm{cor}}^{3}}{12\pi i\omega}\int_{\phi=0}^{2\pi}\int_{\theta=0}^{\pi}\frac{\sin\theta}{\Delta_{0}(\theta,\phi,\omega)} \\
&\times\left[\tau_{\mathrm{u}}(L_{\mathrm{cor}},\theta,\phi,\omega)\,\underline{\underline{u}}_{0}(\theta,\phi,\omega)\right. \\
&\left.+\tau_{v}(\theta,\phi,\omega)\,\underline{\underline{v}}_{0}(\theta,\phi,\omega)\right]\,d\theta\,d\phi,
\end{aligned}
$$

$$(7.70)$$

where the 3×3 dyadic functions

$$
\begin{aligned}
\underline{\underline{u}}_0(\theta,\phi,\omega) &= \left\{ 2\,\underline{\underline{\epsilon}}^{\,L}_{comp}(\omega) - \left[\mathrm{tr}\,\underline{\underline{\epsilon}}^{\,L}_{comp}(\omega) \right] \underline{\underline{I}} \right\} \cdot \underline{\underline{\Omega}}(\theta,\phi) \\
&\quad - \left\{ \mathrm{tr}\left[\underline{\underline{\epsilon}}^{\,L}_{comp}(\omega) \cdot \underline{\underline{\Omega}}(\theta,\phi) \right] \right\} \underline{\underline{I}} \\
&\quad - \frac{1}{\mathrm{tr}\left[\underline{\underline{\epsilon}}^{\,L}_{comp}(\omega) \cdot \underline{\underline{\Omega}}(\theta,\phi) \right]} \left(\mathrm{tr}\left\{ \left[\mathrm{adj}\,\underline{\underline{\epsilon}}^{\,L}_{comp}(\omega) \right] \cdot \underline{\underline{\Omega}}(\theta,\phi) \right\} \right. \\
&\quad \left. - \left[\mathrm{tr}\,\mathrm{adj}\,\underline{\underline{\epsilon}}^{\,L}_{comp}(\omega) \right] \left[\mathrm{tr}\,\underline{\underline{\Omega}}(\theta,\phi) \right] \right) \underline{\underline{\Omega}}(\theta,\phi) \\
\underline{\underline{v}}_0(\theta,\phi,\omega) &= \mathrm{adj}\,\underline{\underline{\epsilon}}^{\,L}_{comp}(\omega) - \frac{\det\underline{\underline{\epsilon}}^{\,L}_{comp}(\omega)}{\mathrm{tr}\left[\underline{\underline{\epsilon}}^{\,L}_{comp}(\omega) \cdot \underline{\underline{\Omega}}(\theta,\phi) \right]}\,\underline{\underline{\Omega}}(\theta,\phi)
\end{aligned}
\tag{7.71}
$$

and scalar functions[1]

$$
\left.
\begin{aligned}
\tau_u(L_{cor},\theta,\phi,\omega) &= \frac{3\left(\kappa_{0+}-\kappa_{0-}\right)}{2L_{cor}} + i\left[\kappa_{0+}^{\frac{3}{2}} - \kappa_{0-}^{\frac{3}{2}} \right] \\
\tau_v(\theta,\phi,\omega) &= i\omega^2\mu_0\left(\kappa_{0+}^{\frac{1}{2}} - \kappa_{0-}^{\frac{1}{2}} \right) \\
\Delta_0(\theta,\phi,\omega) &= \left(t_{B0}^2 - 4t_{A0}\,t_{C0} \right)^{1/2} \\
\kappa_{0\pm} &= \mu_0\omega^2 \frac{-t_{B0} \pm \Delta_0(\theta,\phi,\omega)}{2t_{C0}}
\end{aligned}
\right\},
\tag{7.72}
$$

with

$$
\left.
\begin{aligned}
t_{A0} &= \det\underline{\underline{\epsilon}}^{\,L}_{comp}(\omega) \\
t_{B0} &= \left\{ \mathrm{tr}\left[\mathrm{adj}\,\underline{\underline{\epsilon}}^{\,L}_{comp}(\omega) \right] \cdot \underline{\underline{\Omega}}(\theta,\phi) \right\} - \left[\mathrm{tr}\,\mathrm{adj}\,\underline{\underline{\epsilon}}^{\,L}_{comp}(\omega) \right] \mathrm{tr}\,\underline{\underline{\Omega}}(\theta,\phi) \\
t_{C0} &= \left\{ \mathrm{tr}\left[\underline{\underline{\epsilon}}^{\,L}_{comp}(\omega) \cdot \underline{\underline{\Omega}}(\theta,\phi) \right] \right\} \mathrm{tr}\,\underline{\underline{\Omega}}(\theta,\phi)
\end{aligned}
\right\}.
\tag{7.73}
$$

---

[1]For compact presentation, the dependencies of $\kappa_{0\pm} \equiv \kappa_{0\pm}(\theta,\phi,\omega)$, $t_{A0} \equiv t_{A0}(\omega)$, $t_{B0} \equiv t_{B0}(\theta,\phi,\omega)$ and $t_{C0} \equiv t_{C0}(\theta,\phi,\omega)$ are implicit.

The corresponding weakly nonlinear part of ${}^{\mathrm{ee}}\underline{\underline{P}}_{\Lambda_{\mathrm{step}}}(L_{\mathrm{cor}}, \omega)$ has the integral representation

$$
{}^{\mathrm{ee}}\underline{\underline{P}}_{\Lambda_{\mathrm{step}}}^{\mathrm{NL}}(L_{\mathrm{cor}}, \omega) = \frac{f_{\mathrm{a}} f_{\mathrm{b}} L_{\mathrm{cor}}^{3}}{12\pi i \omega} \int_{\phi=0}^{2\pi} \int_{\theta=0}^{\pi} \frac{\sin\theta}{\Delta_0(\theta, \phi, \omega)} \left\{ \tau_{\mathrm{u}}(L_{\mathrm{cor}}, \theta, \phi, \omega) \right.
$$
$$
\times \left[\underline{\underline{u}}_1(\theta, \phi, \omega) - \frac{\Delta_1(\theta, \phi, \omega)}{\Delta_0(\theta, \phi, \omega)} \underline{\underline{u}}_0(\theta, \phi, \omega) \right]
$$
$$
+ \tau_{\mathrm{v}}(L_{\mathrm{cor}}, \theta, \phi, \omega) \left[\underline{\underline{v}}_1(\theta, \phi, \omega) - \frac{\Delta_1(\theta, \phi, \omega)}{\Delta_0(\theta, \phi, \omega)} \underline{\underline{v}}_0(\theta, \phi, \omega) \right]
$$
$$
+ \frac{3}{2} \left[\left(\frac{1}{L_{\mathrm{cor}}} + i\kappa_{0+}^{\frac{1}{2}} \right) \kappa_{1+} - \left(\frac{1}{L_{\mathrm{cor}}} + i\kappa_{0-}^{\frac{1}{2}} \right) \kappa_{1-} \right]
$$
$$
\times \underline{\underline{u}}_0(\theta, \phi, \omega) + \frac{i}{2} \left(\frac{\kappa_{1+}}{\kappa_{0+}^{\frac{1}{2}}} - \frac{\kappa_{1-}}{\kappa_{0-}^{\frac{1}{2}}} \right) \underline{\underline{v}}_0(\theta, \phi, \omega) \left. \vphantom{\frac{\kappa}{\kappa}} \right\} d\theta \, d\phi,
$$

$$(7.74)$$

where the 3×3 dyadic functions

$$
\underline{\underline{u}}_1(\theta, \phi, \omega) = \left\{ 2\underline{\underline{\epsilon}}_{\mathrm{comp}}^{\mathrm{NL}}(\omega) - \left[\frac{t_{\mathrm{B}1} t_{\mathrm{C}0} - t_{\mathrm{B}0} t_{\mathrm{C}1}}{t_{\mathrm{C}0} \left\{ \mathrm{tr} \left[\underline{\underline{\epsilon}}_{\mathrm{comp}}^{\mathrm{L}}(\omega) \bullet \underline{\underline{\Omega}}(\theta, \phi) \right] \right\}} \right. \right.
$$
$$
\left. + \mathrm{tr}\, \underline{\underline{\epsilon}}_{\mathrm{comp}}^{\mathrm{NL}}(\omega) \right] \underline{\underline{I}} \right\} \bullet \underline{\underline{\Omega}}(\theta, \phi) - \left\{ \mathrm{tr} \left[\underline{\underline{\epsilon}}_{\mathrm{comp}}^{\mathrm{NL}}(\omega) \bullet \underline{\underline{\Omega}}(\theta, \phi) \right] \right\} \underline{\underline{I}} \Big\}
$$
$$
\underline{\underline{v}}_1(\theta, \phi, \omega) = \underline{\underline{\Psi}}(\omega) - \frac{t_{\mathrm{B}1} t_{\mathrm{C}0} - t_{\mathrm{B}0} t_{\mathrm{C}1}}{t_{\mathrm{C}0} \left\{ \mathrm{tr} \left[\underline{\underline{\epsilon}}_{\mathrm{comp}}^{\mathrm{L}}(\omega) \bullet \underline{\underline{\Omega}}(\theta, \phi) \right] \right\}} \underline{\underline{\Omega}}(\theta, \phi)
$$

$$(7.75)$$

with diagonal dyadic

$$
\underline{\underline{\Psi}}(\omega) = \begin{pmatrix} \Psi^x(\omega) & 0 & 0 \\ 0 & \Psi^y(\omega) & 0 \\ 0 & 0 & \Psi^z(\omega) \end{pmatrix},
$$

$$(7.76)$$

having components

$$
\left. \begin{array}{l}
\Psi^x(\omega) = \epsilon_{\mathrm{comp}}^{y,\mathrm{NL}}(\omega)\, \epsilon_{\mathrm{comp}}^{z,\mathrm{L}}(\omega) + \epsilon_{\mathrm{comp}}^{y,\mathrm{L}}(\omega)\, \epsilon_{\mathrm{comp}}^{z,\mathrm{NL}}(\omega) \\[4pt]
\Psi^y(\omega) = \epsilon_{\mathrm{comp}}^{z,\mathrm{NL}}(\omega)\, \epsilon_{\mathrm{comp}}^{x,\mathrm{L}}(\omega) + \epsilon_{\mathrm{comp}}^{z,\mathrm{L}}(\omega)\, \epsilon_{\mathrm{comp}}^{x,\mathrm{NL}}(\omega) \\[4pt]
\Psi^z(\omega) = \epsilon_{\mathrm{comp}}^{x,\mathrm{NL}}(\omega)\, \epsilon_{\mathrm{comp}}^{y,\mathrm{L}}(\omega) + \epsilon_{\mathrm{comp}}^{x,\mathrm{L}}(\omega)\, \epsilon_{\mathrm{comp}}^{y,\mathrm{NL}}(\omega)
\end{array} \right\} ;
$$

$$(7.77)$$

and the scalar functions[2]

$$\left.\begin{aligned}
\Delta_1(\theta,\phi,\omega) &= \frac{t_{\mathrm{B0}}t_{\mathrm{B1}} - 2\left(t_{\mathrm{A1}}t_{\mathrm{C0}} + t_{\mathrm{A0}}t_{\mathrm{C1}}\right)}{\Delta_0(\theta,\phi,\omega)}\\[2mm]
\kappa_{1\pm} &= \frac{\omega^2\left[-t_{\mathrm{B1}} \pm \Delta_1(\theta,\phi,\omega)\right] - 2t_{\mathrm{C1}}\,\kappa_{0\pm}}{2\,t_{\mathrm{C0}}}
\end{aligned}\right\}, \qquad (7.78)$$

wherein

$$\left.\begin{aligned}
t_{\mathrm{A1}} &= \epsilon^{x,\mathrm{NL}}_{\mathrm{comp}}(\omega)\,\epsilon^{y,\mathrm{L}}_{\mathrm{comp}}(\omega)\,\epsilon^{z,\mathrm{L}}_{\mathrm{comp}}(\omega) + \epsilon^{x,\mathrm{L}}_{\mathrm{comp}}(\omega)\,\epsilon^{y,\mathrm{NL}}_{\mathrm{comp}}(\omega)\,\epsilon^{z,\mathrm{L}}_{\mathrm{comp}}(\omega)\\
&\quad + \epsilon^{x,\mathrm{L}}_{\mathrm{comp}}(\omega)\,\epsilon^{y,\mathrm{L}}_{\mathrm{comp}}(\omega)\,\epsilon^{z,\mathrm{NL}}_{\mathrm{comp}}(\omega)\\
t_{\mathrm{B1}} &= \left\{\mathrm{tr}\left[\underline{\underline{\Psi}}(\omega)\bullet\underline{\underline{\Omega}}(\theta,\phi)\right]\right\} - \left[\mathrm{tr}\,\underline{\underline{\Psi}}(\omega)\right]\left[\mathrm{tr}\,\underline{\underline{\Omega}}(\theta,\phi)\right]\\
t_{\mathrm{C1}} &= \left\{\mathrm{tr}\left[\underline{\underline{\epsilon}}^{\mathrm{NL}}_{\mathrm{comp}}(\omega)\bullet\underline{\underline{\Omega}}(\theta,\phi)\right]\right\}\mathrm{tr}\,\underline{\underline{\Omega}}(\theta,\phi)
\end{aligned}\right\}. \qquad (7.79)$$

Only the components $\left[\underline{\underline{\Sigma}}^{[2]}(L_{\mathrm{cor}},\omega)\right]_{mn}$ with $m,n \in \{1,2,3\}$ of the second–order mass operator are needed for the anisotropic dielectric scenario considered here. These form the 3×3 dyadic $^{\mathrm{ee}}\underline{\underline{\Sigma}}^{[2]}(L_{\mathrm{cor}},\omega)$ such that $\left[\underline{\underline{\Sigma}}^{[2]}(L_{\mathrm{cor}},\omega)\right]_{mn} = \left[^{\mathrm{ee}}\underline{\underline{\Sigma}}^{[2]}(L_{\mathrm{cor}},\omega)\right]_{mn}$ for $m,n \in \{1,2,3\}$. Under the assumption of weak nonlinearity, the dyadic $^{\mathrm{ee}}\underline{\underline{\Sigma}}^{[2]}(L_{\mathrm{cor}},\omega)$ may be regarded as the sum

$$^{\mathrm{ee}}\underline{\underline{\Sigma}}^{[2]}(L_{\mathrm{cor}},\omega) = {}^{\mathrm{ee}}\underline{\underline{\Sigma}}^{[2],\mathrm{L}}(L_{\mathrm{cor}},\omega) + {}^{\mathrm{ee}}\underline{\underline{\Sigma}}^{[2],\mathrm{NL}}(L_{\mathrm{cor}},\omega)\,|\underline{E}_{\mathrm{HCM}}(\omega)|^2, \qquad (7.80)$$

with the linear part being

$$^{\mathrm{ee}}\underline{\underline{\Sigma}}^{[2],\mathrm{L}}(L_{\mathrm{cor}},\omega) = -\omega^2\left[\underline{\underline{\alpha}}^{\mathrm{L}}_{\mathrm{a/comp}}(\omega) - \underline{\underline{\alpha}}^{\mathrm{L}}_{\mathrm{b/comp}}(\omega)\right] \bullet {}^{\mathrm{ee}}\underline{\underline{P}}^{\mathrm{L}}_{\Lambda_{\mathrm{step}}}(L_{\mathrm{cor}},\omega)$$
$$\bullet \left[\underline{\underline{\alpha}}^{\mathrm{L}}_{\mathrm{a/comp}}(\omega) - \underline{\underline{\alpha}}^{\mathrm{L}}_{\mathrm{b/comp}}(\omega)\right] \qquad (7.81)$$

and weakly nonlinear part being

$$^{\mathrm{ee}}\underline{\underline{\Sigma}}^{[2],\mathrm{NL}}(L_{\mathrm{cor}},\omega) = -\omega^2\Big\{2\left[\underline{\underline{\alpha}}^{\mathrm{L}}_{\mathrm{a/comp}}(\omega) - \underline{\underline{\alpha}}^{\mathrm{L}}_{\mathrm{b/comp}}(\omega)\right] \bullet {}^{\mathrm{ee}}\underline{\underline{P}}^{\mathrm{L}}_{\Lambda_{\mathrm{step}}}(L_{\mathrm{cor}},\omega)$$
$$\bullet \left[\underline{\underline{\alpha}}^{\mathrm{NL}}_{\mathrm{a/comp}}(\omega) - \underline{\underline{\alpha}}^{\mathrm{NL}}_{\mathrm{b/comp}}(\omega)\right]$$
$$+\left[\underline{\underline{\alpha}}^{\mathrm{L}}_{\mathrm{a/comp}}(\omega) - \underline{\underline{\alpha}}^{\mathrm{L}}_{\mathrm{b/comp}}(\omega)\right]$$
$$\bullet {}^{\mathrm{ee}}\underline{\underline{P}}^{\mathrm{NL}}_{\Lambda_{\mathrm{step}}}(L_{\mathrm{cor}},\omega) \bullet \left[\underline{\underline{\alpha}}^{\mathrm{L}}_{\mathrm{a/comp}}(\omega) - \underline{\underline{\alpha}}^{\mathrm{L}}_{\mathrm{b/comp}}(\omega)\right]\Big\}.$$
$$(7.82)$$

---

[2]For compact presentation, the dependencies of $\kappa_{1\pm} \equiv \kappa_{1\pm}(\theta,\phi,\omega)$, $t_{\mathrm{A1}} \equiv t_{\mathrm{A1}}(\omega)$, $t_{\mathrm{B1}} \equiv t_{\mathrm{B1}}(\theta,\phi,\omega)$ and $t_{\mathrm{C1}} \equiv t_{\mathrm{C1}}(\theta,\phi,\omega)$ are implicit.

Finally, the second–order SPFT estimate of the HCM permittivity dyadic, namely

$$\underline{\underline{\epsilon}}^{[2]}_{\mathrm{SPFT}}(\omega) = \underline{\underline{\epsilon}}^{[2],\mathrm{L}}_{\mathrm{SPFT}}(\omega) + \underline{\underline{\epsilon}}^{[2],\mathrm{NL}}_{\mathrm{SPFT}}(\omega)\, |\,\underline{E}_{\mathrm{HCM}}(\omega)\,|^2, \qquad (7.83)$$

may be specified in terms of the foregoing expressions in this section. The anisotropic dielectric specialization of Eq. (6.14) with $n = 2$ delivers the linear part as

$$\underline{\underline{\epsilon}}^{[2],\mathrm{L}}_{\mathrm{SPFT}}(\omega) = \underline{\underline{\epsilon}}^{\mathrm{L}}_{\mathrm{comp}}(\omega) - \frac{1}{i\omega}\,\left[\underline{\underline{e}}^{\mathrm{L}}(\omega)\right]^{-1} \bullet {}^{\mathrm{ee}}\underline{\underline{\Sigma}}^{[2],\mathrm{L}}(L_{\mathrm{cor}},\omega) \qquad (7.84)$$

and the weakly nonlinear part as

$$\underline{\underline{\epsilon}}^{[2],\mathrm{NL}}_{\mathrm{SPFT}}(\omega) = \underline{\underline{\epsilon}}^{\mathrm{NL}}_{\mathrm{comp}}(\omega) - \frac{1}{i\omega}\,\Big\{ \left[\underline{\underline{e}}^{\mathrm{L}}(\omega)\right]^{-1} \bullet {}^{\mathrm{ee}}\underline{\underline{\Sigma}}^{[2],\mathrm{NL}}(L_{\mathrm{cor}},\omega)$$

$$+\mathcal{N}_{\underline{\underline{\epsilon}}}\left\{\underline{\underline{e}}(\omega)\right\} \bullet {}^{\mathrm{ee}}\underline{\underline{\Sigma}}^{[2],\mathrm{L}}(L_{\mathrm{cor}},\omega)\Big\}, \qquad (7.85)$$

where the 3×3 dyadic $\underline{\underline{e}}(\omega)$ is the analogue of the 6×6 dyadic $\underline{\underline{\mathbf{e}}}(\omega)$ introduced in Eq. (7.49). The linear and weakly nonlinear parts of $\underline{\underline{e}}(\omega)$, namely $\underline{\underline{e}}^{\mathrm{L}}(\omega)$ and $\underline{\underline{e}}^{\mathrm{NL}}(\omega)$, respectively, are as specified in Eqs. (7.50) but with ${}^{\mathrm{ee}}\underline{\underline{\Sigma}}^{[2],\mathrm{L},\mathrm{NL}}(L_{\mathrm{cor}},\omega)$ replacing $\underline{\underline{\Sigma}}^{[2],\mathrm{L},\mathrm{NL}}(L_{\mathrm{cor}},\omega)$, and ${}^{\mathrm{ee}}\underline{\underline{D}}^{0,\mathrm{L},\mathrm{NL}}_{\mathrm{U/comp}}(\omega)$ replacing $\underline{\underline{D}}^{0,\mathrm{L},\mathrm{NL}}_{\mathrm{I/comp}}(\omega)$ therein.

The second–order SPFT estimate represented by Eq. (7.83) implicitly uses the depolarization dyadic ${}^{\mathrm{ee}}\underline{\underline{D}}^0(\omega)$ which is independent of the size of the constituent particles. The sizes of the constituent particles may be taken into account by using ${}^{\mathrm{ee}}\underline{\underline{D}}(\rho,\omega_{\mathrm{NL}})$ in lieu of ${}^{\mathrm{ee}}\underline{\underline{D}}^0(\omega)$, where ρ is a measure of the average linear dimensions of the particles, as discussed in Sec. 6.5. Such an extension to the weakly nonlinear SPFT is available elsewhere [19].

7.3 Nonlinearity enhancement

By the process of homogenization, nonlinear properties may be enhanced. That is, HCMs may be conceptualized which exhibit stronger nonlinear properties than their constituent mediums. Such HCMs represent examples of metamaterials, as discussed in Sec. 6.7. There are possible technological applications of nonlinearity enhancement in the areas of optical computing, real–time holography, optical correlators and phase conjugators [27].

Nonlinearity enhancement via homogenization has been explored especially in the case of HCMs arising from constituent mediums with intensity–dependent permittivities, as prescribed by Eq. (7.15) [14, 28]. Specifically, in representative numerical calculations for weakly nonlinear isotropic dielectric HCMs using the lowest–order SPFT [14], the magnitude of the HCM nonlinear permittivity was observed to exceed that of the constituent mediums by as much as approximately 50%. Greater degrees of enhancement can be achieved in the corresponding anisotropic dielectric scenario [17]. According to the second–order SPFT, the degree of enhancement diminishes slightly as the correlation length increases, for isotropic dielectric [14, 15], isotropic chiral [16] and anisotropic dielectric [18] HCMs. Furthermore, the degree of enhancement is diminished by taking into account the nonzero size of the constituent particles [19].

7.4 Quantum electrodynamic vacuum

Hitherto in this book, vacuum has been considered from the classical perspective, as a linear medium characterized by the permittivity scalar ϵ_0 and permeability scalar μ_0. Under the influence of a strong gravitational field, classical vacuum may be described as an effectively bianisotropic medium but its linearity is preserved, as described in Sec. 2.4.2. We close by departing from this classical viewpoint. Within the realm of quantum electrodynamics (QED), vacuum is a nonlinear medium [29]. However, in the presence of a sufficiently strong, quasistatic, magnetic field, the QED vacuum can be linearized. For optical fields, the linearized representation of QED vacuum is provided by a uniaxial dielectric–magnetic medium which is spatially nonhomogeneous and is described by constitutive parameters that depend on the magnitude and direction of the quasistatic magnetic field.

To be specific, let us consider matter–free space in the presence of a quasistatic magnetic field $\underline{B}_{qs}(\underline{r}) = B_{qs}(\underline{r})\,\hat{u}$ directed along the unit vector $\hat{u}$. The optical fields $\tilde{\underline{F}}_{opt}(\underline{r},t)$ $(F = E, B, D, H)$ in this space satisfy the source–free Maxwell postulates

$$
\left.
\begin{aligned}
\nabla \times \tilde{\underline{E}}_{opt}(\underline{r},t) + \frac{\partial}{\partial t}\,\tilde{\underline{B}}_{opt}(\underline{r},t) = \underline{0} \\[2mm]
\nabla \times \tilde{\underline{H}}_{opt}(\underline{r},t) - \frac{\partial}{\partial t}\,\tilde{\underline{D}}_{opt}(\underline{r},t) = \underline{0} \\[2mm]
\nabla \cdot \tilde{\underline{D}}_{opt}(\underline{r},t) = 0 \\[2mm]
\nabla \cdot \tilde{\underline{B}}_{opt}(\underline{r},t) = 0
\end{aligned}
\right\},
\qquad (7.86)
$$

provided that $|\tilde{\underline{B}}_{\text{opt}}(\underline{r},t)| \ll |\underline{B}_{\text{qs}}(\underline{r})| \ \forall \underline{r}$. According to the linearized version of QED, the optical fields appearing in Eqs. (7.86) obey the constitutive equations [30]

$$
\left.
\begin{aligned}
\tilde{\underline{D}}_{\text{opt}}(\underline{r},t) &= \underline{\underline{\epsilon}}_{\text{QV}}(\underline{r}) \cdot \tilde{\underline{E}}_{\text{opt}}(\underline{r},t) \\
\tilde{\underline{H}}_{\text{opt}}(\underline{r},t) &= \underline{\underline{\nu}}_{\text{QV}}(\underline{r}) \cdot \tilde{\underline{B}}_{\text{opt}}(\underline{r},t)
\end{aligned}
\right\},
\tag{7.87}
$$

where the permittivity dyadic has the uniaxial form

$$
\underline{\underline{\epsilon}}_{\text{QV}}(\underline{r}) = \epsilon_0 \left\{ \left[1 - \frac{8}{\mu_0}\Theta_{\text{QV}}B_{\text{qs}}^2(\underline{r}) \right] \underline{\underline{I}} + \frac{28}{\mu_0}\Theta_{\text{QV}}B_{\text{qs}}^2(\underline{r})\,\hat{\underline{u}}\,\hat{\underline{u}} \right\}
\tag{7.88}
$$

and the impermeability dyadic has the uniaxial form

$$
\underline{\underline{\nu}}_{\text{QV}}(\underline{r}) = \frac{1}{\mu_0} \left\{ \left[1 - \frac{8}{\mu_0}\Theta_{\text{QV}}B_{\text{qs}}^2(\underline{r}) \right] \underline{\underline{I}} - \frac{16}{\mu_0}\Theta_{\text{QV}}B_{\text{qs}}^2(\underline{r})\,\hat{\underline{u}}\,\hat{\underline{u}} \right\}.
\tag{7.89}
$$

Herein the scalar parameter

$$
\Theta_{\text{QV}} = \frac{q_{\text{el}}^4 \hbar}{45(4\pi\epsilon_0)^2 m_{\text{el}}^4 c_0^7} = 8.3229 \times 10^{-32} \ \text{kg}^{-1} \ \text{m} \ \text{s}^2,
\tag{7.90}
$$

with the electronic charge $q_{\text{el}} = 1.6022 \times 10^{-19}$ C, the electronic mass $m_{\text{el}} = 9.1096 \times 10^{-31}$ kg, and the reduced Planck constant $\hbar = 1.0546 \times 10^{-34}$ J s. Thus, the QED vacuum appears to the optical field as a spatiotemporally local, spatially nonhomogeneous, temporally unvarying, uniaxial dielectric–magnetic medium. The derivation of this result is based on the Heisenberg–Euler effective Lagrangian of the electromagnetic field [31, 32]. Notice that by setting $\hbar = 0$ the QED vacuum is converted to the classical vacuum and the influence of the quasistatic field on the optical field disappears.

The planewave properties of QED vacuum are as reported in Sec. 4.5.1 for uniaxial dielectric–magnetic mediums; i.e., QED vacuum is a generally birefringent medium [30]. Also, the corresponding dyadic Green functions are as provided in Sec. 5.2.2 — from these, an integral equation for light scattering in QED vacuum may be deduced [33].

References

[1] N. Bloembergen, *Nonlinear optics*, *4th ed*, World Scientific, Singapore, 1996.

[2] B.D.H. Tellegen, Interaction between radio–waves?, *Nature* **131** (1933), 840.

[3] S.C. Bloch, Heterodyne generation by electromagnetic fields in gaseous plasma, *Brit J Appl Phys* **16** (1965), 1853–1860.

[4] R.W. Boyd, *Nonlinear optics*, *3rd ed*, Academic Press, London, UK, 2008.

[5] A. Lakhtakia and W.S. Weiglhofer, Maxwell Garnett formalism for weakly nonlinear, bianisotropic, dilute, particulate composite media, *Int J Electron* **87** (2000), 1401–1408.

[6] T. Kobayashi, Introduction to nonlinear optical materials, *Nonlin Opt* **1** (1991), 91–117.

[7] S. Kielich and A. Piekara, Frequency– and spatially variable electric and magnetic polarizations induced in nonlinear media by electromagnetic fields, *Acta Physica Polo* **29** (1966), 875–898.

[8] P.W. Atkins and L.D. Barron, Quantum field theory of optical birefringence phenomena I. Linear and nonlinear optical rotation, *Proc R Soc Lond A* **304** (1968), 303–317.

[9] W.S. Weiglhofer and A. Lakhtakia, Mediation of nonlinear polarization by the magnetic field in a composite medium with a chiral component, *Microw Opt Technol Lett* **13** (1996), 285–287.

[10] D. Stroud and P.M. Hui, Nonlinear susceptibilities of granular matter, *Phys Rev B* **37** (1988), 8719–8724.

[11] J.E. Sipe and R.W. Boyd, Nonlinear susceptibility of composite optical materials in the Maxwell Garnett model, *Phys Rev A* **46** (1992), 1614–1629.

[12] A. Lakhtakia and W.S. Weiglhofer, Maxwell Garnett approach for nonlinear dilute particulate composites with bi–isotropic host media, *Int J Electron* **80** (1996), 665–676.

[13] A. Lakhtakia and W.S. Weiglhofer, Maxwell Garnett formalism for cubically nonlinear, gyrotropic, composite media, *Int J Electron* **84** (2000), 285–294.

[14] A. Lakhtakia, Application of strong permittivity fluctuation theory for isotropic, cubically nonlinear, composite mediums, *Opt Commun* **192** (2001), 145–151 (2001).

[15] T.G. Mackay, A. Lakhtakia and W.S. Weiglhofer, Homogenisation of isotropic, cubically nonlinear, composite mediums by the strong–permittivity–fluctuation theory: third–order considerations, *Opt Commun* **204** (2002), 219–228.

[16] T.G. Mackay, A. Lakhtakia and W.S. Weiglhofer, The strong–property–fluctuation theory for cubically nonlinear, isotropic chiral composite mediums, *Electromagnetics* **23** (2003), 455–479.

[17] M.N. Lakhtakia and A. Lakhtakia, Anisotropic composite materials with intensity–dependent permittivity tensor: the Bruggeman approach, *Electromagnetics* **21** (2001), 129–138.

[18] T.G. Mackay, Geometrically derived anisotropy in cubically nonlinear dielectric composites, *J Phys D: Appl Phys* **36** (2003), 583–591.

[19] J. Cui and T.G. Mackay, Depolarization regions of nonzero volume for anisotropic, cubically nonlinear, homogenized nanocomposites, *J Nanophoton* **1** (2007), 013506.

[20] H.E. Brandt (ed), *Selected papers on nonlinear optics*, SPIE Optical Engineering Press, Bellingham, WA, USA, 1991.

[21] E. Cumberbatch, Self–focusing in non–linear optics, *IMA J Appl Maths* **6** (1970), 250-262.

[22] O. Levy, Y. Yagil and D.J. Bergman, Field–induced tuning of the optical properties of nonlinear composites near resonance, *J Appl Phys* **76** (1994), 1431–1435.

[23] A. Lakhtakia, Extended Maxwell Garnett formulae for weakly nonlinear dielectric-in-dielectric composites, *Optik* **103** (1996), 85–87.

[24] X.C. Zeng, D.J. Bergman, P.M. Hui and D. Stroud, Effective–medium theory for weakly nonlinear composites, *Phys Rev B* **38** (1988), 10970–10973.

[25] G.Ya. Slepyan, S.A. Maksimenko, F.G. Bass and A. Lakhtakia, Nonlinear electromagnetics in chiral media: self–action of waves, *Phys Rev E* **52** (1995), 1049–1058.

[26] G.Ya. Slepyan, A. Lakhtakia and S.A. Maksimenko, Bruggeman and Maxwell Garnett models of a chiral composite with weak cubic nonlinearities, *Microw Opt Technol Lett* **12** (1996), 342–246.

[27] H.B. Liao, R.F. Xiao, H. Wang, K.S. Wong and G.K.L. Wong, Large third–order optical nonlinearity in Au:TiO_2 composite films measured on a femtosecond time scale, *Appl Phys Lett* **72** (1998), 1817–1819.

[28] R.W. Boyd, R.J. Gehr, G.L. Fischer and J.E. Sipe, Nonlinear optical properties of nanocomposite materials, *Pure Appl Opt* **5** (1996), 505–512.

[29] J.D. Jackson, *Classical electrodynamics, 3rd ed*, Wiley, New York, NY, USA, 1999.

[30] S.L. Adler, 1971. Photon splitting and photon dispersion in a strong magnetic field. *Ann Phys (NY)* **67** (1971), 599–647.

[31] W. Heisenberg and H. Euler, Folgerungen aus der Diracschen Theorie des Positrons, *Z Phys* **98** (1936), 714–732.

[32] J. Schwinger, On gauge invariance and vacuum polarization, *Phys Rev* **82** (1951), 664–679.

[33] A. Lakhtakia and T.G. Mackay, Integral equation for scattering of light by a strong magnetostatic field in vacuum, *Electromagnetics* **27** (2007), 341–354.

Appendix A

Dyadic Notation and Analysis

A brief introduction to dyadics and their properties is provided in this Appendix. For further details about dyadics, especially in the context of electromagnetics, the reader is referred elsewhere [1, 2].

A dyadic is a linear mapping from one n–vector to another. For 3–vectors, the mapping can be straightforwardly described in terms of Cartesian vectors. Thus, a 3×3 dyadic $\underline{\underline{D}}$ is a linear mapping from 3–vector $\underline{a}$ to 3–vector $\underline{b}$:

$$\underline{b} = \underline{\underline{M}} \cdot \underline{a} \, .$$

The identity dyadic $\underline{\underline{I}}$ is such that $\underline{a} \cdot \underline{\underline{I}} = \underline{\underline{I}} \cdot \underline{a} = \underline{a}$; likewise, the null dyadic $\underline{\underline{0}}$ is defined so that $\underline{a} \cdot \underline{\underline{0}} = \underline{\underline{0}} \cdot \underline{a}$ equals the null vector $\underline{0}$.

These properties lead to the idea of a *dyad* that is composed of two vectors, i.e., $\underline{\underline{d}} = \underline{p}\,\underline{q}$. It follows that $\underline{\underline{d}} \cdot \underline{a} = \underline{p}\,(\underline{q} \cdot \underline{a})$ and $\underline{a} \cdot \underline{\underline{d}} = (\underline{a} \cdot \underline{p})\,\underline{q}$ are vectors, whereas $\underline{\underline{d}} \times \underline{a} = \underline{p}(\underline{q} \times \underline{a})$ and $\underline{a} \times \underline{\underline{d}} = (\underline{a} \times \underline{p})\underline{q}$ are dyads. The transpose of a dyad $\underline{p}\,\underline{q}$ is the dyad $\underline{q}\,\underline{p}$. Sometimes, dyads are called bivectors.

A dyadic is not necessarily a dyad. The general representation of a dyadic is the sum

$$\underline{\underline{M}} = \sum_{\ell=1,2,\ldots} \gamma^{(\ell)}\,\underline{p}^{(\ell)}\underline{q}^{(\ell)} \, ,$$

where $\gamma^{(\ell)}$ are some scalar coefficients, while $\underline{p}^{(\ell)}$ and $\underline{q}^{(\ell)}$ are vectors.

All vectors can be written using matrix notation. Thus, the vector $\underline{p} = \hat{x}\,p_x + \hat{y}\,p_y + \hat{z}\,p_z$ in a Cartesian coordinate system is equivalent to the column vector

$$\underline{p} \equiv \begin{pmatrix} p_x \\ p_y \\ p_z \end{pmatrix}$$

of size 3, and the dyad $\underline{d} = \underline{p}\,\underline{q}$ is equivalent to the 3×3 matrix

$$\underline{d} \equiv \begin{pmatrix} p_x q_x & p_x q_y & p_x q_z \\ p_y q_x & p_y q_y & p_y q_z \\ p_z q_x & p_z q_y & p_z q_z \end{pmatrix}.$$

Hence, most dyadics in electromagnetics can be written as 3×3 matrixes. The identity dyadic $\underline{\underline{I}}$ is equivalent to the 3×3 identity matrix, and the null dyadic $\underline{\underline{0}}$ to the 3×3 null matrix. The usual algebra of matrixes can thus be used for dyadics as well.

The trace of a dyadic is the sum of the diagonal elements in its matrix representation. Likewise, the determinant of a dyadic is the same as the determinant of its equivalent matrix. A dyadic can be transposed in the same way as a matrix. If it is nonsingular, a dyadic can be inverted.

The antisymmetric dyadic

$$\underline{p} \times \underline{\underline{I}} = \underline{\underline{I}} \times \underline{p} \equiv \begin{pmatrix} 0 & -p_z & p_y \\ p_z & 0 & -p_x \\ -p_y & p_x & 0 \end{pmatrix}$$

is often useful to denote gyrotropic electromagnetic properties that are characteristic of ferrites and plasmas. The simplest antisymmetric dyadic is $\underline{\hat{u}} \times \underline{\underline{I}}$, where $\underline{\hat{u}}$ is any vector of unit magnitude. The trace of any antisymmetric dyadic is zero.

Even vector differential operators can be thought of as dyadics. Thus, the *curl* operator is written as $\nabla \times \underline{\underline{I}}$ and the *divergence* operator as $\nabla \cdot \underline{\underline{I}}$, with

$$\nabla = \underline{\hat{x}}\frac{\partial}{\partial x} + \underline{\hat{y}}\frac{\partial}{\partial y} + \underline{\hat{z}}\frac{\partial}{\partial z}$$

written as a vector in a Cartesian coordinate system.

References

[1] H.C. Chen, *Theory of electromagnetic waves*, McGraw–Hill, New York, NY, USA, 1983.
[2] J. Van Bladel, *Electromagnetic fields*, Hemisphere, Washington, DC, USA, 1985.

Epilogue

प्यास कुछ और भी भड़का दी झलक दिखला के,
तुझको पर्दा रुख़-ए-रौशन से हटाना होगा।
– कैफ़ी आज़मी (१९५८)

You have intensified my thirst by a fleeting glimpse,
Now you shall have to unveil your radiant face.
– Kaifi Azmi (1958)

Index

achirality, 44
aciculate molecules, 47
active medium, 28, 108
algebraic multiplicity, 116
amphoteric refraction, 118
attenuation, 83, 164

Böttcher's equation, 161
Beltrami
    current densities, 129
    fields, 89, 100
birefringence, 85, 89
    circular, 90
Born series, 162
boundary conditions
    frequency domain, 10, 86
    time domain, 4
bounds
    Bergman–Milton, 171
    Hashin–Shtrikman, 158, 171
    Wiener, 171
Bravais lattice, 37
Brillouin scattering, 184
Bruggeman formalism, 160, 181

causal dispersion relations, 22
causality, 20, 25
charge density
    macroscopic, 2
    microscopic, 1
chiral sculptured thin film, 47, 102,
    104

chiroferrite, 44
chiroplasma, 44
column vector, 205
comparison medium, 162
    weakly nonlinear, anisotropic
        dielectric, 192
    weakly nonlinear, isotropic chiral,
        186
    weakly nonlinear, isotropic
        dielectric, 184
complementary function, 126
constitutive relations, 7
    Boys–Post, 9
    Tellegen, 9, 84
correlation length, 164
counterposition, 117
covariance function
    Gaussian, 185
    step, 164
    three–point, 165
    two–point, 164
crystal optics, 81, 92
crystal symmetry
    glide plane reflection, 78
    improper rotation, 56
    inversion, 56
    proper rotation, 56
    reflection, 56
    screw axis rotation, 78
    translation, 56, 78
crystallography, 57
cubic crystal system, 56, 57

About the Authors

Tom G. Mackay is a graduate of the Universities of Edinburgh, Glasgow and Strathclyde. He started off his career as a bioengineer, at Glasgow Royal Infirmary in 1988. Gradually, he gravitated towards applied mathematics and completed his Ph.D. in that subject under the supervision of Werner S. Weiglhofer (Department of Mathematics, University of Glasgow) in 2001. Since then he has been based in the School of Mathematics at the University of Edinburgh where he is now a Reader. He is also an Adjunct Professor in the Department of Engineering Science and Mechanics at the Pennsylvania State University. In 2006/07 he held a *Royal Society of Edinburgh/Scottish Executive Support Research Fellowship*. For the past ten years, the electromagnetic theory of complex mediums and homogenization have been the chief focuses of his research.

Akhlesh Lakhtakia is the Charles Godfrey Binder (Endowed) Professor of Engineering Science and Mechanics at the Pennsylvania State University. He received his B.Tech. (1979) and D.Sc. (2006) degrees in Electronics Engineering from Banaras Hindu University, and his M.S. (1981) and Ph.D. (1983) degrees in Electrical Engineering from the University of Utah. He has published more than 625 journal articles; contributed 18 chapters to research books and encyclopedias; edited, co-edited, authored or co-authored 12 books and 8 conference proceedings; and reviewed for 110 journals. He was the Editor-in-Chief of the international journal *Speculations in Science and Technology* from 1993 to 1995, and is the founding Editor-in-Chief of the online *Journal of Nanophotonics* published by SPIE from 2007. He is a Fellow of the Optical Society of America, SPIE, and the Institute of Physics (UK). At Penn State, he was awarded the PSES Outstanding Research Award in 1996, the Faculty Scholar Medal in Engineering in 2005, and the PSES Premier Research Award in 2008. His current research interests lie in the electromagnetics of complex mediums, sculptured thin films, and nanotechnology.